D0743036

Dynamics of Lotic Ecosystems

Dynamics of Lotic Ecosystems

Edited by

Thomas D. Fontaine, III
Savannah River Ecology Laboratory
University of Georgia

Steven M. Bartell
Environmental Sciences Division
Oak Ridge National Laboratory

ANN ARBOR SCIENCE
THE BUTTERWORTH GROUP

Copyright © 1983 by Ann Arbor Science Publishers
230 Collingwood, P.O. Box 1425, Ann Arbor, Michigan 48106

Library of Congress Catalog Card Number 82-048641
ISBN 0-250-40612-8

Manufactured in the United States of America
All Rights Reserved

Butterworths, Ltd., Borough Green, Sevenoaks
Kent TN15 8PH, England

PREFACE

Increases in the number of stream ecologists, technical articles, and frequency of special symposia on lotic ecosystems reflect a growing interest in the structure and function of streams and rivers. Different perspectives in the past stimulated a variety of specializations and methods directed at improving and enlarging our understanding of lotic systems. Naturalists detailed the unique morphological and behavioral adaptations displayed by plants and animals that inhabit riffles and pools of quickly flowing streams. On a different scale, ecologists have endeavored to quantify interactions among groups of organisms and their physical and chemical environment that contribute to the spatial and temporal heterogeneity so characteristic of lotic habitats. Dictated by societal needs, engineers have focused on means to estimate the capacity of larger rivers to process and dilute industrial and domestic wastes.

While progress has been made in these areas of specialization and others, we believe that communication across borderlines of separate disciplines, the kind of communication that promotes synthesis of ideas and stimulates useful generalizations or theory, has been all too infrequent in the study of lotic systems. Accordingly, we convened a symposium that would provide a format for the assembly of individuals from academia, industry and government to explore common interests, discuss current understanding, and outline general principles that govern the structure and function of flowing water systems. We encouraged participants to generalize and extrapolate on the basis of their own research, including theoretical work, modeling, experiments and field observations, as well as the research of others. As a result, topics of solicited and contributed papers within this volume transcend orders of magnitude in space and time, from organisms to watersheds, from diel cycles to processes that span decades. The domain of research discussed includes energy flow, material cycling, potential effects of management alternatives on water quality, and elucidation of physical, chemical and biological processes that order assemblages of organisms in lotic habitats.

The variety of research interests and experience that was represented at the symposium stimulated provocative and sometimes controversial discussion that is difficult to capture in book form. A detailed set of general principles did not readily emerge from four days of interaction. Nevertheless, we believe that participants left the symposium with new or modified conceptual frameworks on which to build future research. For these individuals, this volume preserves many of the ideas and observations discussed in Augusta. For others, we hope that this book imparts at least some of the excitement of the symposium and provides a useful reference for future research concerning the dynamics of lotic systems.

Steven M. Bartell   Thomas D. Fontaine, III

ACKNOWLEDGMENTS

We owe a debt of gratitude to the entire staff of the Savannah River Ecology Laboratory for assistance in organization and execution of the symposium. We gratefully acknowledge the cooperation and financial support of the United States Department of Energy. Michael H. Smith and Robert I. Nestor provided advice and encouragement that we appreciate. We especially thank Nancy Barber, symposium secretary, for answering the seemingly endless correspondence associated with the symposium and the proceedings, and for limitless help in conducting all phases of the meetings. We are particularly grateful to the following people who critically reviewed the manuscripts for this volume:

J. D. Allan, University of Maryland, College Park, MD

T. F. H. Allen, Department of Botany, University of Wisconsin, Madison, WI

P. M. Allred, Savannah River Ecology Laboratory, Aiken, SC

T. G. Bahr, New Mexico State University, Las Cruces, NM

J. R. Barnes, Brigham Young University, Provo, UT

P. C. Baumann, Columbus, OH

E. F. Benfield, Virginia Polytechnic Institute & State University, Blacksburg, VA

A. C. Benke, Georgia Institute of Technology, Atlanta, GA

R. E. Bilby, Weyerhauser, Inc., Bellingham, WA

C. E. Boyd, Auburn University, Auburn, AL

M. M. Brinson, East Carolina University, Greenville, NC

L. A. Burns, Environmental Protection Agency, Athens, GA

D. R. Cameron, Agriculture Canada Research Station, Saskatchewan, CANADA

S. R. Carpenter, Notre Dame University, Notre Dame, IN

K. C. Cummins, Oregon State University, Corvallis, OR

D. L. DeAngelis, Oak Ridge National Laboratory, Oak Ridge, TN

R. C. Eckhart, University of Maine, Orono, ME

J. W. Elwood, Oak Ridge National Laboratory, Oak Ridge, TN

K. C. Ewel, University of Florida, Gainesville, FL

S. G. Fisher, Arizona State University, Tempe, AZ

G. L. Godshalk, University of Southern Mississippi, Hattiesburg, MS

B. Hannon, University of Illinois, Urbana, IL

R. L. Henry, III, Savannah River Ecology Laboratory, Aiken, SC

R. A. Herendeen, University of Illinois, Urbana, IL

R. A. Hough, Wayne State University, Detroit, MI

H. B. N. Hynes, University of Waterloo, Waterloo, Ontario, CANADA

R. H. Karlson, University of Delaware, Newark, DE

N. K. Kaushik, University of Guelph, Guelph, Ontario, CANADA

W. M. Kemp, University of Maryland, College Park, MD

D. R. Keeney, University of Wisconsin-Madison, Madison, WI

R. L. Knight, University of Florida, Gainesville, FL

W. P. McCafferty, Purdue University, West Lafayette, IN

C. D. McIntire, Oregon State University, Corvallis, OR

R. J. Mackay, University of Toronto, Toronto, Ontario, CANADA

J. J. Messer, Utah State University, Logan, UT

G. S. Minshall, Idaho State University, Pocatello, ID

W. J. Mitsch, University of Louisville, Louisville, KY

R. J. Naiman, Woods Hole Oceanographic Institution, Woods Hole, MA

J. D. Newbold, Oak Ridge National Laboratory, Oak Ridge, TN

B. L. Peckarsky, Cornell University, Ithaca, NY

W. H. Resh, University of California-Berkeley, Berkeley, CA

A. L. Sheldon, University of Montana, Missoula, MT

K. F. Suberkropp, Kellogg Biological Station, Hickory Corners, MI

J. E. Titus, State University of New York at Binghamton, Binghamton, NY

J. H. Thorp, III, Savannah River Ecology Laboratory, Aiken, SC

F. J. Triska, U.S. Geological Survey, Menlo Park, CA

R. L. Todd, University of Georgia, Athens, GA

J. B. Wallace, University of Georgia, Athens, GA

C. E. Warren, Oregon State University, Corvallis, OR

J. V. Ward, Colorado State University, Fort Collins, CO

K. E. Willson, Ontario Ministry of the Environment, Toronto, Ontario, CANADA

L. A. Yarbro, University of North Carolina, Chapel Hill, NC

Fontaine **Bartell**

Thomas D. Fontaine, III is a systems ecologist at the University of Georgia's Savannah River Ecology Laboratory. He received both an MS and PhD in Environmental Engineering Sciences from the University of Florida, where he developed and implemented numerical models for assessing the effects of aquatic plant management alternatives on energy flow and nutrient cycling in lake ecosystems. Dr. Fontaine is currently involved in developing models for predicting heavy metal speciation and toxicity in lentic and lotic systems as affected by a variety of physical, chemical and biological conditions. In addition, he is exploring the use of operations research techniques for attaining optimal solutions to environmental–economic problems.

Steven M. Bartell is an aquatic systems ecologist in the Environmental Sciences Division at Oak Ridge National Laboratory, Oak Ridge, Tennessee. He received his PhD in Oceanography and Limnology from the University of Wisconsin–Madison. Dr. Bartell's research interests have focused on relationships between selective planktivory and phosphorus cycling in pelagic systems, phytoplankton nutrient dynamics, and multivariate analysis of phytoplankton community structure. He is currently active in development of models that predict fates and effects of aromatic hydrocarbons in lotic and lentic ecosystems, and development of probabilistic methods for evaluation of ecological risk posed by toxicants in aquatic systems.

CONTENTS

Section 1
System Concepts

Section 2
Synthesis Studies

Section 3
Dynamics and Control of System Components

SECTION 1

SYSTEM CONCEPTS

There are fundamental conceptual and methodological problems in ecology, and stream ecology is no exception. Despite decades of addressing "system level" problems, ecologists have yet to rigorously define the criteria that are necessary and sufficient to delineate "ecosystems." Difficulties in boundary identification, and variations in spatial and temporal scale continue to plague the study and comparison of different ecosystems. Do ecosystems really exist outside the context of their investigation? If ecosystems are purely heuristic devices, like "populations" and "communities," ecologists may have been misled by (1) assuming *a priori* that ecosystems have any general properties or interesting characteristics and (2) invalid and ill-founded comparisons of ecosystems, operationally defined. While these opinions might be unduly pessimistic, much work must be focused on methods for useful system definition and valid systems comparison. The papers contained in this section address some of these problems in the discipline of stream ecology.

Elwood et al. present a mathematical treatment of the nutrient spiraling concept that can be used to describe the relative importance of physical, chemical and biological factors in shaping observed temporal and spatial patterns of material transport, retention and utilization. They propose the operational term "spiraling length" as the longitudinal distance required for an individual molecule to complete one inorganic–organic cycle and suggest that field measurements thereof offer a stream length-independent method for legitimately comparing different streams. Elwood et al. hypothesize that spiraling lengths are shorter and cycling is more efficient in natural streams than in perturbed streams.

Ward and Stanford complement the ideas of Elwood et al. and address important questions regarding how impoundments act as perturbations to natural streams. They offer a conceptual framework, the serial discontinuity hypothesis, that may be a useful addition to the nutrient spiraling concept and river continuum hypothesis that emerged from the International Biological Program. Of central importance to their hypothesis is the concept of "discontinuity distance," the degree to which a particular community type or functional attribute may be displaced upstream or downstream by a perturbation.

Similar to Elwood et al. in their attempt to define global measures for comparing stream systems, Hughes and Omernik propose an alternative to the commonly used stream order method for classifying streams. The authors advocate the use of more fundamental hydrologic characteristics, especially mean annual discharge per unit watershed area.

Mathematical models have become increasingly useful for exploration of interrelationships among biotic and abiotic components of lotic systems. McIntire, and Wlosinski and Minshall address the difficult subject of conceptualization and aggregation in stream simulation models. McIntire emphasizes the importance of process aggregation in stream models and suggests that such models can realistically simulate system level behavior without specifically modeling population processes. He argues for the development of field techniques that can be used to quantify process level fluxes. Wlosinski and Minshall present four versions of a stream model, each version differing in its level of aggregation. Interestingly, most accurate simulations were obtained when components were aggregated in a way similar to the process level aggregation suggested by McIntire.

The work of civil and environmental engineers concerning hydrodynamics, transport of materials, and processing of wastes in streams and rivers has received limited attention from basic stream ecologists. Joeres and Tetrick's description of both hydrological and biological processes identifies a modeling approach that might prove useful to more biologically inclined stream ecologists. Their work suggests an interface for cooperative work among biologists, chemists and engineers to increase understanding of basic lotic system structure and function, as well as develop necessary tools for intelligent management of lotic resources.

1. RESOURCE SPIRALING: AN OPERATIONAL PARADIGM FOR ANALYZING LOTIC ECOSYSTEMS

J. W. Elwood, J. D. Newbold, R. V. O'Neill, and W. Van Winkle
Environmental Sciences Division
Oak Ridge National Laboratory
Oak Ridge, Tennessee

ABSTRACT

Spiraling is defined as the spatially dependent cycling of nutrients and the processing (i.e., oxidation and conditioning) of organic matter in lotic ecosystems. This conceptual view of resource use provides a framework for describing and measuring the temporal and spatial dynamics of nutrients and organic carbon in streams and rivers. In addition, structural and functional aspects of stream populations and communities which enhance the retention and use (i.e., spiraling) of nutrients and carbon can be interpreted in terms of ecosystem productivity and stability. Nutrient spiraling length and organic carbon turnover length (defined, respectively, as the average distance required for one complete cycle of a nutrient and the average distance between the entry of organic carbon into the system and its oxidation) are indices of the use of these resources supplied from the surrounding watershed relative to their rate of downstream transport. Unlike processing efficiency, however, both spiraling and turnover lengths are independent of the length of the stream over which they are measured, and thus they can be legitimately compared among streams and stream reaches of different sizes. We suggest that nutrient spiraling in most undisturbed streams is predominantly a biotic process involving the sorption, retention, and turnover of nutrients by microbes associated with inorganic sediments and detritus on the stream

3

bottom. If the spiraling length is shortened, the productivity of lotic eco-systems can increase. The nutrient supply can be stabilized by storing nutrients, thereby damping variation resulting from temporal and spatial variations in lateral inputs. The concept provides a useful operational and analytical paradigm for dealing with the linkages between metabolic processes involved in resource retention, use, and turnover and between transport processes, both of which affect resource supply in stream ecosystems.

> All land represents a downhill flow of nutrients from the hills to the sea. This flow has a rolling motion. Plants and animals suck nutrients out of the soil and air and pump them upward through the food chains; the gravity of death spills them back into the soil and air. Mineral nutrients, between their successive trips through this circuit, tend to be washed downhill. Without the impounding action of soils and lakes, plants and animals would have to follow their salts to the coast line.
>
> *A. Leopold, 1941*

THE SPIRALING CONCEPT

The idea that nutrients move downhill as they cycle through ecosystems is neither new nor restricted to streams. Leopold (1941) elaborated this view in an elegant essay, stressing the importance of biota in slowing the downhill flow of nutrients from mountains to sea. Ecologists since that time have been well aware of the phenomenon, and a paradigm of nutrient "cycling" has developed in the scientific literature. This paradigm includes the notion that all ecosystems are open or "leaky." The impor-tance of nutrient retention in ecosystems is thus a widely accepted aspect of the concept of nutrient cycling. It is interesting to note, however, that Leopold (1941) emphasized downward movement within ecosystems rather than the loss of nutrients from them. It is this distinction that the spiraling concept addresses.

The term spiraling was introduced into the "ecological lexicon" by Webster (1975) to describe the combined process of cycling and downhill transport in ecosystems (i.e., the "rolling motion" referred to by Leopold). Associated with each passage through a nutrient cycle is a finite downhill displacement that stretches the cycle into a continuous spiral. The spiraling concept allows us to visualize a nutrient cycle as simultaneously closed and open. It is closed in the sense that a nutrient atom may pass through the same trophic level or chemical state (or ecosystem compart-ment) many times during its residence in a stream or stream system, and it is open in the sense that the completion of a cycle does not occur in place but involves some downhill displacement before the cycle is closed. The concept of cycling in closed and open systems is compared with the concept of spiraling in an open, spatially distributed system in Figure 1. In the closed ecosystem (Figure 1a), the nutrient cycle continues indefinitely. The rate at which nutrients are used in such systems depends entirely on the rate at which they are turned over or regenerated. Because the amount of nutrient in the closed system is fixed, the primary phenomena of interest are the pathways of cycling and the rate of cycling of nutrients in such systems. This closed-system approach could be used either for studying laboratory microcosms and global nutrient cycling where the nutrient supply is fixed or for studying systems in which cycling is very rapid relative to inputs and losses.

Because most ecosystems are not closed, however, it is necessary to consider nutrient transport into, through, and out of the systems [i.e., to treat them as open systems (Figure 1b)]. In an open system a nutrient atom has a characteristic residence time, cycling a finite number of times before being exported across some arbitrarily defined boundary of the ecosystem. The standing stock of nutrients in the open system is not fixed,

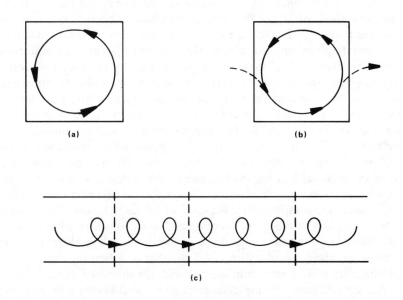

(a)   (b)

(c)

Figure 1. Schematic representations of (a) cycling in closed systems (e.g., microcosms), (b) cycling in open systems (e.g., lakes), and (c) spiraling in open systems (e.g., streams). Solid lines around system a are functionally defined boundaries, whereas those in b and c are arbitrarily defined operational boundaries. The dashed vertical lines in c represent arbitrary operational boundaries of a stream reach.

but depends on the retentiveness of the ecosystem. In this case the rate at which nutrients are used depends not only on the rate of recycling but also on the number of cycles an atom completes before leaving the system or being sequestered in an unavailable form (e.g., precipitated or adsorbed and sedimented).

Spiraling in open systems (Figure 1c) differs from the first two cycling models because it has a spatial dimension, representing the direction of dominant downhill transport. As a nutrient atom completes a cycle, it undergoes a finite downhill displacement. The basic distinction between spiraling (Figure 1c) and the open-cycle model (Figure 1b) is that the "openness" of the spiraling system represents downhill movement within the system rather than import or export across ecosystem boundaries, as in the open-cycle model. The upstream and downstream boundaries of the spiraling model, in turn, can be regarded as incidental and may be arbitrarily defined. The broken lines in Figure 1c show two such arbitrarily bounded, adjacent ecosystems. In one a nutrient atom cycles twice, whereas in the other ecosystem, which is longer, the nutrient atom cycles three times. If these two arbitrarily defined ecosystems were viewed as nonspatial systems, of the type represented in Figure 1b, we might conclude that the longer system was more retentive or less leaky than the shorter system. Although this conclusion would be correct, it would also miss the point, which is that the two systems are identical except in length. The spiraling concept emphasizes that transport occurs in conjunction with the cycle rather than as an alternative to it. This conceptual approach to viewing nutrient cycling in open systems also deemphasizes the view of an ecosystem as a spatially bounded unit (e.g., a lake, a watershed, or a reach of stream). In the spiraling context ecosystem boundaries become incidental; cycling continues indefinitely as the nutrient moves along a spatial axis. Thus applying the spiraling concept to lotic ecosystems, where unidirectional downhill transport dominates both the physical and biotic character of the system, focuses not on the number of cycles completed within the system, but on the spatial distance over which one complete cycle occurs (i.e., on the tightness of the spiral). The shorter this distance, the more times a nutrient is used within a reach of stream. An important implication of the short cycling distance is that, for two identical streams with equal supplies of a limiting nutrient, the productivity will be greater in the one with the shortest cycling distance.

As the foregoing example suggests, the spiraling concept does not represent a radical departure from traditional approaches to ecosystem analyses; instead, it offers a different perspective. Although the concept is applicable to all ecosystems, our purpose is to suggest the potential utility of and need for a spiraling approach in analyzing lotic ecosystems.

UTILITY OF THE SPIRALING CONCEPT

The spiraling concept has three basic facets that make it useful in the analysis of stream ecosystem. First, and simplest, it is a useful approach to measuring, reporting, and conceptualizing nutrient and carbon dynamics (e.g., use, cycling, and transport) in lotic ecosystems where continuous, unidirectional transport of materials is a major defining feature. The spiraling concept is an intuitive analytic framework for quantifying the relationships between the hydrodynamic transport of nutrients and organic matter and the ecological processing (e.g., uptake, remineralization, oxidation, etc.) of these materials. To this end, we present spiraling length as an index of nutrient spiraling and turnover length as an index of organic carbon spiraling. Second, the spiraling concept focuses on the interactions of hydrologic transport phenomena with ecological processes that control rates of nutrient cycling and organic carbon oxidation. Identifying these interactions may be of critical importance to our understanding of stream ecosystem structure and function. The ultimate utility of the spiraling concept, however, depends on the importance of such interactions. Finally, the spiraling concept emphasizes the closely linked nature of spatial and temporal dynamics of a stream ecosystem. Responses to perturbations and to environmental variations occur locally, but they are also propagated downstream. To fully understand such dynamics, and, for example, to address questions of ecosystem stability, we must construct models that represent spatial and temporal dynamics simultaneously. These three facets of spiraling are considered in the following sections.

The role of nutrient cycling in streams has generally not been regarded as important. This can be traced to the assumption that, except in rare cases, streams are not nutrient-limited systems (e.g., Hynes, 1969). The observation of Ruttner (1926) that running water continually bathes organisms in nutrients and oxygen and removes waste products led many stream ecologists to assume that nutrient supplies in streams were not limiting. Thus attention was focused on adaptations to the stream's physical environment rather than on addressing questions of nutrient dynamics in what appeared to be a nutritionally rich environment. Also, biotic processes on the stream bottom and transport processes in streams generally are considered separately, with the result that issues of nutrient cycling and nutrient supply down the longitudinal gradient of streams have remained obscure.

There is little doubt, however, that spiraling of nutrients and carbon in streams and rivers does occur. Spiraling of nutrients in streams has been demonstrated by radiotracer studies (e.g., Ball and Hooper, 1963; Elwood and Nelson, 1972; Newbold et al., 1981), and the effects of this spiraling have been widely observed [e.g., algal blooms below discharges of organic wastes (Hynes, 1960)].

On the other hand, a substantial amount of progress has been made in understanding streams, particularly their trophic structure, without considering the spatially dependent aspects of resource supply. Although the focus of work in streams has been on trophic pathways (e.g., Minshall, 1967; King and Ball, 1967; Cummins et al., 1973), most work has been done without reference to longitudinal transport of nutrient and energy sources. In most cases, it was assumed either that transport phenomena were unimportant quantitatively or that effects of transport were irrelevant because of longitudinal homogeneity. The few reported investigations relating transport-dependent aspects of trophic pathways to rates of longitudinal transport in streams have concerned removal of seston by filter feeders (e.g., Ladle et al., 1972; Wallace et al., 1977; McCullough et al., 1979). Similarly, studies on the processing and fate of organic materials (*inter alios* Kaushik and Hynes, 1971; Petersen and Cummins, 1974; Iversen, 1973; 1975; Hynes et al., 1974; Sedell et al., 1975; Anderson and Grafius, 1975; Triska and Sedell, 1976) have dealt only with transport-independent aspects. The idea that downstream areas in the river continuum are dependent on and structured to capitalize on reduced carbon substrates from upstream (Vannote et al., 1980) emphasizes, however, the importance of carbon spiraling in lotic systems.

Although both transport and in-stream processing of organic matter have been examined in organic carbon budgets of streams (Fisher and Likens, 1973), the work on nutrient dynamics in streams and whole watersheds has emphasized transport to the exclusion of in-stream retention and recycling (e.g., Hobbie and Likens, 1973; Dillon and Kirchner 1975). This exclusion of lateral exchange processes and longitudinal recycling appears to have been based on the assumption that biogeochemical sinks are transient and unimportant to nutrient dynamics. It is generally true for conservative nutrients, such as phosphorus, that long-term downstream losses will equal watershed inputs upstream, but the stream ecosystems can alter the timing, form, and magnitude of nutrient delivery downstream (Kaushik et al., 1975; Peters, 1978; Hill, 1979; Rigler, 1979; Meyer and Likens, 1979). The implications of these alterations raise intriguing questions concerning their effects on the dynamics of stream ecosystems. In addition, for nitrogen or other nonconservative nutrients, failure to account for some in-stream processes, such as denitrification, volatilization (of ammonia), and nitrogen fixation, can result in significant errors in mass balances for a watershed and/or reaches of a stream.

As limnologists have long noted, the continuous flow of water and the transport of materials in streams would appear to make streams physiologically rich environments. Nutrients and carbon are transported downstream in large quantities, and turbulent mixing, by reducing diffusion gradients, brings the nutrients into contact with substrates on the stream bottom (i.e., algae and bacteria), making nutrients available for sorption and use; yet, paradoxically, streams are also depauperate. Although total quantities of transport in streams draining undisturbed watersheds are high, concentrations of dissolved nutrients, such as nitrogen and phosphorus, which are generally found to limit aquatic productivity, tend to be low (Omernik, 1977), with concentrations of soluble phosphorus and nitrogen comparable to those in nutrient-limited oligotrophic lakes. Thus nutrient supply may play a major role in limiting the productivity of lotic ecosystems.

Both algae and heterotrophic microbes that decompose nutrient-poor detritus are generally assumed to satisfy their own requirements for essential elements by absorbing and assimilating dissolved mineral nutrients. Thus microbial decomposition and primary production rates should be related to the supply of dissolved nutrients. Egglishaw (1968; 1972) found that decomposition rates within and among Scottish streams were correlated with nutrient levels, and Stockner and Shortreed (1978) reported that increases in nitrogen and phosphorus levels in stream water increased algal biomass in a rain forest stream. Our results from Walker Branch (Elwood et al., 1981b; Newbold et al., in press), a woodland stream in eastern Tennessee, demonstrated that decomposition rates and algal biomass are phosphorus limited but not nitrogen limited. If streams are nutrient limited, then their productivity is dependent on lateral inputs of nutrients from the surrounding watershed (e.g., groundwater) and on nutrients made available from upstream through recycling processes in the stream ecosystem. Because biomass and productivity are related to nutrient supply, spiraling of nutrients may be a central issue in understanding the productivity of these systems.

If productivity is stoichiometrically related to the use of nutrients, then productivity (P), of organisms on the stream bottom (algae, bacteria, and fungi), which depends on the uptake of soluble nutrients, is a function of the nutrient concentration in water (C) and of the standing stock of available (i.e., recyclable) nutrient on the stream bottom (N):

$$P = f(C, N) \tag{1}$$

In the absence of lateral inputs of available nutrients throughout the length of a stream, the nutrient concentration in stream water and the standing stock of nutrient on the stream bottom at a given location will be controlled by spiraling of nutrients that have entered the system from the surrounding watershed upstream. Under this condition an index of spiraling which incorporates both the rate of nutrient utilization and its regeneration back to the original available form will be related to productivity.

AN INDEX OF NUTRIENT SPIRALING

In this section we discuss the two phenomena underlying the spiraling concept, present two alternative derivations and interpretations of an index of nutrient spiraling, and generalize this index for systems consisting of multiple biotic and abiotic compartments exchanging nutrients with water.

Spiraling of nutrients in streams combines two phenomena: (1) soluble and particulate materials are retained (i.e., sorbed and ingested) by biota on the stream bottom so that their downstream transport is slowed, and (2) materials are reused (i.e., recycled) as they move downstream. Consider, for example, a stream with no lateral inputs or losses of nutrients and in a steady state. Nutrients are distributed among ecosystem components (e.g., water, detritus, and aufwuchs), each subject to a different rate of downstream displacement. Conservation of matter in a steady-state system implies that total downstream flux, F_T (g/s), of the nutrient is constant with respect to both time and distance and that the total amount of nutrient passing downstream is equal to upstream inputs.

The first derivation and interpretation of an index of nutrient spiraling is based on the average downstream velocity and cycling rate of nutrient in all compartments of a stream ecosystem. In a stream of unit width, the total downstream flux of nutrient, F_T (g/s), is related to the standing stock of available nutrient per unit length of stream, N (g/m), and the average downstream velocity of the nutrient, v (m/s):

$$F_T = Nv \tag{2}$$

Equation 2 provides a measure of the retention aspect of spiraling. As nutrient uptake from water occurs and the standing stock of nutrient (and biomass) on the stream bottom increases, the average downstream velocity of the nutrient decreases. This decrease is a result of the loss of nutrient from a compartment with a high downstream velocity (i.e., water) to compartments with a comparatively low downstream velocity (i.e., algae and bacteria associated with sediments on the stream bottom).

Average downstream velocity can be viewed as an average for all compartments (including the water) weighted by the standing stock of nutrient in each.

The second aspect of spiraling can be expressed in terms of the rate at which nutrients in the stream are cycled. If a proportion, k (s^{-1}), of the standing stock is cycled per unit time, then the use of nutrients, U (g m^{-1} s^{-1}), or the amount of nutrient beginning (or completing) a cycle per unit time is:

$$U = kN \qquad (3)$$

A nutrient atom, on the average, completes a cycle in $T = 1/k$ sec. Thus Equation 3 describes the cycling aspect of spiraling in terms of nutrient utlization and stock of available (recyclable) nutrients in the stream.

Combining Equations 2 and 3 yields a ratio that combines both the retention and cycling aspects of spiraling and is in units of length:

$$v/k = vT = F_T/U = S \qquad (4)$$

$$(m/s)/s^{-1} = (m/s) \ (s) = (g/s)/(g \ m^{-1} \ s^{-1}) = (m)$$

The resulting parameter, S, represents the average downstream distance (velocity × time) traveled by a nutrient atom as it completes a cycle. We refer to this distance as the spiraling length. Note that rapid cycling (high k) combined with a high resistance to downstream transport (low v) will yield a short spiraling length. The number of times a nutrient is expected to cycle in a reach of stream of length X is simply $n = X/S$. Thus spiraling can be represented by S, without the need to reference an experimental reach of some arbitrary length.

The second derivation and interpretation of an index of spiraling involves downstream fluxes and exchange fluxes of nutrients. The spiraling index can be developed more explicitly by considering a two-compartment system consisting of microbes growing on substrates (denoted the particulate compartment P) and water (W) flowing downstream over these substrates (Figure 2). Spiraling length can be partitioned into expected travel distances in each of these compartments. These distances can be calculated as ratios of downstream fluxes in the water (F_W) or particulate (F_P) compartment to the exchange fluxes from water to particulates (R_W) or particulates to water (R_P). Thus the expected distance downstream that a nutrient atom will travel in the water compartment before being taken up by the particulate compartment can be calculated in an analogous manner as $S_W = F_W/R_W$, whereas the

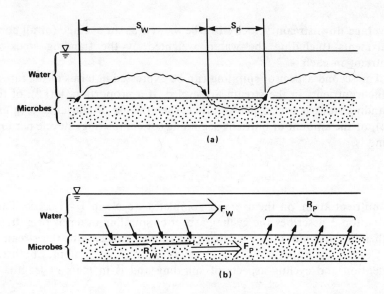

Figure 2. Nutrient spiraling in a stream ecosystem. (a) The components of spiraling length (S) of a nutrient. The nutrient atom travels an average longitudinal distance, S_W, in the water compartment (W) before being taken up by particulates, plus an average distance, S_P, in the particulate compartment (P) before returning to the water. (b) Downstream nutrient fluxes (g/s) in water (F_W) and in the particulate compartment (F_P) and exchange fluxes (g s^{-1} m^{-1}) of nutrients from water to particulates (R_W) and from particulates back to water (R_P).

expected travel distance in the particulate compartment can be calculated in an analogous manner as $S_P = F_P/R_P$. The first distance, S_W, is designated the uptake length, and S_P is designated the turnover length. If uptake and turnover of nutrient are the result of biotic mechanisms, then S_W can be viewed as an index of the effectiveness of stream biota in scavenging nutrients from the water column, whereas S_P is an index of the ability of the biota to avoid downstream transport and to recycle nutrients to the water.

Assuming the stream is in steady state, with no lateral inputs or losses, implies that $R_W = R_P$, and thus $S_P = F_P/R_W$. The sum of these two lengths:

$$S = S_W + S_P = (F_W/R_W) + (F_P/R_P) =$$

$$(F_W + F_P)/R_W = F_T/R_W \qquad (5)$$

is the spiraling length. The shorter the spiraling length, the more times the nutrient can be reused within a given reach of stream. Note that under these assumptions, the spiraling length is a ratio of the total nutrient supply being transported downstream to the rate of nutrient utilization or cycling by biota.

Although the last interpretation of spiraling length is based on the steady-state assumption, the fluxes used to calculate this index can be measured in non-steady-state systems. Thus the ability to estimate spiraling length is not dependent on satisfying the steady-state assumption.

For systems consisting of multiple biotic and abiotic compartments exchanging nutrients with water, the spiraling length concept is easily generalized in a manner that illustrates the connections between the two alternative derivations just discussed. The downstream flux (F_i) of a nutrient in compartment i can be represented as

$$F_i = N_i v_i \tag{6}$$

where N_i is the standing stock of nutrients in compartment i per unit length of stream bottom (g/m), and v_i is the average downstream velocity of compartment i. Assuming first-order kinetics, the turnover rate of nutrients in compartment i is k_i (s^{-1}); thus the mean residence time of a nutrient atom in compartment i at steady state is $T_i = 1/k_i$. The release flux of nutrient from compartment i (R_i), which is the multiple compartment analog of the utilization rate (U) in Equation 3, is

$$R_i = k_i N_i = N_i/T_i \tag{7}$$

During its residence time (T_i), the atom travels downstream a distance (as in Equation 4):

$$S_i = V_i/k_i = v_i T_i \tag{8}$$

Substituting for v_i from Equation 6 and for T_i from Equation 7,

$$S_i = (F_i/N_i)(N_i/R_i) = F_i/R_i \tag{9}$$

The total spiraling length (S) of the system then is the sum of the S_i's for the various compartments, weighted by the proportion of the total nutrient flux passing through each compartment:

$$S = \sum_i (b_i S_i) = \sum_i (b_i F_i/R_i) \tag{10}$$

where b_i is the weighting factor (i.e., the probability that an atom beginning a cycle will enter compartment i before completing a cycle). For a system in which all recycling involves passage through the water compartment (as shown in Figure 2 for a two-compartment system), b_i for water is unity and for the other compartments is R_i/R_W.

AN INDEX OF CARBON SPIRALING

Carbon spiraling is somewhat more difficult to conceptualize than spiraling of a conservative nutrient. When carbon is released from the biota as CO_2, it does not necessarily continue to spiral downstream; instead it may be lost to the atmosphere. Moreover, the rate of supply of inorganic carbon adjusts itself (via exchange with the atmosphere) in response to utilization rate, unlike the supply of a conservative nutrient such as phosphorus, which is fixed entirely by rates of delivery from the watershed. For these reasons, the spiraling concept is not well suited for analysis of the inorganic phase of the carbon cycle.

The processing and transport of organic carbon, on the other hand, is quite amenable to the spiraling approach. Organic carbon may either be fixed within the stream or enter the stream from terrestrial sources. From the time it enters the stream or is fixed in the stream as organic carbon, a carbon atom in organic form travels some distance downstream, perhaps passing through a number of ecosystem compartments, before being oxidized. We refer to this distance as the carbon turnover length S_c and define it as the average or expected downstream distance traveled by a carbon atom during its residence in the stream in a fixed or reduced form (Figure 3). The carbon turnover length, therefore, refers to only a

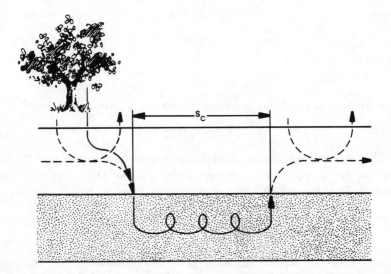

Figure 3. Carbon spiraling in a stream ecosystem. Dashed lines represent flows of CO_2, and solid lines represent flows of reduced carbon (detritus and algae). The turnover length for organic carbon in a stream, S_c, is defined as the expected travel distance between its entry into the system as allochthonous detritus or through fixation (of CO_2) in the stream and its exit from the system by oxidation to CO_2 or emergence.

discontinuous portion of the complete spiral of a carbon atom, which would also involve downstream passage in an inorganic form. Our definition for carbon turnover length is equivalent to our previous definition for the turnover length (S_p) of a conservative nutrient. We can, therefore, express S_c in terms of rates and fluxes of organic carbon, using the relationships derived earlier.

For simplicity, in the subsequent discussion we consider organic matter as consisting of only one compartment. The implications of lumping are the same as those discussed earlier for nutrient spiraling. Specifically, if we restrict ourselves to steady-state (or long-term average) measurements and uniform conditions longitudinally, we can validly use weighted estimates for rates and fluxes to determine the turnover length. Using the assumption of a single compartment, we can write:

$$S_c = v_c/k_c = v_c T_c = F_c/R_c \qquad (11)$$

in which the relationships are taken from Equations 8 and 9. In this case v_c (m/s) is the mean downstream velocity of transport for all organic carbon in the stream, and k_c (s^{-1}) is the rate of carbon respiration. If we let B represent the standing stock of organic carbon in the stream (g/m^2), then

$$F_c = v_c B \qquad (12)$$

and the flux, R_c (g m^{-1} s^{-1}), of carbon lost to respiration is:

$$R_c = k_c B \qquad (13)$$

Note that, if we are considering a lumped compartment, the nonrespiratory losses of carbon from individual subcompartments are balanced by gains to other compartments and thus have no net effect on k_c or R_c.

In a manner like spiraling length, the carbon turnover length is a basic measure of the rate at which the ecosystem uses carbon relative to the rate at which it is transported downstream. If S_c is short, organic carbon inputs are respired near the location of entry or fixation. If it is long, organic carbon is lost from the local stream reach to be oxidized or deposited somewhere downstream, perhaps in a lake, reservoir, or estuary, rather than in the lotic ecosystem itself.

As an index, the carbon turnover length may provide a new and useful measure for examining and comparing organic matter budgets for streams. Fisher and Likens (1973) defined "ecosystem efficiency" for a stream as the ratio of carbon respired within the stream to the total input of organic carbon. Pointing out some difficulties with the use of efficiency as an index of carbon processing when comparing streams of different sizes, Fisher (1977) proposed a system metabolism index (SMI). The SMI for a stream is the ratio of respiration in the system to the inputs occurring along the stream in excess of that which would maintain a longitudinally constant concentration of organic matter in transport. For example, in a reach with no tributary or groundwater inputs, the SMI would have a value of 1.00 if respiration and lateral inputs were equal. When the SMI is less than 1, respiration is insufficient to account for the accrual of inputs along the stream length, and, at steady state, concentrations of organic matter increase downstream. Unlike efficiency, SMI can be calculated independent of reach length and can be interpreted easily for a stream of any size, but it does have the major drawback that it essentially disregards the role of hydrologic transport.

We now show that the carbon turnover length provides a link between ecosystem efficiency and SMI, and allows a consistent interpretation for all three indexes.

For simplicity, we neglect the role of tributary and groundwater inputs to a stream and assume a stream of unit width. The efficiency (E) at steady state for a reach of stream x meters long is:

$$E = \frac{\text{Respiration}}{\text{Inputs}} = \frac{R_c x}{F_c + 1} \tag{14}$$

where I (g m^{-2} s^{-1}) is the rate at which organic matter enters along the length of the stream or is fixed by primary production within the stream, F_c is the downstream flux of organic carbon measured at the upstream end of the study reach, and R_c is the flux of carbon lost to respiration in the stream, as defined previously. Since we are neglecting downstream increases in flow, the system metabolism index is simply

$$SMI = R_c / 1 \tag{15}$$

Substituting for I and dividing numerator and denominator by R_c, we obtain:

$$E = \frac{x}{(F_c / R_c) + (x / SMI)} = \frac{x}{S_c + (x / SMI)} \tag{16}$$

since the carbon turnover length (S_c) equals F_c / R_c.

According to this result, the efficiency of a reach depends only on the turnover length, the length of the reach, and SMI. The result is approximate, however, because it is based on the assumption that streamflow as well as I, R, S_c, and SMI are constant throughout the reach.

IMPLICATIONS AND MECHANISMS OF SPIRALING

Spiraling of resources (nutrients and reduced carbon substrates) in streams can be viewed from two perspectives: (1) its impact on both the temporal and spatial quantity and quality of nutrient and energy supplies in stream ecosystems, and (2) the interactions of biotic processes that affect the recycling of nutrients and the oxidation of organic matter with transport phenomena. The first perspective focuses on the temporal and spatial heterogeneity of nutrient and energy inputs to and losses from lotic systems. Nutrient and energy resources in streams are quite variable and often unstable and unpredictable, with both pulsed or episodic inputs and losses resulting from seasonal and hydrologic changes. Spiraling of these resources, some of which may be limiting, thus represents a means by which lotic ecosystems can increase and stabilize their nutrient and energy supply, thereby increasing their average productivity. Soluble and particulate nutrients and energy (in the form of reduced carbon substrates) released at one point in a stream become available as resources at points farther down the longitudinal gradient.

The second perspective focuses on structural and functional mechanisms of stream organisms that enhance the spiraling of resources. Mechanisms that shorten the spiraling length of nutrients and the turnover length of organic carbon in streams can be considered adaptive because they increase the use of nutrient and energy supplies in the stream community.

Most autotrophic and heterotrophic production in streams occurs on the stream bottom. Thus a stream would have to shorten its spiraling length to increase nutrient utilization, and thereby increase productivity. It could do this by increasing the uptake rate of available nutrients in water (R_W), which would reduce the uptake length ($S_W = F_W/R_W$). A second mechanism would be to reduce the downstream velocity of nutrient in the particulate phase or increase the release rate of nutrient from particulates (R_P). In Walker Branch over 90% of the standing stock of phosphorus in the system is associated with detritus (fine and coarse particulate matter) on the stream bottom. Increasing the cycling rate of phosphorus associated with this detritus reduces the spiraling length of phosphorus and increases the productivity of this phosphorus-limited stream (Elwood et al., 1981b).

Spiraling of nutrients tied up in detritus could occur by shredding and physical fragmentation of detritus; this increases the leaching of nutrients (e.g., Cowen and Lee, 1973). There is evidence that macroinvertebrate shredders increase the loss of nutrients from decomposing detritus

(Woodall, 1972) and make them more available to organisms downstream (Short and Maslin, 1977). Fisher and Likens (1973) estimated that macroinvertebrate detritivores in Bear Brook used only 1 to 2% of the available energy, but Webster and Patten (1979) calculated that detritivores in a small forest stream in North Carolina annually ingested 80% of the leaf litter input. Thus biological fragmentation of coarse particulate organic matter and leaching of nutrients associated with the detritus will enhance the cycling of nutrients associated with these substrates.

Mineralizing nutrients tied up in microbial biomass (bacteria, algae, and fungi) associated with substrates on the stream bottom is another mechanism for increasing the available nutrient supply and reducing the spiraling length. Macfadyen (1961) hypothesized that, since microbial populations represent bottlenecks to nutrient and energy flow, grazing on microbes will increase the availability of these resources. Johannes (1965) suggested that grazing of detritus by protozoans in marine ecosystems was important in the generation of phosphorus. He concluded that the effect of these consumers on detritus mineralization was direct, resulting from the mineralization of nutrients tied up in microbial biomass. Several investigators (Barsdate et al., 1974; Fenchel and Harrison, 1975; Fenchel, 1977; Fenchel and Jorgensen, 1977; Lopez et al., 1977) have shown that the decomposition of detritus is enhanced in the presence of protozoan grazers and is often phosphorus limited. In soil systems with amoebal grazers, significantly more carbon, nitrogen, and phosphorus were shown to be mineralized than in soil systems without grazers (Elliott et al., 1979). Fenchel and Harrison (1975) found that bacterial densities were lower but bacterial uptake and excretion of phosphorus (per unit of bacterium) were higher in grazed systems. Grazers, however, accounted for only 4 to 5% of the regeneration of phosphorus. They concluded, therefore, that the higher uptake and excretion of phosphorus by bacteria in the grazed detrital systems was an indirect effect of grazers. Three possible explanations for the higher bacterial growth rates in the presence of grazers include (1) bioturbation (i.e., turbulence caused by physical activity of consumers) which increases microturbulence at the cell surface, thereby reducing diffusion gradients; (2) increasing the available surface area for nutrient exchange by preventing (through grazing) microbial cell density on sediment particles from becoming self-limiting; and (3) production by grazers of exudates, other than phosphorus and nitrogen, which stimulate microbial growth. Whatever the exact mechanism, these results indicate that consumers play an important role in controlling nutrient dynamics and that, in the absence of consumers, nutrients would be tied up in microbial biomass, resulting in a reduction in decomposition of organic carbon.

Other adaptations that would increase the supply of nutrients and shorten the spiraling length are: (1) the production of enzymes, such as alkaline phosphatase, which release nutrients sequestered in soluble and particulate organic fractions and make them available to algae and bacteria, and (2) production of mucilaginous exudates that trap nutrients in the soluble and particulate phase and prevent the diffusion of enzymes and products of enzymatic degradation away from the cell surface (Zobell, 1943). Alkaline phosphatase production by algae, bacteria, and fungi in soil and planktonic systems is well documented (e.g., Torriani, 1960; Henley, 1973). The high levels of alkaline phosphatase measured in aufwuchs and detritus samples from Walker Branch (Sayler et al., 1979) suggest that microbes in stream ecosystems may be dependent on the enzymatic hydrolysis of organic phosphorus associated with detritus for their phosphorus supply. Costerton et al. (1974) and Geesey et al. (1977) suggested that the reason the majority of bacteria found in rivers are of the gram-negative type is that they retain their degradative enzymes in a protective association with the cell wall. Thus the products of enzymatic degradation are immediately available to the cell's transport system and are not transported away from the cell wall by the current. Other adaptations that would shorten spiraling length (S) include the development of growth forms with high surface area to volume ratios, which enhance the scavenging of nutrients from water. Filamentous algae and bacteria, which are able to withstand relatively high current velocities, can be viewed as a morphological adaptation directed at increasing the possible uptake rate of nutrients.

Another mechanism for stabilizing the nutrient availability in stream ecosystems and increasing productivity is to store nutrients in sites that are less vulnerable to downstream transport (i.e., compartments with low v). Nutrient uptake onto stream sediments occurs via two primary mechanisms—abiotic adsorption onto organic and inorganic substrates and absorption by attached living microorganisms. Results of laboratory experiments using fine particulate organic matter and inorganic substrates from Walker Branch, Tennessee, indicate that most of the uptake of PO_4-P is biotic (Elwood et al., 1981a). The uptake length of phosphorus (S_W) appears to vary seasonally, ranging from approximately 6 m in late fall after leaf drop to over 180 m in summer (Table 1). Thus uptake length is shortest in late autumn when the standing crop of detritus and the surface area for microbial colonization are at their annual maximum. If the major site of nutrient uptake is microbes associated with stream sediments, then uptake length will be a function of microbial biomass, which, in turn, is a function of the colonizable surface area (Sanders, 1967; Hargrave and Phillips, 1977). The greater retention time of phosphorus in Walker Branch in late autumn in comparison with that during summer and early fall (Table 1) suggests that the turnover of phosphorus sorbed by substrates on the stream bottom is much slower during the later fall period. Phosphorus associated with leaf detritus collected after a November release of $^{32}PO_4$ appears to exhibit an asymptotic decline in activity (Fig. 4), with approximately 70% of the ^{32}P having a turnover time of 15 d. There appears to be little or no turnover of the remaining 30% in station 1 samples, however. This implies that a substantial fraction of the phosphorus sorbed from water by detritus is sequestered by microbes in storage sites with a very slow turnover rate. Leaf detritus collected after the ^{32}P release in July, however, did not exhibit this slow turnover compartment (Newbold et al., 1981). Slow turnover rates in late fall may be a result of the increased nutrient supply from leaf detritus. If this sequestered phosphorus is available to microbes, it could provide an important stabilizing factor during a later period when the demand is high but the supply is reduced.

Table 1.   Seasonal Changes in the Uptake Length (S_W) and Expected Residence Time (T) of Phosphorus in a 120-m Reach of Walker Branch*

Date	S_W, m	Residence Time, d
June–July (6/30/69)	115	128
July–August (7/5/78)	164	16
September–October (9/14/70)	105	22
November–December (11/10/69)	6	217

*The dates in parentheses are the days when the uptake length was measured. ^{32}P loss was determined after measurement of the uptake length to estimate the residence time of recycled phosphorus in this reach of stream.

The presence of an ecosystem component that exhibits a very slow turnover rate has been identified as an important contributor to system stability (O'Neill et al., 1975; Webster et al., 1975). Such a component, usually nonliving or in inactive organic matter, can be found in a wide range of ecosystem types (Reichle et al., 1975). This slow component has turnover rates between 10 and 200 times slower than the autotrophic components of the system (O'Neill et al., 1975).

The slow component remains relatively unchanged in response to short-term perturbations in energy input to the system, thereby increasing ecosystem resistance to disturbance (Webster et al., 1975). For terrestrial ecosystems, the slow component supports 2 to 1000 times more heterotrophic biomass than living plant materials (O'Neill and Reichle, 1980). The stability of this energy base appears to more than compensate for the increased difficulty in extracting energy from this source. Although most

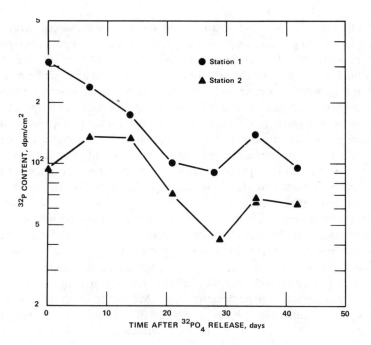

Figure 4. Mean concentration of ^{32}P associated with leaf detritus at two locations downstream as a function of time after a 30-min release of carrier-free ^{32}PO$_4$ to Walker Branch. Station 1 was located 15 m downstream from the point of release, and Station 2 was located 120 m downstream.

of this theoretical work has focused on terrestrial ecosystems, the stabilizing influence of the slow component has also been pointed out for lakes (Clesceri et al., 1977), marshes (Pomeroy, 1975), and aquatic microcosms (Giddings and Eddlemon, 1977).

Even more important than its role as a stable energy base is the contribution of the slow component in retention and recycling of nutrients. This role becomes more important as nutrient limitations increase. It can be shown for forest ecosystems (O'Neill and Reichle, 1980) that as energy and water limitations become less important, the role of the slow component increases. The lack of identification of a slow component and the assumption of no recycling make stream resiliency appear anomalous when compared with other ecosystem types (Webster et al., 1975). Therefore the identification of this large, slowly accumulating pool of phosphorus on leaf detritus in the stream and the recycling of nutrients have important implications for our understanding of the stability properties of stream ecosystems.

Until recently, our understanding of the behavior of ecosystems has been limited to primary responses of specific system components (species and populations). To a large extent this is still true of lotic systems, a fact attributable, in part, to the lack of a conceptual and analytic framework for dealing with the openess of these systems, which have both a temporal and a spatial component associated with nutrient cycling and processing of energy. The spiraling concept is an approach for dealing with the added spatial dimension of nutrient and energy processing in lotic systems, thus providing a means of comparing the dynamics of different lotic systems on both temporal and spatial bases and contrasting the dynamics of lotic systems with other aquatic and terrestrial ecosystems that may or may not be spatially distributed.

It has been argued either that streams and rivers should not be viewed as ecosystems or that the ecosystem concept should be redefined for lotic systems because of (1) the absence of "in-place" cycling of nutrients (Hynes, 1970), and (2) the lack of longitudinal interdependencies of biological communities (Rzoska, 1978). The underlying issue in both arguments appears to be whether homeostatic, self-regulatory mechanisms to optimize the use and exchange of nutrients and energy could evolve in lotic systems without positive and negative feedbacks that result from closed cycling of nutrients in biotic communities. The facts that spatially dependent nutrient cycling in streams and rivers does occur and that most of it, at least for phosphorus, involves biotic mechanisms and pathways mean that stream communities are longitudinally coupled. Furthermore, given the fact that streams can be nutrient limited, we can argue that lotic ecosystems are longitudinally interdependent, in the sense that energy processing in downstream areas depends on the retention and cycling of nutrients by communities in upstream areas. Since an increase in the use and retention of nutrients upstream translates into a negative effect on downstream communities, negative feedbacks in the cycling of nutrients do exist in lotic ecosystems. By applying the spiraling concept, we can begin to examine many complex community- and system-level interactions in streams and rivers to better understand their longitudinal interdependencies and the mechanisms in stream communities for stabilizing the supply of nutrients and energy resources.

ACKNOWLEDGMENTS

This research was supported by the National Science Foundation's Ecosystem System Studies Program under Interagency Agreement No. DEB–78–03012 with the U. S. Department of Energy under contract

W–7405–eng–26 with Union Carbide Corporation. We appreciate the constructive review of this manuscript by S. G. Hildebrand, P. J. Mulholland, and an anonymous reviewer. This paper is Publication No. 1826, Environmental Sciences Division, Oak Ridge National Laboratory.

REFERENCES

Anderson, N. H., and E. Grafius, 1975, Utilization and Processing of Allochthonous Material by Stream Trichoptera, *Verh. Internat. Verein. Limnol.*, 19: 3083-3088.

Ball, R. C., and F. F. Hooper, 1963, Translocation of Phosphorus in a Trout Stream Ecosystem, in V. Schultz and A. W. Klement, Jr. (Eds.), *Radioecology*, pp. 217-228, Reinhold Publishing Company, New York.

Barsdate, R. J., R. T. Prentki, and T. Fenchel, 1974, Phosphorus Cycle of Model Ecosystems: Significance for Decomposer Food Chains and Effect of Bacteria Grazers, *Oikos*, 25: 239-251.

Clesceri, L. S., C. W. Boylan, and R. A. Park, 1977, *Microdynamics of Detritus Formation and Decomposition and Its role as a Stabilizing Influence in Freshwater Lakes*, FWI Report 77-12, Rensselaer Polytechnic Institute, Troy, NY.

Costerton, J. W., J. M. Ingram, and K. J. Cheng, 1974, Structure and Function of the Cell Envelope of Gram-Negative Bacteria, *Bact. Rev.*, 38: 87-110.

Cowen, W. F., and G. F. Lee, 1973, Leaves as Sources of Phosphorus, *Environ. Sci. Technol.*, 7: 853-854.

Cummins, K. W., R. C. Peterson, F. O. Howard, J. C. Wuycheck, and V. I. Holt, 1973, The Utilization of Leaf Litter by Stream Detrivores, *Ecology*, 54: 336-345.

Dillon, P. J., and W. B. Kirchner, 1975, The Effects of Geology and Land Use on the Export of Phosphorus from Watersheds, *Water Res.*, 9: 135-148.

Egglishaw, H. J., 1968, The Quantitative Relationship Between Bottom Fauna and Plant Detritus in Streams of Different Calcium Concentrations, *J. Appl. Ecol.*, 5: 731-740.

————, 1972, An Experimental Study of the Breakdown of Cellulose in Fast-Flowing Streams, in V. M. Santolini and J. W. Hopton (Eds.), *Detritus and Its Role in Aquatic Ecosystems, Mem. Ist. Ital. Idrobiol.* 29, Suppl., pp. 405-429, Pallanza, Italy.

Elliott, E. T., C. V. Cole, D. C. Coleman, R. V. Anderson, H. W. Hunts, and J. F. McClellan, 1979, Amoebal Growth in Soil Microcosms: A Model System of C, N, and P Trophic Dynamics, *Int. J. Environ. Stud.*, 13: 169-174.

Elwood, J. W., and D. J. Nelson, 1972, Periphyton Production and Grazing Rates in a Stream Measured with a [32]P Material Balance Method, *Oikos,* 23: 295-303.

_____, J. D. Newbold, R. V. O'Neill, R. W. Stark, and P. T. Singley, 1981a, The Role of Microbes Associated with Organic and Inorganic Substrates in Phosphorus Spiralling in a Woodland Stream, *Verh. Internat. Verein. Limnol.,* 21: 818-824.

_____, J. D. Newbold, A. F. Trimble, and R. W. Stark, 1981b, The Limiting Role of Phosphorus in a Woodland Stream Ecosystem: Effects of P Enrichment on Leaf Decomposition and Primary Producers, *Ecology,* 62: 146-158.

Fenchel, R. M., and B. B. Jorgensen, 1977, Detritus Food Chains of Aquatic Ecosystems: The Role of Bacteria, In M. Alexander (Ed.), *Advances in Microbial Ecology,* Vol. 1, pp. 1-58, Plenum Press, New York.

_____, and P. Harrison, 1975, The Significance of Bacterial Grazing and Mineral Cycling for the Decomposition of Particulate Detritus, in J. M. Anderson and A. Macfadyen (Eds.), *The Role of Terrestrial and Aquatic Organisms in Decomposition Processes,* British Ecological Society Symposium 17, pp. 285-299, Blackwell Scientific Publications, Oxford, England.

Fenchel, T., 1977, The Significance of Bactivorous Protozoa in the Microbial Community of Detrital Particles, in J. Cairns, Jr. (Ed.), *Aquatic Microbial Communities,* pp. 529-544, Garland Publishing, Inc., New York.

Fisher, S. G., 1977, Organic Matter Processing by a Stream-Segment Ecosystem: Fort River, Massachusetts, U.S.A., *Int. Rev. Gesamten Hydrobiol.,* 62: 701-727.

_____, and G. E. Likens, 1973, Energy Flow in Bear Brook, New Hampshire: An Integrative Approach to Stream Ecosystem Metabolism, *Ecol. Monogr.,* 43: 421-439.

Geesey, G. G., W. T. Richardson, H. G. Yocmans, R. T. Irvin, and J. W. Costerton, 1977, Microscopic Examination of Natural Sessile Bacterial Populations from an Alpine Stream, *Can. J. Microbiol.,* 23: 1733-1736.

Giddings, J. M., and G. K. Eddlemon, 1977, The Effects of Microcosm Size and Substrate type on Aquatic Microcosm Behavior and Arsenic Transport, *Arch. Environ. Contam. Toxicol.,* 6: 491-505.

Hargrave, B. T., and G. A. Phillips, 1977, Oxygen Uptake of Microbial Communities on Solid Surfaces, in J. Cairns, Jr. (Ed.), *Aquatic Microbial Communities,* pp. 545-587, Garland Publishing, Inc., New York.

Henley, F. P., 1973, Inorganic Nutrient Uptake and Deficiency in Algae, *CRC Crit. Rev. Microbiol.,* 3: 69-113.

Hill, A. R., 1979, Denitrification in the Nitrogen Budget of a River Ecosystem, *Nature,* 281: 291-292.

Hobbie, J. E., and G. E. Likens, 1973, Output of Phosphorus, Dissolved Organic Carbon, and Fine Particulate Carbon from Hubbard Brook Watersheds, *Limnol. Oceanogr.,* 18: 734-742.

Hynes, H. B. N., 1960, *The Biology of Polluted Waters,* Liverpool University Press, Liverpool, England.

_____, 1969, The Enrichment of Streams, in *Eutrophication: Causes, Consequences, Correctives,* pp. 188-196, National Academy Sciences, Washington, D. C.

_____, 1970, *The Ecology of Running Waters,* Liverpool University Press, Liverpool, England.

_____, N. K. Kaushik, M. A. Lock, D. L. Lush, Z. S. J. Stocker, R. R. Wallace, and D. D. Williams, 1974, Benthos and Allochthonous Organic Matter in Streams, *J. Fish. Res. Board Can.,* 31: 545-553.

Iversen, T. M., 1973, Decomposition of Autumn-Shed Beech Leaves in a Spring Brook and Its Significance for the Fauna, *Arch. Hydrobiol.,* 72: 305-312.

_____, 1975, Disappearance of Autumn-Shed Beech Leaves Placed in Bags in Small Streams, *Verh. Internat. Verein. Limnol.,* 19: 1687-1692.

Johannes, R. E., 1965, Influence of Marine Protozoa on Nutrient Regeneration, *Limnol. Oceanogr.,* 10: 443-450.

Kaushik, N. K., and H. B. N. Hynes, 1971, The Fate of Dead Leaves That Fall into Streams, *Arch. Hydrobiol.,* 68: 465-515.

_____, J. B. Robinson, P. Sain, H. R. Whiteley, and W. Stammers, 1975, A Quentitative Study of Nitrogen Loss from Water of a Small, Spring-Fed Stream, in *Water Pollution Research Canada,* Proceedings of the 10th Canadian Symposium, pp. 100-117, Institute for Environmental Studies, University of Toronto, Toronto.

King, D. C., and R. C. Ball, 1967, Comparative Energetics of a Polluted Stream, *Limnol. Oceanogr.,* 12: 27-33.

Ladle, M., J. A. B. Bass, and W. R. Jenkins, 1972, Studies on Production and Food Consumption by the Larval Simuliidae (Diptera) of a Chalk Stream, *Hydrobiologia,* 39: 429-448.

Leopold, A., 1941, Lakes in Relation to Terrestrial Life Patterns, in *A Symposium on Hydrobiology,* pp. 17-22, University of Wisconsin Press, Madison.

Lopez, G. R., J. S. Levinton, and L. B. Slobodkin, 1977, The Effect of Grazing by the Detritivore (*Orchestia grillus*) on *Spartina* Litter and Its Associated Microbial Community, *Oecologia,* 30: 111-127.

McCullough, D. A., G. W. Minshall, and C. E. Cushing, 1979, Bioenergetics of Lotic Filter-Feeding Insects *Simulium* spp. (Diptera) and *Hydropsyche occidentalis* (Trichoptera) and Their Function in Controlling Organic Transport in Streams, *Ecology,* 60: 585-596.

Macfadyen, A., 1961, Metabolism of Soil Invertebrates in Relation to Soil Fertility, *Ann. Appl. Biol.,* 49: 215-218.

Meyer, J. L., and G. E. Likens, 1979, Transport and Transformation of Phosphorus in a Forest Stream Ecosystem, *Ecology,* 60: 1255-1269.

Minshall, G. W., 1967, Role of Allochthonous Detritus in the Trophic Structure of a Woodland Spring Brook Community, *Ecology,* 48: 139-149.

_____, J. W. Elwood, M. S. Schulze, R. W. Stark, and J. C. Barmier, Continuous Ammonium Enrichment of a Woodland Stream: Uptake Kinetics, Leaf Decomposition, and Nitrification, *Freshw. Biol.,* in press.

Newbold, J. D., J. W. Elwood, R. V. O'Neill, and W. Van Winkle, 1981, Measuring Nutrient Spiralling in Streams, *Can. J. Fish. Aquat. Sci.,* 38: 860-863.

Omernik, J. M., 1977, *Non-Point Source Stream Nutrient Level Relationships: A Nationwide Survey,* Report EPA-60013-77-105, Ecological Research Series, Environmental Research Laboratory, Office of Research and Development, U.S. Environmental Protection Agency, Corvallis, OR.

O'Neill, R. V., W. F. Harris, B. S. Ausmus, and D. E. Reichle, 1975, A Theoretical Basis for Ecosystem Analysis with Particular Reference to Element Cycling, in F. G. Howell, J. B. Gentry, and M. H. Smith (Eds.), *Mineral Cycling in Southeastern Ecosystem,* ERDA Symposium Series CONF-740513, pp. 28-40, NTIS, Springfield, VA.

_____, and D. E. Reichle, 1980, Dimensions of Ecosystem Theory, in R. Waring and J. Franklin (Eds.), *Forests: Fresh Perspectives from Ecosystem Analysis,* pp. 11-26, Oregon State University Press, Corvallis.

Peters, R. H., 1978, Concentrations and Kinetics of Phosphorus Fractions in Water from Streams Entering Lake Memphremagog, *J. Fish. Res. Board Can.,* 35: 315-328.

Petersen, R. C., and K. W. Cummins, 1974, Leaf processing in a Woodland Stream, *Freshwater Biol.,* 4: 343-368.

Pomeroy, L. R., 1975, Mineral Cycling in Marine Ecosystems, in F. G. Howell, J. B. Gentry, and M. H. Smith (Eds.), *Mineral Cycling in Southeastern Ecosystems,* ERDA Symposium Series CONF-740513, pp. 209-223, NTIS, Springfield, VA.

Reichle, D. E., R. V. O'Neill, and W. F. Harris, 1975, Principles of Energy and Material Exchange in Ecosystems, in W. H. Van Dobben, W. H. and R. H. Low-McConnell (Eds.), *Unifying Concepts in Ecology,* pp. 27-43, Dr. W. Junk, Publishers, The Hague.

Rigler, F. H., 1979, The Export of Phosphorus from Dartmoor Catchments: A Model to Explain Variations of Phosphorus Concentrations in Streamwater, *J. Mar. Biol. Assoc. U.K.,* 59: 659-687.

Ruttner, A. J., 1926, Bermerkungen uber den Sauerstoffgehalt der Gewasser und dessen respiratorischen Wert, *Naturwissenschaften,* 14: 1237-1239.

Rzoska, J., 1978, *On the Nature of Rivers, With Case Histories of Nile, Zaire, and Amazon,* Dr. W. Junk, Publishers, The Hague.

Sanders, W. M., III, 1967, The Growth and Development of Attached Stream Bacteria. Part I. Theoretical Growth Kinetics of Attached Stream Bacteria, *Water Resour. Res.,* 3: 81-87.

Sayler, G. S., M. Puziss, and M. Silver, 1979, Alkaline Phosphatase Assay for Freshwater Sediments: Application to Perturbed Sediment Systems, *Appl. Environ. Microbiol.,* 38: 922-927.

Sedell, J. R., F. J. Triska, and N. S. Triska, 1975, The Processing of Conifer and Hardwood Leaves in Two Coniferous Forest Streams. I. Weight Loss and Associated Invertebrates, *Verh. Internat. Verein. Limnol.,* 19: 1617-1627.

Short, R. A., and P. E. Maslin, 1977, Processing of Leaf Litter by a Stream Detritivore: Effect on Nutrient Availability to Collectors, *Ecology,* 58: 935-938.

Stockner, J. G., and K. R. S. Shortreed, 1978, Enhancement of Autotrophic Production by Nutrient Addition in a Coastal Rainforest Stream on Vancouver Island, *J. Fish. Res. Board Can.,* 35: 28-34.

Torriani, A., 1960, Influence of Inorganic Phosphate in the Formation of Phosphatases by *Echerichia coli, Biochim. Biophys. Acta,* 38: 460-469.

Triska, F. J., and J. R. Sedell, 1976, Decomposition of Four Species of Leaf Litter in Response to Nitrate Manipulation, *Ecology,* 57: 783-792.

Vannote, R. L., G. W. Minshall, K. W. Cummins, J. R. Sedell, and C. E. Cushing, 1980, The River Continuum Concept, *Can. J. Fish. Aquat. Sci.,* 37: 130-137.

Wallace, J. B., J. R. Webster, and W. R. Woodall, 1977, The Role of Filter Feeders in Flowing Waters, *Arch. Hydrobiol.,* 75: 506-532.

Webster, J. R., 1975, Analysis of Potassium and Calcium Dynamics in Stream Ecosystems on Three Southern Appalachian Watersheds of Contrasting Vegetation, Ph.D. Thesis, University of Georgia, Athens.

———, and B. C. Patten, 1979, Effects of Watershed Perturbation on Stream Potassium and Calcium Dynamics, *Ecol. Monogr.,* 19: 51-72.

———, J. B. Waide, and B. C. Patten, 1975, Nutrient Cycling and the Stability of Ecosystems, in F. G. Howell, J. B. Gentry, and M. H. Smith (Eds.), *Mineral Cycling in Southeastern Ecosystems,* ERDA Symposium Series CONF-740513, pp. 1-27, NTIS, Springfield, VA.

Woodall, W. R., 1972, Nutrient Pathways in Small Mountain Streams, Ph.D. Thesis, University of Georgia, Athens.

Zobell, C. E., 1943, The Effect of Solid Surfaces on Bacterial Activity, *J. Bact.,* 46: 39-56.

2. THE SERIAL DISCONTINUITY CONCEPT OF LOTIC ECOSYSTEMS

James V. Ward

Department of Zoology and Entomology
Colorado State University
Fort Collins, Colorado

Jack A. Stanford

University of Montana Biological Station
Bigfork, Montana

ABSTRACT

Recent theoretical concepts of lotic ecosystems deal primarily with origins and fates of organic resources and inorganic nutrients as prescribed by the stream continuum and nutrient spiraling concepts. These concepts are based on gradient analysis in which stream systems are, of necessity, viewed as uninterrupted continua. Few riverine ecosystems, however, remain free-flowing over their entire course. Rather, regulation by dams has typically resulted in an alternating series of lentic and lotic reaches. The serial discontinuity concept is an attempt to attain a broad theoretical perspective of regulated lotic ecosystems. Discontinuity distance (DD), defined as the longitudinal shift of a given parameter by stream regulation, may be positive (downstream shift), negative (upstream shift), or near zero. The direction and intensity of DD vary as functions of the specific parameter examined and the position of the dam(s) along the longitudinal stream profile. The serial discontinuity concept can be applied to physical parameters (e.g., temperature summation) and biological phenomena at the population (e.g., species abundance patterns), community (e.g., biotic diversity), or ecosystem

29

levels (e.g., Photosynthesis/Respiration). Regulated streams, according to the serial discontinuity concept, are viewed as large-scale experimental systems in which disruptions in continuum processes and nutrient spirals create conditions amenable to testing and developing basic theories of stream ecology.

INTRODUCTION

During the past decade two important theoretical concepts of lotic ecosystems have emerged. The river continuum concept (Vannote et al., 1980) describes the gradient of physical conditions and resulting biotic responses from the headwaters to the mouths of river systems. The nutrient spiraling concept (Webster, 1975; Wallace et al., 1977; Webster and Patten, 1979; Elwood et al., this volume) is concerned with the unidirectional and biologically mediated recycling (spiraling) of nutrients, including fixed carbon, along the river continuum. These concepts are based on gradient analysis (*sensu* Whittaker, 1967) in which stream systems are, of necessity, viewed as uninterrupted continua. Few riverine ecosystems, however, remain free-flowing over their entire course. Rather, regulation by dams has typically resulted in an alternating series of lentic and lotic reaches.

The effects of impoundments on lotic reaches immediately downstream from dams were recently summarized by Ward and Stanford (1979a). The serial discontinuity concept presented here is an attempt to attain a broad theoretical perspective of regulated lotic systems over the entire longitudinal stream profile. The concept treats physical parameters (e.g., thermal regima) and biological phenomena at the population (e.g., species abundance patterns), community (e.g., biotic diversity), and ecosystem levels (e.g., photosynthesis/respiration). According to this concept, regulated streams are viewed as large-scale experimental systems in which disruptions in continuum processes and nutrient spirals create conditions amenable to testing and developing basic theories of stream ecology.

THE SERIAL DISCONTINUITY CONCEPT

The serial discontinuity concept, at this initial stage of its derivation, contains the following presuppositions: (1) The river continuum and nutrient spiraling hypotheses are conceptually sound and their underlying assumptions are valid. (2) The watershed is free of pollution and other disturbance, except impoundment. (3) The remaining lotic reaches were not disturbed during reservoir construction (e.g., riparian vegetation and substrate were not modified). (4) Unless otherwise stated, the impoundments are assumed to be deep-release storage reservoirs, which thermally stratify and which do not release oxygen-deficient or gas-supersaturated waters. We intend to present the hypothesized ramifications of modifying thermal and flow regimes by impoundment as major disruptions of continuum processes, without additional complicating factors.

In Figures 1 and 2 the solid lines represent hypothetical curves of various parameters as functions of distance along the uninterrupted stream continuum. These idealized curves were derived from data contained in various sources (especially, Vannote et al., 1980; Cummins, 1975; 1977; 1979). According to such conceptualizations, natural headwater streams are characterized as heavily canopied, light-limited heterotrophic systems with low-amplitude thermal regimes and coarse substrates. Of course, not all headwater streams are canopied by terrestrial vegetation (Minshall, 1978), nor do they all receive substantial groundwater inputs to moderate temperature and flow patterns. The fact that the majority of research on natural streams in North America has been conducted in the eastern deciduous forest has fostered such generalizations since many undisturbed lotic ecosystems for which the most intensive data are available do indeed exhibit these general characteristics.

The dashed lines in Figures 1 and 2, which synthesize our present understanding of regulated streams, indicate hypothesized modifications of those parameters when dams are placed on upper, middle, and lower reaches. Impoundments are viewed as theoretical dimensionless points on the longitudinal stream profile (i.e., only the modification of the downstream lotic ecosystem is shown; the limnological dynamics within the reservoir are not). It must be stressed that the curves presented are highly idealized. The vertical axes are intentionally presented without scales. It is probably not possible to quantify precisely any of the parameters along the entire stream continuum. The best data are available for stream orders 1 through 5. Much additional research is needed to confirm or refute even the relative changes postulated here as resulting from impoundment. Further refinements will be required to account for geographical differences and synergistic effects engendered by other disturbances to watersheds.

The differential effects of dam position on parameter modification are illustrated in Figures 1 and 2. We postulate that a parameter that may be greatly modified in the lotic reach below a dam placed at one point on the longitudinal stream profile may be little affected by impounding a different reach. A major impoundment at any position on a river system

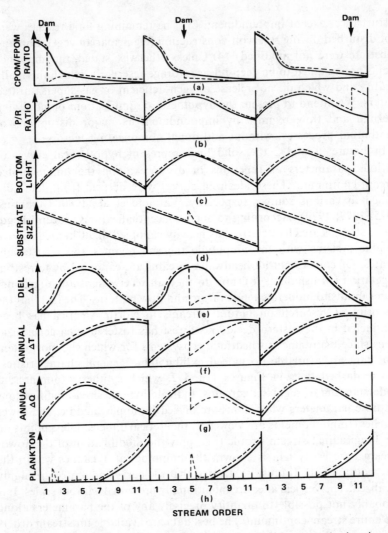

Figure 1. Relative changes in various parameters as a function of stream order, based on our interpretation of natural stream continua theory (solid lines) and postulated effects (dashed lines) of damming headwaters (left column), middle reaches (center column), and lower reaches (right column) of a river system. See text for further explanation.

will directly and indirectly affect all ecological aspects of the downstream lotic ecosystem at some level of resolution. Some, however, will be more severely influenced than others, and gross measurements may be little affected in some instances. For example, a headwater dam (Figure 1a) will greatly depress the ratio of coarse particulate to fine particulate

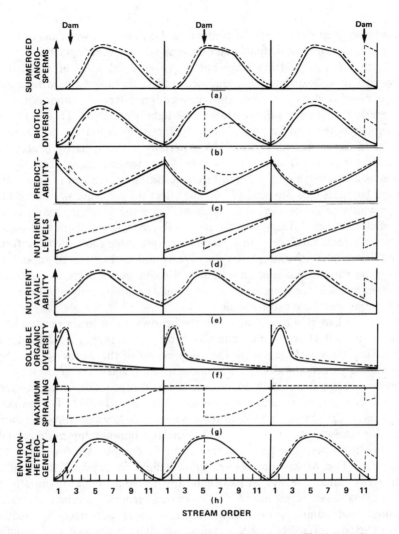

Figure 2. Relative changes in additional parameters (see Fig. 1 legend).

organic matter (CPOM/FPOM) below the impoundment because in-stream transport of detritus is blocked, whereas impounding the lower reaches of a river system will have little effect on the size composition of detritus. Invertebrate functional feeding groups will reflect changes in the CPOM/FPOM ratios (Short and Ward, 1980). Because of the importance of direct allochthonous inputs (primarily leaf litter) in the energy budgets of the upper reaches of forested streams, a headwater dam would be expected to modify functional composition most severely (shredders would be greatly reduced). In contrast, a dam on the lower reaches may not greatly alter the trophic relationships of the receiving stream.

Light intensity at the stream bottom (Figure 1c) will not be greatly changed by a headwater impoundment unless the downstream riparian vegetation is disrupted, but the clarification effect of impounding a formerly turbid river will greatly increase light penetration. This accounts for the hypothesized increase in photosynthesis-to-respiration ratio below dams in lower reaches (Figure 1b), especially if water clarity is accompanied by increased substrate stability. Clear water released from dams creates a hydrodynamic disequilibrium resulting in removal of fine sediment particles (see Simons, 1979). The effect on substrate size composition would be greatest in lower reaches (Figure 1d) where the majority of particles are small. Damming the Brazos River in Texas changed a sand-bottom stream into one with a predominantly rubble substrate (Stanford and Ward, 1979). Although few data are available, the effect of impoundment on the composition and quality of sedimentary detritus of downstream reaches (see Webster et al., 1979) may be biologically significant at all points on the stream profile.

It is unlikely that damming either the headwaters or lower reaches will have much effect on the maximum diel temperature range (diel ΔT) of the receiving stream (Figure 1e). The suppression of the diel range will be considerable below a dam in the middle reaches, where the greatest daily thermal range is normally exhibited, and will profoundly influence the biotic community structure (Ward and Stanford, 1979b). The annual thermal range (Figure 1f) would remain similar below a deep-release dam in the headwaters but would be reduced by impounding middle, and especially lower, reaches. Interpretation is complicated by shifts in latitude. The Mississippi River, for example, is exposed to quite a different climatic regime at its origin in Minnesota than at its mouth in Louisiana. Since water near $4°C$ is discharged from deep-release dams in winter and summer, the annual thermal amplitude may be greatly constricted. Surface-release reservoirs, especially if shallow, may significantly modify the receiving stream biota by raising summer temperatures (Fraley, 1979). Other thermal modifications engendered by stream regulation and their biological implications are discussed in detail in Ward (1976a) and Ward and Stanford (1979b).

In a pattern not unlike temperature, the flow regime of natural streams often exhibits maximum variation in middle reaches. The relative flow constancy of headwaters is attributed to their spring-fed nature and the moderation of precipitation by terrestrial watershed processes, whereas the discharge pattern of large rivers is dampened by the cumulative variations of many tributaries (Hynes, 1970). The middle reaches are most influenced by local meteorological events and hence exhibit the most variable and unpredictable flow regimes.

Storage reservoirs may moderate flow fluctuations (ΔQ) in the middle reaches (Figure 1g) by storing water during spates and major runoff periods and releasing additional water during periods of normally low flow. Storage reservoirs have a lesser influence on the already relatively constant flow regime of headwaters and lower reaches. The downstream flow regime is largely a function of the purpose of the reservoir. See Ward (1976b) for a discussion of the types of flow regulation and effects on stream biota.

True plankton communities (Figure 1h) occur only in the lower reaches of river systems except below impoundments and natural lakes. Plankton released from surface-release impoundments typically result in greatly enhanced populations of filter-feeding stream invertebrates immediately below the dam (Simmons and Voshell, 1978; Ward and Short, 1978). Lentic plankton are rapidly eliminated downstream, with concomitant major shifts in the invertebrate functional feeding-group composition. Plankton are also released from deep-release dams, but apparently not in sufficient numbers or with sufficient temporal predictability to greatly influence the trophic structure of the receiving stream biota (Ward, 1975).

Impoundment will not allow development of submerged angiosperms (Figure 2a) in the headwaters unless the canopy is disrupted, nor will major changes occur in the middle reaches where dense populations normally occur. In contrast, damming lower reaches may greatly enhance rooted aquatic plants, since the high nutrient levels can be used in the clear water and more stable substrate below the dams. High-gradient mountain streams lacking angiosperms may develop dense beds in regulated sections (Ward, 1976c). Attached algae exhibit a similar response (see Lowe, 1979).

It is postulated that biotic diversity, relative to unregulated lotic systems, will be modified irrespective of the position of the impoundment along the stream profile (Figure 2b). Regulation of the headwaters will suppress the biotic diversity in the receiving stream, primarily because of the disruption of detrital transport and the spiraling of nutrients and organic matter. The severe reduction of biotic diversity induced by damming middle reaches has been attributed primarily to the altered thermal regime. For example, it has been suggested (Ward, 1976a; Ward and Stanford, 1979b; Vannote et al., 1980) that daily variation in temperature, which is suppressed below dams in middle reaches, is partly responsible for maximizing species diversity in natural lotic systems by providing a wide range of thermal optima, even though suboptimal conditions occur over a portion of the diel cycle for each species.

Although few biological data are available for lower reaches, enhanced environmental heterogeneity below dams would likely lead to an increase in biotic diversity. The only known record of enhanced zoobenthic diversity below a dam occurs in a river that exhibited increased substrate and thermal heterogeneity in the regulated section (Ward and Stanford, 1979b).

The high biotic diversity in the middle reaches of natural streams may result not only from spatial heterogeneity but also from temporal heterogeneity (i.e., low predictability, Figure 2c). Lind (1971), for example, contrasted the relatively constant supply of organic matter discharged from a reservoir in Texas with the seasonal variations in organic transport which typify unregulated streams. Within limits, low predictability (high temporal heterogeneity) in flow and temperature regimes (and other factors) may enhance species packing of lotic organisms by several mechanisms (Patrick, 1970; Ward, 1976a); increased predictability resulting from impoundment may contribute to the reduced biotic diversity of regulated streams in middle reaches.

Although total nutrient levels (Figure 2d) generally increase along the stream continuum (Cummins, 1977), availability (Figure 2e) is probably greatest in middle reaches where the light regime and substrate are most suitable for plant growth. If the residence time of headwaters is increased, deep-release impoundments could conceivably raise nutrient levels downstream, but availability would not be altered. In middle and lower reaches, reservoirs generally act as nutrient sinks, but nitrate may be greater in outflowing than inflowing water (see Soltero et al., 1973). Greater nutrient availability induced by impoundment in lower reaches, because of increased clarity and substrate stability, may compensate for reduced levels in the receiving stream.

The relative diversity of soluble organic compounds (Figure 2f) is highest in headwaters of natural streams (Vannote et al., 1980), and impoundment of upper reaches would likely exert the greatest effect on this parameter. If residence time is increased, limnological phenomena within the reservoir (including biotic uptake and transformations) may reduce the chemical diversity below a headwater dam. Such a conclusion must, however, remain highly speculative at this time.

The homeostatic feedback mechanisms which control in situ nutrient cycling in autotrophic ecosystems (the "circular causal systems" of Hutchinson, 1948) are not directly applicable to stream ecosystems because of the unidirectional movement of water. If we view the stream on a spatio-temporal scale, however, a storage–cycle–release phenomenon, termed "nutrient spiraling" (Webster, 1975; Wallace et al., 1977; Webster and Patten, 1979, Elwood et al., this volume), becomes apparent. As stated by Cummins (1979), "communities in each successive stream order are dependent upon the inefficiency or 'leakage' from the preceding orders."

Vannote et al. (1980) emphasized the adjustments made by the biotic community along the river continuum which "are structured to process materials . . . thereby minimizing the variance in system structure and function." They further "propose that biological communities, developed in natural streams in dynamic equilibrium, assume processing strategies involving minimum energy loss. . . ," which is equated with maximum spiraling. From this we have deduced that maximum spiraling is maintained throughout the natural stream continuum (Figure 2g) by biotic adjustments to continually changing physical conditions. We propose that the disruption of nutrient spiraling by impoundment will be severely manifested in upper and middle reaches but less severely altered in large rivers where dissolved and particulate matter entering the reservoir will not differ greatly from that passed through the dam. There is some indication that the "food quality" of detritus may vary as a function of stream order (Naiman and Sedell, 1979). It is probable that limnological phenomena within reservoirs alter the food quality (as well as the amount and the chemical and size composition) of detritus, but no data are available.

Finally, we propose that environmental heterogeneity (viewed broadly to include both spatial and temporal components) exhibits a pattern along the stream profile (Figure 2h) which is similar to biotic diversity (see, e.g., Hedrick et al., 1976) and that the response to regulation will also be similar.

Multiple Impoundment

Figures 1 and 2 consider the differential effects on a given parameter of single dams positioned in the headwaters, the middle, or the lower reaches of a river system. Many river systems, however, are alternating series of lentic and lotic reaches because of multiple impoundment. For example, eleven main-stem dams have been constructed on the Snake River, a tributary of the Columbia (Robinson, 1978). The few data available on cumulative effects of multiple impoundment (e.g., Denisova, 1971) deal primarily with physico-chemical changes within the reservoirs. Virtually nothing is known regarding cumulative effects of multiple impoundment on the remaining lotic segments. If, for example, a factor modified by an upstream impoundment has not been returned to normal levels before reaching the next reservoir, will the interaction be neutral, cumulative, or ameliorative, and to what extent? Not only are precise answers to such questions generally unknown but the questions themselves have rarely been asked.

We have developed a hypothetical framework to visualize the basin-wide effects of impoundment (Figure 3) in an attempt to attain a broad theoretical perspective of regulated lotic ecosystems. The approach is applicable to smaller drainage basins (e.g., fifth-order systems) or portions of larger watersheds. The framework may also be used as a sub-model to investigate the effects of individual dams in a series.

Two components are apparent (see Figure 3). Discontinuity distance (DD), defined as the longitudinal shift of a given parameter by stream regulation, has a length variable (X), which is the displacement of the parameter in stream-order units (kilometers may be more useful than stream-order units, especially in xeric regions). An upstream shift is indicated by X_{neg} and a downstream shift by X_{pos}; X_o indicates that no major longitudinal shift is apparent. The theoretical example in Figure 3 shows a downstream shift of the parameter A maximum by five stream orders (DD = +5).

The second component is parameter intensity (PI), defined as the difference in absolute parameter units between the natural and the

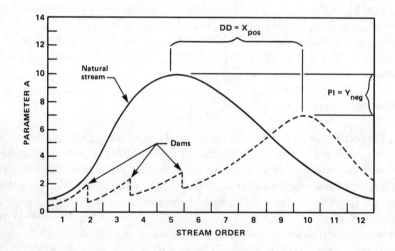

Figure 3. Theoretical framework for conceptualizing the influence of impoundment on ecological parameters in a river system. Discontinuity distance (DD) is the downstream (positive) or upstream (negative) shift of a parameter a given distance (X) due to stream regulation. PI is a measure of the difference in the parameter intensity attributed to stream regulation. See text for further explanation.

regulated lotic system. Parameter intensity can be elevated (Y_{pos}), depressed (Y_{neg}), or unchanged (Y_o) in comparison with the natural lotic system. For parameter A (Figure 3), PI is –3.

We contend that the conceptual framework exemplified in Fig. 3 provides a structure for designing and interpreting research on regulated lotic ecosystems. It is intentionally simplistic and general in this initial formulation, not only as a reflection of our inability to quantify precisely the effects of regulation but also so that it may be applicable to a variety of physical and biological parameters at different levels. For example, regulation of a river system may cause the distribution pattern of a lotic species to shift three stream orders downstream where mean annual abundance is depressed from 3000 to 2000 organisms/m^2 (DD = +3 and PI = –1000 organisms/m^2), or the position of the maximum photosynthesis-to-respiration ratio may not shift longitudinally but may be elevated from 1.1 to 1.3 at that point (DD = 0 and PI = +0.2).

One contribution of the model may be simply to focus conceptually on the basin-wide ramifications of stream regulation. The temporal shift of maximum temperatures in the Columbia River as additional dams were constructed (Jaske and Goebel, 1967), for example, could have been predicted and the biological implications better understood given a comprehensive view of the watershed. From a preliminary analysis of the distribution of hydropsychid caddisflies over nearly 300 km of a river with four mainstream impoundments (Stanford and Ward, 1981), it appears that the thermal regime and particulate organic carbon dynamics have undergone major downstream shifts since regulation of the river system. A simple mathematical model incorporating a variable number of reaches and impoundments should be initially developed and tested to examine trends in single parameters before attempting to apply this framework to the interactions of multiple factors. We hesitate to speculate further on the potential utility of the model without the results of experimental field research designed specifically within this framework. The ultimate goal of such a scheme is to stimulate research leading to the causal relationships essential to a fuller understanding of basic and applied aspects of stream ecology, and its success should be judged by that criterion.

ACKNOWLEDGMENTS

We wish to thank K. W. Cummins and J. R. Webster for suggestions regarding the manuscript. We gratefully acknowledge the support provided for our plenary presentation by the Savannah River Ecology

Laboratory and the Institute of Ecology, University of Georgia. This manuscript was prepared while J. V. Ward was supported by the Colorado Experiment Station.

REFERENCES

Cummins, K. W., 1975, The Ecology of Running Waters; Theory and Practice, in D. B. Baker et al. (Eds.), *Proceedings of the Sandusky River Basin Symposium,* pp. 277–293, International Joint Commission on the Great Lakes, Heidelburg College, Tiffin, OH.

———, 1977, From Headwater Streams to Rivers. *Am. Biol. Teacher,* 39: 305-312.

———, 1979, The Natural Stream Ecosystem, in J. V. Ward and J. A. Stanford (Eds.), *The Ecology of Regulated Streams,* pp. 7-24, Plenum Press, New York.

Denisova, A. I., 1971, The Effect of the Position of a Reservoir in a Cascade on Its Hydrochemistry, *Hydrobiol. J.,* 7: 8-15.

Fraley, J. J., 1979, Effects of Elevated Stream Temperatures Below a Shallow Reservoir on a Cold Water Macroinvertebrate Fauna, in: J. V. Ward and J. A. Stanford (Eds.), *The Ecology of Regulated Streams,* pp. 257-272, Plenum Press, New York.

Hedrick, P. W., M. E. Ginevan, and E. P. Ewing, 1976, Genetic Polymorphism in Heterogeneous Environments, *Annu. Rev. Ecol. Systemat.,* 7: 1-32.

Hutchinson, G. E., 1948, Circular Causal Systems in Ecology, *Ann. N. Y. Acad. Sci.,* 50: 221-246.

Hynes, H. B. N., 1970, *The Ecology of Running Waters,* University of Toronto Press, Toronto.

Jaske, J. T., and J. G. Goebel, 1967, Effects of Dam Construction on Temperature of the Columbia River, *J. Am. Water Works Assoc.,* 59: 935-942.

Lind, O. T., 1971, The Organic Matter Budget of a Central Texas Reservoir, in G. E. Hall (Ed.), *Reservoir Fisheries and Limnology,* pp. 193-202, Special Publication No. 8, American Fisheries Society, Washington, D.C.

Lowe, R. L., 1979, Phytobenthic Ecology and Regulated Streams, in J. V. Ward and J. A. Stanford (Eds.), *The Ecology of Regulated Streams,* pp. 25-34, Plenum Press, New York.

Minshall, G. W., 1978, Autotrophy in Stream Ecosystems, *BioScience,* 28: 767-771.

Naiman, R. J., and J. R. Sedell, 1979, Benthic Organic matter as a Function of Stream Order in Oregon, *Arch. Hydrobiol.,* 87: 404-422.

Patrick, R., 1970, Benthic Stream Communities, *Am. Sci.,* 58: 546-549.

Robinson, W. L., 1978, The Columbia: A River System Under Siege, *Ore. Wildl.,* 33: 3-7.

Short, R. A., and J. V. Ward, 1980, Leaf Litter Processing in a Regulated Rocky Mountain Stream, *J. Fish. Aquat. Sci.,* 37: 123-127.

Simons, D. B., 1979, Effects of Stream Regulation on Channel Morphology, in J. V. Ward and J. A. Stanford (Eds.), *The Ecology of Regulated Streams*, pp. 95-111, Plenum Press, New York.

Simmons, G. M., Jr., and J. R. Voshell, Jr., 1978, Pre- and Post-Impoundment Benthic Macroinvertebrate Communities of the North Anna River, in J. Cairns, E. F. Benfield, and J. R. Webster (Eds.), *Current Perspectives on River-Reservoir Ecosystems*, pp. 45-61, North American Benthological Society, Roanoke, VA.

Soltero, R. A., J. C. Wright, and A. A. Horpestad, 1973, Effects of Impoundment on the Water Quality of the Bighorn River, *Water Res.*, 7: 343-354.

Stanford, J. A., and J. V. Ward, 1979, Stream Regulation in North America, in J. V. Ward and J. A. Stanford (Eds.), *The Ecology of Regulated Streams*, pp. 215-236, Plenum Press, New York.

Stanford, J. A., and J. V. Ward, 1981, Preliminary Interpretations of the Distribution of Hydropsychidae in a Regulated River, in G. P. Moretti (Ed.), *Proceedings of the Third International Symposium on Trichoptera* pp. 323-328, Dr. W. Junk, Publishers, The Hague.

Vannote, R. L., G. W. Minshall, K. W. Cummins, J. R. Sedell, and C. E. Cushing, 1980, The River Continuum Concept, *Can. J. Fish. Aquat. Sci.*, 37: 130-137.

Wallace, J. B., J. R. Webster, and W. R. Woodall, 1977, The Role of Filter Feeders in Flowing Waters, *Arch. Hydrobiol.*, 79: 506-532.

Ward, J. V., 1975, Downstream Fate of Zooplankton from a Hypolimnial Release Mountain Reservoir, *Verh. Internat. Verein. Limnol.*, 19: 1798-1804.

_____, 1976a, Effects of Thermal Constancy and Seasonal Temperature Displacement on Community Structure of Stream Macroinvertebrates, in G. W. Esch and R. W. McFarlane (Eds.), *Thermal Ecology II*, ERDA Symposium Series CONF-750425, pp. 302-307, NTIS, Springfield, VA.

_____, 1976b, Effects of Flow Patterns Below Large Dams on Stream Benthos: A Review, in J. F. Orsborn and C. H. Allman (Eds.), *Instream Flow Needs Symposium*, Vol. II, pp. 235-253, American Fisheries Society, Bethesda, MD.

_____, 1976c, Comparative Limnology of Differentially Regulated Sections of a Colorado Mountain River, *Arch. Hydrobiol.*, 78: 319-342.

_____, and R. A. Short, 1978, Macroinvertebrate Community Structure of Four Special Lotic Habitats in Colorado, U.S.A., *Verh. Internat. Verein. Limnol.*, 20: 1382-1387.

_____, and J. A. Stanford (Eds.), 1979a, *The Ecology of Regulated Streams*, Plenum Press, New York.

_____, and J. A. Stanford, 1979b, Ecological Factors Controlling Stream Zoobenthos with Emphasis on Thermal Modification of Regulated Streams, in J. V. Ward and J. A. Stanford (Eds.), *The Ecology of Regulated Streams*, pp. 35-55, Plenum Press, New York.

Webster, J. R., 1975, Analysis of Potassium and Calcium Dynamics in Stream Ecosystems on Three Southern Appalachian Watersheds of Contrasting Vegetation, Ph.D. Thesis, University of Georgia, Athens.

Webster, J. R., and B. C. Patten, 1979, Effects of Watershed Perturbation on Stream Potassium and Calcium Dynamics, *Ecol. Monogr.*, 49: 51-72.

Webster, J. R., E. F. Benfield, and J. Cairns, Jr., 1979, Model Predictions of Effects of Impoundment on Particulate Organic Matter Transport in a River System, in J. V. Ward and J. A. Stanford (Eds.), *The Ecology of Regulated Streams,* pp. 339-364, Plenum Press, New York.

Whittaker, R. H., 1967, Gradient Analysis of Vegetation, *Biol. Rev.,* 42: 207-264.

3. A CONCEPTUAL FRAMEWORK FOR PROCESS STUDIES IN LOTIC ECOSYSTEMS

C. David McIntire
  Department of Botany and Plant Pathology
  Oregon State University
  Corvallis, Oregon

ABSTRACT

This paper presents an approach to lotic ecosystem studies based on the concept of relevant dynamic processes. A process is defined as a systematic series of actions relevant to the dynamics of the system as it is modeled. Each process is expressed as a set of variables, and state variables represent either biomass or process capacity, depending on whether or not qualitative aspects within a process are represented. Process dynamics in lotic ecosystems are discussed relative to (1) the potential to expand process capacity, (2) process production, (3) realized growth of process capacity, and (4) process regulation. A set of variables are defined which aid in the investigation of mechanisms regulating process dynamics in lotic ecosystems. Process models are discussed in relation to their role in hypotheses generation, data assessment, and the setting of research priorities. Specific hypotheses relative to process dynamics in lotic ecosystems are proposed and related to the behavior of a lotic ecosystem model.

43

INTRODUCTION

Some of the more recent theoretical and applied problems in lotic ecology require innovative approaches at levels of resolution incompatible with population studies. Theoretical concepts dealing with lotic processes that involve many populations collectively are beginning to emerge, but such concepts have not yet stimulated the development of the new field methodology necessary for parameter estimation. As one historically interested in aquatic plants, I have found that it is relatively easy to promote a holistic treatment of autotrophic processes (e.g., primary production), particularly among my colleagues with zoological interests, even though such processes may involve an assemblage of 200 or more taxa. The mere suggestion, however, that certain problems in lotic ecology justify a holistic treatment of such macroconsumer processes as grazing, shredding, and collecting sometimes brings a strong negative response manifest through both theoretical and methodological concerns. In other words, individuals primarily interested in population interactions may feel uncomfortable with levels of resolution at which such details are integrated out. Moreover, there is a strong tendency to express dynamics of high level processes as the summation of lower level processes or to model the total function of a group of diverse taxa in the behavioral form of a single taxon.* Yet, theoretically, there is no less justification for defining and investigating relevant macroconsumer processes above the population level than for the study of primary production as a total ecosystem process. Therefore, the problem must be primarily methodological, in addition to an understandable skepticism for the abstract.

Lotic ecosystems have been investigated from the trophic level viewpoint (e.g., Odum, 1957; Warren et al., 1964) and more recently within the conceptual framework of paraspecies or functional groups (Cummins, 1974; Boling et al., 1975). These approaches involve the classification of taxonomic entities into groups of organisms considered to be similar to each other, usually with respect to trophic function. The dynamics of a trophic level or paraspecies are treated analytically (and conceptually) as a summation of activities associated with the constituent taxa of each functional group. The paraspecies approach provides a refinement that allows a multistage representation of life history phenomena within a structure that can always be mapped into a trophic response. In this

---

*For further discussion, see Overton's reference to the so-called "aggregation problem" (Overton, 1977) and Zeigler (1976).

paper, I present an alternative approach based on the concept of relevant dynamic processes. Process definition and relevancy are determined by objective questions and methodological limitations. This approach ignores the dynamics of individual taxonomic entities and emphasizes total higher (coarser) level functions above the species level. In other words, there is no attempt to aggregate species into compartments of a lumped model as discussed by Zeigler (1976) or to simulate high level behavior from lower level representations (Wlosinski and Minshall, this volume). Instead, the process perspective examines behaviors that are unique and relevant to the particular level of organization under consideration, behaviors that often provide a perspective different from that of a lumped version of a base model of interacting species. Parameter estimation ideally is based on direct measurements of the relevant set of processes. Examples of such measurements include estimates of primary production in the field and the laboratory (McIntire and Phinney, 1965; Odum, 1956) and the more recent investigations of the process of shredding by leaf-pack studies (Sedell et al., 1975).

THE PROCESS CONCEPT

Ecosystem Processes and Process Capacity

The ecological literature often refers to various physical and biological processes in contexts that are usually intuitively understandable without a formal, theoretical structure. For analytical purposes, however, it is desirable to formalize the process concept by defining terms and establishing a system for diagramming relationships. The definitions apply only to biological processes; physical processes are treated conceptually as driving or control functions.

> Definition: A process is a systematic series of actions relevant to the dynamics
> of the system as it is conceptualized (or modeled).

The process concept is illustrated in Figure 1. This diagram is compatible with three primary consumer processes found in lotic ecosystems—grazing, shredding, and collecting (McIntire and Colby, 1978). These processes correspond to resource utilization of periphyton, large particulate organic matter (LPOM), and fine particulate organic matter (FPOM) and to functional groups designated as grazers (or scrapers), shredders, and collectors in recent literature (Cummins, 1974; 1975). In this case, a process (the entire contents of the circle) is

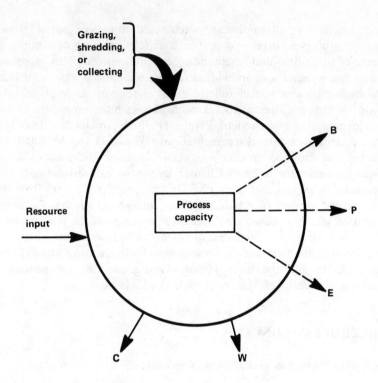

Figure 1. Schematic representation of the process concept illustrating a state variable, the process capacity, and relevant input and output variables. Components of the process capacity are (1) quantitative, biomass, and (2) qualitative, genetic information. Other variables are: C, cost of processing; W, waste discharge; B, loss to decomposition; P, loss to predation; and E, export.

elaborated into one state variable (the process capacity) and six variables representing inputs and outputs associated with the process. Theoretically, process capacity includes two components—a quantitative aspect, which is the biomass at any instant of time involved in the process, and a qualitative aspect, which relates to taxonomic composition and physiological state. In the stream model developed by McIntire and Colby (1978), qualitative changes within a process were not represented, and the state variables were reported as biomass. Other variables associated with grazing, shredding, and collecting include resource consumption (resource input), cost of processing (C), waste discharge (W), loss to decomposition (B), loss to predation (P), and export (E). These variables are disucssed in the following section.

At this point, the concept of process capacity needs further clarification. The potential performance of a unit of biomass associated with a process relative to a given set of inputs may change with temporal and spatial changes in the taxonomic composition and physiological state of that biomass; or, stated from a population perspective, one combination of populations (or parts of populations) involved in a particular process may do things at different rates (i.e., may require a different parameterization) than a different combination involved in the same process. Process capacity expresses both quantitative and qualitative aspects of the process state and, therefore, represents a unit that is time and spatially invariant relative to its relationships with other components of the system. As yet, we have not invented an approach to the direct or indirect estimation of process capacity, and most current models represent only the quantitative aspects (biomass) within a process. The problem is to develop a set of rules that will map process biomass (a variable that changes qualitatively) into process capacity (a corresponding variable that maintains a constant functional potential). Regardless of how state variables are expressed (biomass or capacity), the process convention ignores taxonomic categories and is, therefore, different from the paraspecies and trophic level approaches, which classify taxonomic entities into ecologically similar groups. Even in rare instances where relevant functional attributes of an ecosystem correspond exactly to the functional dynamics of paraspecies or other species aggregations, the process viewpoint places the emphasis on the capacity of a system or subsystem to process inputs, whereas the functional-group approach usually is more concerned with state variable dynamics, i.e., the biomass or numerical abundance of organisms in each functional group.

Ecosystems can be conceptualized as hierarchical systems of biological processes with physical and chemical processes expressed as driving or control variables. Depending on the resolution levels of interest, the various biological processes are the systems, subsystems, and suprasystems under consideration. This view of ecological systems is consistent with FLEX, a general ecosystem-model paradigm developed by Overton (1972; 1975) and based on the general systems theory of Klir (1969). Figure 2 illustrates two higher level processes relevant to lotic systems. Herbivory has two subsystems, primary production and grazing, whereas detritivory is partitioned into the processes of shredding, collecting, and detrital decomposition. Herbivory and detritivory are analogous consumer processes, differing only with respect to whether or not resources are generated from within the system (autochthonous production) or from allochthonous inputs. At the next higher level, herbivory and detritivory are subsystems of the process of primary consumption. At any

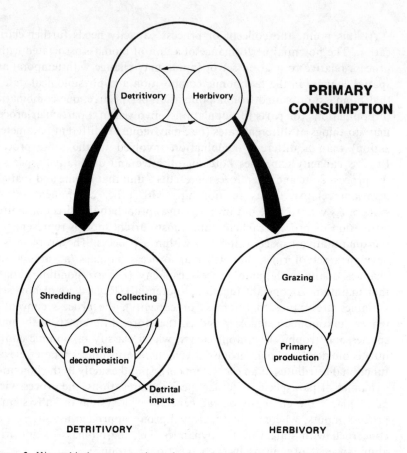

Figure 2. Hierarchical representation of selected lotic processes showing primary consumption and its subsystems.

particular level, a process can be elaborated into the relevant variables, interaction functions, and parameters. For example, in the model of McIntire and Colby (1978), herbivory and detritivory have two and seven state variables, respectively.

Some Useful Variables

Process dynamics in ecosystems can be examined relative to (1) the potential to expand process capacity, (2) process production, (3) the realized growth of process capacity, and (4) process regulation. In the discussion that follows, process capacity is stressed, but the concepts obviously apply to process biomass if the qualitative component of capacity is ignored.

Consumer Processes

The potential to expand the capacity of a consumer process at time k is given by

$$g_0(k) = [S(k)]^{-1} [aD(k) - C(k)] \qquad (1)$$

where $S(k)$ = the state variable value at time k; i.e., the process capacity or biomass if quality is ignored
$D(k)$ = the process demand at time k
$C(k)$ = the cost of processing at time k
a = an efficiency parameter

Process demand (D) represents resource consumption by a process when resources are in unlimited supply. The cost of processing (C) is the metabolic loss of energy during processing. The efficiency parameter (a) is the proportion of resource intake that is incorporated into process capacity. The variable g_0 is a specific growth rate, the potential to expand process capacity per unit capacity in the absence of resource limitation and negative effects from other processes. It is analogous to the intrinsic rate of natural increase, a population parameter defined by Birch (1948). Therefore g_0 is theoretically unaffected by density-dependent factors and is a function of physical processes that act in a density-independent way. Obviously g_0 approaches some maximum (say, $g_0$max) as density-independent factors become optimal. If the state variable represents process capacity rather than biomass, $g_0$max is a constant for a given process because, by definition, it is unaffected by density-dependent relationships and qualitative changes within the process. On the other hand, if the state variable is biomass, theoretically $g_0$max will vary with qualitative changes in the process biomass.

Process production for consumer processes is defined as the net elaboration of the process capacity (or biomass) regardless of its fate during the period under consideration. Bioenergetically, it is process assimilation (i.e., the amount of a resource incorporated into process capacity) minus the cost of processing. Therefore, process production is analogous to the concept of secondary production (Ricker, 1958). Process production is derived from the expression

$$g_1 = [S(k)]^{-1} [aR(k) - C(k)] \qquad (2)$$

where $R(k)$ is the realized consumption of resources at time k (i.e., the actual rate at which resources are consumed by the process under existing conditions). It follows that the process production rate at time k is $[aR(k) - C(k)]$ or $g_1(k)S(k)$. Thus g_0 is a specific growth rate with unlimited resources, and g_1 is the analogous rate when resources vary according to system dynamics.

Waste loss associated with processing (W) at time k is:

$$W(k) = R(k) - aR(k) \tag{3}$$

Fecal discharge is the principal biological mechanism accounting for W, and this waste usually represents a resource for another process or is exported from the system. In the model of McIntire and Colby (1978), W from all processes was treated as FPOM, a resource processed by collecting.

The realized growth of process capacity may be obtained after accounting for export and interactions with other processes. The equations are:

$$g_2(k) = [S(k)]^{-1} [aR(k) - C(k) - E(k)] \tag{4}$$

$$g_3(k) = [S(k)]^{-1} [aR(k) - C(k) - E(k) - B(k)] \tag{5}$$

and

$$g_4(k) = [S(k)]^{-1} [aR(k) - C(k) - E(k) - B(k) - P(k)] \tag{6}$$

Here $E(k)$, $B(k)$, and $P(k)$ correspond to export or emergence losses, losses to decomposer processes, and losses to predation, respectively. In the model of McIntire and Colby, B for processes of grazing, shredding, collecting, and invertebrate predation is conceptualized as part of the incremental respiratory loss (i.e., the variable C). If process capacity is gained directly from outside the system (immigration), an additional term (I) must be added to Equation 6 to account for this import. The variable $g_r(k)$ is defined as the actual or realized specific growth rate associated with a process at time k, and $g_r(k)S(k)$ is the realized process growth rate. For grazing, shredding, and collecting, g_r equals g_4, whereas, for vertebrate predation, g_r equals g_3. If a process remains in a steady state relative to the time resolution under consideration, g_r fluctuates around a mean of zero. During a known successional sequence, g_r follows a trajectory that is consistent with the corresponding changes in community structure and function.

Autotrophic Processes

Concepts related to autotrophic process dynamics are analogous to the concepts presented for consumer processes. When light energy, nutrients, and space are not limiting:

$$g_o(k) = [S(k)]^{-1} [P_{gmax}(k) - C(k)] \tag{7}$$

where P_{gmax} is the gross primary production when resources are in unlimited supply. Again, g_o approaches $g_o max$ as density-independent factors become optimal. When resources vary with system dynamics,

$$g_1(k) = [S(k)]^{-1} [P_g(k) - C(k)] \qquad (8)$$

where P_g is the realized gross primary production. If the process represents the function of autotrophic organisms only, the rate of net primary production at time k is $g_1(k)S(k)$. For aquatic ecosystems, however, it is often convenient to include the activities of tightly coupled heterotrophic microorganisms within the process boundary, as in the case of periphyton. If this is done, $g_1(k)S(k)$ simply represents a net elaboration of autotrophic process capacity—not net primary production, and C, the cost of processing, expresses the integrated metabolic losses from the activities of both autotrophic and heterotrophic organisms. Expressions analogous to Eqs. 4, 5, and 6 for autotrophic processes are:

$$g_2(k) = [S(k)]^{-1} [P_g(k) - C(k) - E(k)] \qquad (9)$$

$$g_3(k) = [S(k)]^{-1} [P_g(k) - C(k) - E(k) - B(k)] \qquad (10)$$

$$g_4(k) = [S(k)]^{-1} [P_g(k) - C(k) - E(k) - B(k) - G(k)] \qquad (11)$$

where G is the loss to the process of grazing and the other symbols are the same as previously stated.

Process Regulation

In studies of whole ecosystems, it is often difficult to understand mechanisms accounting for system dynamics from plots of state variables. In other words, values for state variables go up and down, but it is not always intuitively obvious why such variations occur. Likewise, state variable dynamics in ecosystem models are often difficult to interpret; this may seem surprising considering that the model is the investigator's own creation. Therefore, we frequently find ourselves in the frustrating situation of not even being able to understand the behavior of our simple models of the real world, let alone the real world itself.

The g-variables defined above provide a convenient basis for investigating regulatory mechanisms in large ecosystem models. The value of such variables in field research is unexplored. From Equations 1, 2, and 4 to 11:

$g_o - g_1$ is the regulating effect of resource limitation
$g_1 - g_2$ is the regulating effect of export losses
$g_2 - g_3$ is the regulating effect of decomposer processes
$g_3 - g_4$ is the regulating effect of predator processes

To analyze state variable dynamics, we simply plot g_o, g_r, and all intermediate $g_i[i = 1,2,\ldots,g_{r-1}]$ against time and examine the areas between curves relative to a plot of the corresponding state variable. If the model provides a satisfactory representation of the system of interest, such plots can serve as a basis for generating hypotheses by providing good insight into regulatory mechanisms.

PROCESS MODELS

Process models can be developed to optimize predictions without concern for whether or not internal structures are reasonable representations of the detailed mechanisms that generate the emergent, higher level behavior of interest. Such models might provide useful forecasts without telling us very much about how the system works. Alternatively, we may develop a process model for scientific purposes—to generate hypotheses, to synthesize concepts and the results of field and laboratory studies, to evaluate a data base, and to set priorities for future research. In this case the model is an iterative tool that provides new insights, and its structure and parameters are updated as new information becomes available through research.

Example

The lotic ecosystem process model of McIntire and Colby (1978) is an example of a model developed for research purposes. Briefly, a stream was conceptualized as a hierarchical system of biological processes (Figures 3 and 4). The ecosystem was considered mechanistically as two coupled subsystems representing the processes of primary consumption and predation. At a finer level of resolution, predation included processes of invertebrate and vertebrate predation, whereas herbivory and detritivory were the subsystems of primary consumption (Figures 2 and 4). Behavioral aspects of the so-called standard run of the McIntire and Colby model are summarized in Tables 1 and 2. This run attempts to simulate process dynamics of a small (first- or second-order), shaded stream site receiving a relatively large input of allochthonous detritus. Energy inputs and hydrologic properties for the standard run were published in Fig. 3 and Table 1 (McIntire and Colby, 1978). On an annual time resolution, gross primary production expressed as organic matter (dry weight) is about 70 g/m^2, and detrital inputs are 480 g/m^2 (Table 1). The model predicts that the mean periphyton biomass is only about 1 g/m^2, whereas fine particulate biomass (FPOM) and large particulate

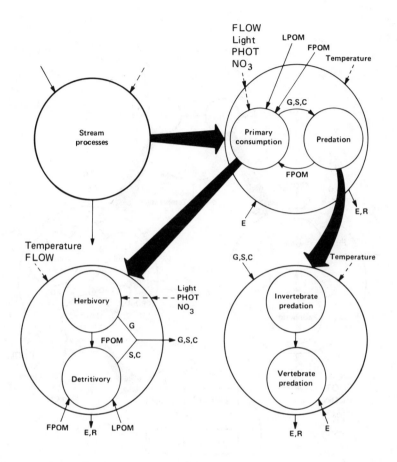

Figure 3. Schematic representation of a lotic ecosystem showing the hierarchical decomposition of the primary consumption and predation subsystems. The symbols refer to inputs and outputs from processes of grazing (G), shredding (S), and collecting (C); large (LPOM) and small particle (FPOM) detritus; export or emergence (E); respiration (R); temperature; stream discharge (FLOW); photoperiod (PHOT); and nutrient concentration (NO_3). (Data from McIntire and Colby, 1978).

biomass (LPOM) are about 10 and 60 g/m² in mean value, respectively. At a steady state primary production is partitioned into metabolic losses (14%), losses to primary consumption (69%), and export (17%); whereas primary consumption accounts for 62 and 49% of FPOM and LPOM,

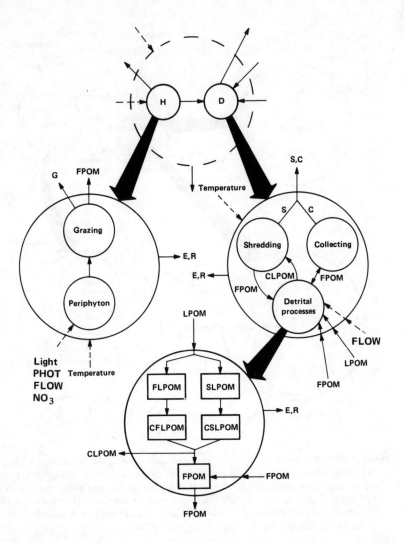

Figure 4. Schematic representation of the mechanistic structure of the herbivory and detritivory subsystems in a lotic ecosystem. The symbols are: CLPOM, conditioned large particle detritus; FLPOM, quickly decomposing material; SLPOM, slowly decomposing material; CFLPOM, quickly decomposing material suitable for consumption by macroconsumers; and CSLPOM, slowly decomposing material suitable for consumption by macroconsumers (other symbols are the same as given in Fig. 3). (Data from McIntire and Colby, 1978).

Table 1. The McIntire and Colby Stream Model*

Property	Algae	FPOM	LPOM
Mean biomass, g/m^2)	0.94	10.48	60.35
Inputs, $g\ m^{-2}\ yr^{-1}$)			
Gross primary production	71.14		
Terrestrial			473.63
Aquatic (feces)		480.49	
Mechanical (from LPOM)		55.08	
Losses, % of inputs			
Periphyton respiration	14		
Primary consumption	69	62	49
Export	17	15	16
Mechanical (to FPOM)			11
Microbial decomposition		23	24

*Standard run: annual dynamics of autotrophic and detrital processes. Acronyms refer to periphyton (Algae), fine particulate organic matter (FPOM), and large particulate organic matter (LPOM).

Table 2. The McIntire and Colby Stream Model*

Property	Graze	Shred	Collect	I Pred	V Pred
Mean biomass, g/m^2	1.08	1.23	2.57	0.41	6.03
Production, $g\ m^{-2}\ yr^{-1}$	3.40	6.36	12.05	0.84	5.48
Turnover, times/yr	3.16	5.17	4.69	2.07	0.91
Assimilation, $g\ m^{-2}\ m^{-1}$	27.05	41.83	70.31	3.60	15.45
Losses, % of assimilation					
Respiration/postmortem					
decomposition	87	85	83	76	65
To v-predation	7	6	7	16	
To i-predation	3	3	4		
Emergence/mortality	2	6	6	7	35

*Standard run; abbreviations are annual dynamics of grazing, Graze; shredding, Shred; collecting, Collect; invertebrate predation, I Pred; and vertebrate predation, V Pred.

respectively. Production and mean biomass associated with grazing are relatively low (3.40 g m^{-2} yr^{-1} and 1.08 g/m^2), whereas collecting exhibits the highest production (12.05 g m^{-2} yr^{-1}) of the macroconsumer processes. Turnover for primary consumer processes range from 3.16 times/yr (grazing) to 5.17 times/yr (shredding); corresponding values for invertebrate predation and vertebrate predation are 2.07 and 0.91 times/yr, respectively. Also, the standard run predicts that from 9 to 11% of the production by the primary consumer processes is lost to predation.

Since the McIntire and Colby model was developed to help increase understanding of mechanisms regulating biological processes in streams—not to predict state variable dynamics of a particular natural ecosystem, model validation was based on ranges of biomass and production values found in the literature (McIntire and Colby, 1978, p. 177). Such comparisons indicate that the values in Tables 1 and 2 are well within the ranges of values reported from various studies of corresponding variables in natural systems. In the next section, the use of process models for hypothesis generation is demonstrated by comparing output from the standard run to output after selected modifications.

Hypothesis Generation

Plots of g variables for the standard run suggest mechanisms that regulate processes of grazing, shredding, collecting, invertebrate predation, and vertebrate predation in small headwater streams (Figures 5 and 6). Grazing is obviously resource limited throughout the year (i.e., the area between the trajectory of g_0 and that of g_1 is relatively large), while shredding is regulated primarily by export (emergence) and predation except for a 1-month period in late spring. The effect of resource supply on the dynamics of collecting is relatively minor in comparison with the regulatory influence of export and predation. Invertebrate predation is resource limited during summer and fall, but predation exerts more influence earlier in the year. Vertebrate predation is resource limited throughout the year, except for a brief period of exponential growth in March and April.

Hypothesis I:

In small, shaded streams receiving high LPOM inputs, grazing is limited primarily by food resources, whereas shredding and collecting are regulated by emergence patterns (export) and predation; invertebrate predation and vertebrate predation are food resource limited for about 7 and 10 months during the year, respectively.

The effect of predation on primary consumer processes can be examined by comparing selected output from the standard run with corresponding output without predation, i.e., with the initial biomass for invertebrate and vertebrate predation set at zero (Table 3). Without predation, production associated with grazing, a resource limited process, is only 23% of that found with predation; mean biomass is about the same for the two cases. For collecting, however, which is a process regulated by

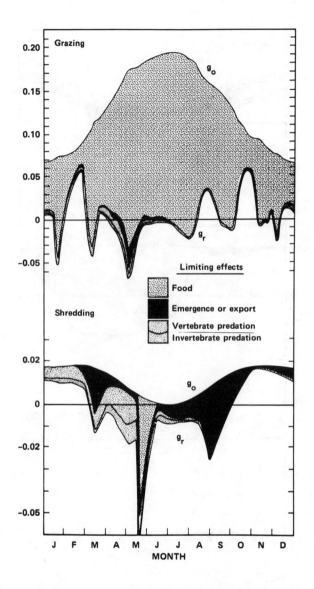

Figure 5. Diagram of g function trajectories representing processes of grazing and shredding for the standard run of the McIntire and Colby Stream Model. The symbol g_o is the process specific growth rate with unlimited resources, and g_r is the realized specific growth rate. (Data from McIntire and Colby, 1978).

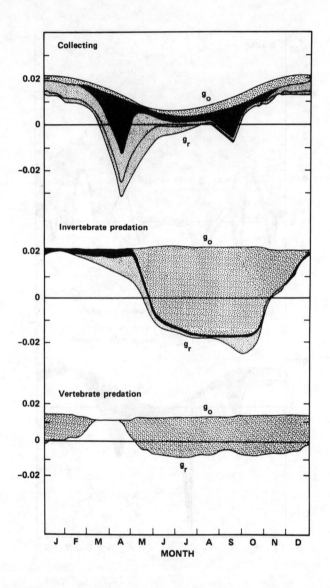

Figure 6. Diagram of g function trajectories representing processes of collecting, inverte-brate predation, and vertebrate predation for the standard run of the McIntire and Colby Stream Model. The symbols g_o and g_r are as given in Fig. 5. (Data from McIntire and Colby, 1978).

Table 3. The McIntire and Colby Stream Model*

Property	Graze		Shred		Collect	
	P	w/o P	P	w/o P	P	w/o P
Mean biomass, g/m^2	1.08	1.13	1.23	1.64	2.57	8.21
Production, g m^{-2} yr^{-1}	3.40	0.78	6.36	3.15	12.05	11.71
Turnover, times/yr	3.16	0.69	5.17	1.92	4.69	1.43
Assimilation, g m^{-2} yr^{-1}	27.05	25.76	41.83	47.30	70.31	195.32
Losses, % of assimilation						
Respiration/postmortem						
decomposition	87	97	85	93	83	94
To v-predation	7	0	6	0	7	0
To i-predation	3	0	3	0	4	0
Emergence/export	2	3	6	7	6	6

*Standard run with (P) and without (w/o P) the process of predation. The abbreviations are annual dynamics of grazing, Graze; shredding, Shred; and collecting, Collect.

Table 4. The McIntire and Colby Stream Model*

Property	Graze		Shred		Collect	
	S	w/o E	S	w/o E	S	w/o E
Mean biomass, g/m^2	1.08	1.08	1.23	1.52	2.57	2.38
Production, g m^{-2} yr^{-1}	3.40	3.90	6.36	8.55	12.05	12.01
Turnover, times/yr	3.16	3.61	5.17	5.63	4.69	5.04
Assimilation, g m^{-2} yr^{-1}	27.05	27.65	41.83	52.42	70.31	64.99

*Standard run with (S) and without (w/o E) LPOM export. The abbreviations are annual dynamics of grazing, Graze; shredding, Shred; and collecting, Collect.

predation and emergence (export), mean biomass is about 3.2 times greater without predation, whereas production without predation is 97% of that with predation. The process of shredding, which is regulated by resources, predation, and emergence, depending on season, exhibits changes in both mean biomass and production when predation is deleted from the run. Mean biomass increases by 33%, and production decreases by about 50%.

Hypothesis II:

If a consumer process is limited by food resources, its annual production tends to increase as pressure from predation increases, whereas the mean biomass associated with the process may change very little.

Hypothesis III:

If a consumer process is regulated primarily by predation and export (or emergence), an increase in predation tends to decrease mean biomass, while production remains about the same.

The rate at which LPOM is dislodged from a stream section and moved downstream is difficult to estimate, although such measurements are often attempted in the field. The standard run predicts that 16% of the annual input of LPOM is exported from a particular stream section; this value is roughly compatible with the energy budget proposed by Petersen and Cummins (1974). The effect of total retention of LPOM on primary consumer processes was investigated by setting LPOM export (not FPOM export) equal to zero and comparing these results to the standard run (Table 4). This manipulation has the greatest impact on shredding, which exhibits a 24% increase in mean biomass and a 34% increase in production. The process of collecting is relatively unaffected by total retention of LPOM; this is surprising considering that FPOM inputs increase by 8%. Since the effect of food resource limitation on collecting is minor (Fig. 6), the slight benefit of an increase in resources is balanced by the negative effect of an increase in predation which is stimulated by more shredding. Total retention of LPOM also is associated with an increase in FPOM export, from 15 to 18% of inputs.

Hypothesis IV:

In small, shaded streams production associated with shredding is more sensitive to the quantity and timing of LPOM inputs than is production associated with collecting or grazing.

One hypothesis in lotic ecology suggests that macroconsumer processes such as grazing, shredding, collecting, and predation play a prominent role in the processing of autochthonous and allochthonous inputs. If macroconsumer processes are deleted from the standard run, periphyton respiration increases form 14 to 42% of annual gross primary production and export increases from 17 to 58% (Table 5). Without grazing, gross primary production and mean periphyton biomass increase from 71 to 534 g m^{-2} yr^{-1} and from 1 to 20 g/m^2, respectively. Without shredding and collecting, the annual mean biomass of LPOM increases from 60 to 191 g/m^2. Under these circumstances, however, 69% of the LPOM input is processed by microbial decomposition, and 31% is exported. In other words, the model indicates that rates of microbial decomposition of

Table 5. The McIntire and Colby Stream Model*

Property	Algae		LPOM	
	S	w/o C	S	w/o C
Mean biomass, g/m^2	0.94	20.09	60.35	191.29
LPOM input, $g\ m^{-2}\ yr^{-1}$			473.63	473.63
GPP, $g\ m^{-2}\ yr^{-1}$	71.14	533.74		
Losses, % of inputs				
Periphyton respiration	14	42		
Microbial decomposition			24	69
Export	17	58	16	31
Primary consumption	69	0	49	0
Mechanical			11	0

*Standard run with (S) and without (w/o C) macroconsumer processes of grazing, shredding, collecting, invertebrate predation, and vertebrate predation. The abbreviations are annual dynamics of periphyton, Algae; large particulate organic matter, LPOM; and rate of gross primary production, GPP.

Table 6. The McIntire and Colby Stream Model*

Variable	BC	H1	H2	FULL
Algae				
Mean biomass g/m^2	0.94	0.96	1.03	1.16
GPP, $g\ m^{-2}\ yr^{-1}$	71.14	105.59	141.09	188.57
Turnover, times/yr	65.00	100.90	126.11	151.53
Graze				
Mean biomass, g/m^2	1.08	1.71	2.26	2.76
Production, $g\ m^{-2}\ yr^{-1}$	3.40	6.53	10.66	18.56
Turnover, times/yr	3.16	3.83	4.71	6.73

*Standard run with four different light energy input schedules. The abbreviations are annual dynamics of periphyton, Algae; grazing, Graze; and the rate of gross primary production, GPP. The light energy inputs are the Berry Creek schedule, BC; a hypothetical schedule, H1; 2× the hypothetical schedule, H2; and a constant saturation intensity of 2600 ft-c.

allochthonous inputs which have been measured in recent field and laboratory studies can account for the degradation of over two-thirds of the total input in the absence of macroconsumer processes.

Hypothesis V:

If macroconsumers are removed from a stream section, microbial activity will process about two-thirds of the allochthonous inputs, and the rate of detrital export will just about double in magnitude; annual gross primary production increases about sevenfold.

Output from the standard run was obtained for four different input schedules to examine the effect of light energy on herbivory. The Berry Creek light schedule (see Fig. 3, McIntire and Colby, 1978) represented a low input level typical of many shaded, low-order woodland streams; this schedule was used to generate all previous output designated as a standard run. A hypothetical light schedule followed a seasonal pattern similar to the Berry Creek schedule (summer shading), but the intensity was considerably higher than the latter at corresponding times. A third schedule was generated by multiplying each value in the hypothetical schedule by two, and a fourth was a constant intensity (2600 fc) above the light saturation value for primary production.

The mechanistic behavior of herbivory is illustrated in Table 6 by the annual dynamics of its two subsystems—periphyton processes (ALGAE) and grazing (GRAZE). The model predicts that periphyton biomass is relatively unaffected by changes in light energy, whereas annual gross primary production increases from about 71 to 189 g/m^2 with increasing light energy. As the productivity of the system increases with increasing light energy, the periphyton biomass turns over more rapidly, as much as once every 2 to 4 days, and the increase in primary production is channeled into the process of grazing rather than being stored as periphyton biomass. Mean biomass and annual production associated with grazing both increase with corresponding increases in primary production. Therefore, relative to state variable dynamics, the increase in light energy is reflected by the primary consumer process rather than the primary producer process; the very small changes in periphyton biomass would be undetectable in the field. In fact, periphyton biomasses in the neighborhood of 1 g/m^2 are usually inconspicuous and require some kind of chlorophyll extraction procedure for estimation.

Hypothesis VI:

Periphyton biomass is a poor predictor of the rate of primary production and the capacity of a stream ecosystem to support grazing.

Hypotheses I to VI relate to the fine resolution dynamics of processes identified in the McIntire and Colby model. The hierarchical structure of the model also permits the generation and examination of hypotheses related to coarse resolution dynamics.

Production for the total stream ecosystem relative to a particular temporal and spatial resolution is given by

$$\text{Prod(S)} = \text{GPP} + A_d - \text{CR} \tag{12}$$

where GPP is the gross primary production, A_d is the assimilation by processes involved in the processing of allochthonous inputs, and CR is the respiratory loss for the entire system. The term A_d is complicated by the fact that FPOM, the resource of collecting, is derived from both autochthonous and allochthonous sources. In Equation 12, A_d does not include assimilation of FPOM derived from autotrophic processes within the system (i.e., FPOM discharged by grazing) since this component is already included in the GPP term. Equations for calculating A_d were presented by McIntire and Colby (1978). For the stream model,

$$Prod(S) = E_g + E_s + E_c + E_i + E_v + E_a + W^* \qquad (13)$$

when the system is in a steady state. The terms E_s, E_c, E_i, E_v, and E_a are emergence and export losses for processes of grazing, shredding, collecting, invertebrate predation, vertebrate predation, and primary production (periphyton), respectively; and W^* represents export of waste (feces) derived from tissue elaborated with the system.

If an assumption is made about microbial assimilation of allochthonous material it is possible to calculate Prod(S) from the stream model output. Here it is assumed that microbial production (excluding periphyton) is 9% of assimilation ($A_b = R_b/0.91$, where A_b and R_b are microbial assimilation and respiration, respectively). The 9% figure was suggested by Odum (1957) and was discussed more recently by Jones (1975). At best, this value is only an educated guess that helps provide an example of total ecosystem behavior relative to different energy inputs.

Table 7 gives the annual total ecosystem production relative to six combinations of light energy and LPOM input. Runs S99, S96, S93, and S90 attempt to represent inputs along a continuum from a small, headwater stream with low light and high LPOM input (S99) to a higher order system with full sunlight and little or no LPOM input (S90). With

Table 7. The McIntire and Colby Stream Model*

Run	Light schedule	LPOM input, $g\ m^{-2}\ yr^{-1}$	Prod(S)–9% $g\ m^{-2}\ yr^{-1}$
S99	BC	947.27	47.38
S96	H1	473.63	37.34
S93	H2	157.88	34.55
S90	Full	0	39.32
S102	Full	947.27	105.17
S87	BC	0	20.97

*Annual total ecosystem production, Prod(S)–9%, predicted by the standard run for different combinations of light energy and LPOM input. Light schedules are the same as those in Table 6.

these inputs, the model predicts that total ecosystem production changes very little along the hypothetical continuum, with a mean Prod(S) value of 39.65 g m^{-2} yr^{-1}. In contrast, runs S102 and S87 represent energy inputs that are atypical of the continuum; S102 has both high light and high LPOM input, and S87 has low light and no LPOM input. The Prod(S) value for S102 is 2.65 times greater than the continuum mean, whereas that for S87 is only 53% of the continuum mean, i.e., greater than three standard deviations less than the mean.

Hypothesis VII:

Annual total ecosystem production per unit area does not change appreciably along the continuum from small, headwater streams to larger rivers.

DISCUSSION

A relatively new approach to the study of lotic ecosystems is based on the concepts of functional groups (Cummins, 1974) and paraspecies (Boling et al., 1975). Since this approach has identified such categories as scrapers (grazers), collectors, shredders, and macropredators, (see Figure 8, Cummins, 1974), which, at least superficially, appear similar to corresponding processes in the McIntire and Colby model, it is necessary to explore differences between the functional group perspective and the process viewpoint presented in this paper. In other words, is the conceptual framework for process studies really different from the concept of functional groups, or is it, in fact, just another way of expressing the same idea?

The functional group and process perspective both attempt to deal with the behavior of a diverse assemblage of species collectively and both depend on a good knowledge of natural history, particularly at the population level of organization. Beyond these similarities, the concepts exhibit some theoretical and operational differences.

For functional groups or paraspecies, the basic unit of interest is a taxonomic entity, usually a species population. This is true because these approaches involve the classification of taxonomic entities. For example, Boling et al. (1975, p. 200) defined a paraspecies "as a collection of one or more taxonomic species which are similar to each other throughout their residence time in the ecosystem," and Cummins (1974) provided a list of taxa that he classified into various functional groups. Moreover, sampling in the field often involves collecting, identifying, sorting, counting, and weighing organisms obtained from intensive study sites. Also, food consumption rates, if considered relevant, are obtained for representative individual taxa and extrapolated to a functional group. Another characteristic of the functional group approach is the emphasis on state variable dynamics, i.e., there is a tendency to compare sites by examining differences in group biomasses.

In contrast to the functional group approach, the basic unit of the process viewpoint is the dynamic process as defined in an earlier section. Again, taxonomic entities are not ignored, as a knowledge of their natural history plays an important role in the identification of processes relevant to the objective questions under consideration. Once the structure is defined, however, emphasis is on the capacity of the system or subsystem to process inputs and on mechanisms that regulate process dynamics. Furthermore, interest shifts from monitoring biomass of groups of species to investigating variables controlling the use of resource categories. Ideally, parameter estimation is based on direct measurement of processes in the field or laboratory rather than on measurements of population parameters. The approach to estimation is determined in part by the levels of resolution relevant to project objectives. For example, for lotic processes, estimation of primary production and the total cost of processing (community respiration) are compatible with the ecosystem level, whereas measurements of grazing, shredding, and collecting are finer level processes. Historically, ecologists have measured total ecosystem processes (e.g., primary production and community respiration) directly for many years, but finer level macroconsumer processes have not been studied directly until very recently; i.e., they were treated instead as the estimated mean behavior of a defined group of taxonomic entities studied at an even finer level.

On theoretical grounds, the process approach could be criticized for ignoring details of population interactions and life histories. This argument suggests that we cannot possibly understand ecosystems without first understanding how the constituent populations interact and that abstractions such as grazing, shredding, and collecting have little basis in reality. In other words, all the action is at the population level of organization so to speak. This argument is similar to the assertion that a knowledge of enzyme systems is necessary before we can understand anything about whole individual organisms. This point of view made life difficult for some ecologists during the spectacular expansion of molecular biology about 15 years ago. In any case, a counter argument suggests that natural history at the population level provides the basis for identifying relevant processes and that individual populations are simply swept up into a coarser level of organization, a level at which their detailed terminology is no longer appropriate. Certainly, resource categories (e.g., LPOM and FPOM) exhibit increases from allochthonous and internal inputs and decreases from the actions of physical and biological activities (processes). Such changes are real and can be monitored in the field. The process approach emphasizes the direct analysis of the capacity of the system to generate these changes and the mechanisms that regulate this capacity. Instead of analyzing and summing the activities of many individual species, the collective manifestation of these activities is investigated, usually by focusing on the dynamics of the corresponding resource category.

State variable measurement and parameter estimation are probably the most serious problems associated with the process viewpoint. For example, there is no standard methodology for measuring grazing, shredding, and collecting directly in the field. Although some investigators may feel that good methods for process measurements will never be available, recent studies of shredding indicate otherwise. One of the most promising approaches to the measurement of micro- and macro-consumer processes is to monitor the disappearance of a resource under controlled conditions in the field or laboratory. Leaf pack studies by Sedell et al. (1975) and Cummins et al. (1980) are notable examples. Lack of suitable methods for the direct measurement of macroconsumer processes is understandable, considering that we have only recently become interested in the capacity of lotic systems to process organic materials. It is important that new concepts stimulate methodological advancement rather than allowing current methodological limitations to restrict or inhibit the development of new concepts.

Interest in the biological processes of lotic ecosystems has been evident at some level of organization for at least 30 years or more. This paper has attempted to provide a little cleaner, more explicit conceptual framework for future process studies and simply represents a first iteration toward that goal. The hypotheses presented in the previous section are examples of tentative, process-oriented contentions that can serve as a basis for future research. Of even more importance, the process approach is well suited for investigating many applied problems, particularly when interest revolves around the capacity of flowing water systems to cope with perturbation. Indeed, advances in stream ecology have been significant during the past decade. Perhaps further development of the process viewpoint will help generate additional insights.

ACKNOWLEDGMENTS

Very special thanks are due my long-time friend and colleague W. S. Overton for providing the theoretical basis for the stream ecosystem model and for stimulating my interest in ecosystem modeling and the process concept. The invaluable programming assistance of J. A. Colby during the development of the stream model is gratefully acknowledged. He also contributed some important ideas toward the development of an approach to the investigation of process regulation.

REFERENCES

Birch, L. C., 1948, The Intrinsic Rate of Natural Increase of an Insect Population, *J. Anim. Ecol.* 17: 15–26.

Boling, R. H., R. C. Petersen, and K. W. Cummins, 1975, in B. C. Patten (Ed.) *Systems Analysis and Simulation in Ecology,* Vol. III, pp. 183-204, Academic Press, Inc., New York.

Cummins, K. W., 1974, Structure and Function of Stream Ecosystems, *Bioscience,* 24: 631-641.

———, 1975. The Ecology of Running Waters: Theory and Practice. in *Proceedings of the Sandusky River Basin Symposium,* International Reference Group on Great Lakes Pollution from Land Use Activities, pp. 277-293, Heidelberg College, Triffen, OH.

Cummins, K. W., G. L. Spengler, G. M. Ward, R. M. Speaker, R. W. Ovink, D. C. Mahan, and R. L. Mattingly, 1980, Processing of Confined and Naturally Entrained Leaf Litter in a Woodland Stream Ecosystem, *Limnol. Oceanogr.,* 25: 952-957.

Jones, J. G., 1975, Heterotrophic Microorganisms and Their Activity, in B. A. Whitton (Ed.), *River Ecology,* Blackwell Scientific Publications, Oxford, England.

Klir, G. J., 1969, An Approach to General Systems Theory, Van Nostrand–Reinhold, Princeton, NJ.

McIntire, C. D., and H. K. Phinney, 1965, Laboratory Studies of Periphyton Production and Community Metabolism in Lotic Environments, *Ecol. Monogr.,* 35: 237-258.

———, and J. A. Colby, 1978, A Hierarchical Model of Lotic Ecosystems, *Ecol. Monogr.,* 48: 167-190.

Odum, H. T., 1956, Primary Production in Flowing Waters, *Limnol. Oceanogr.,* 1: 102-117.

———, 1957, Trophic Structure and Productivity of Silver Springs, Florida, *Ecol. Monogr.,* 27: 55-112.

Overton, W. S., 1972, Toward a General Model Structure for a Forest Ecosystem, in J. F. Franklin, L. J. Dempster, and R. H. Waring (Eds.), *Proceedings of the Symposium on Research on Coniferous Forest Ecosystems,* pp. 37-47, Pacific Northwest Forest and Range Experiment Station, Portland, OR.

———, 1975. The Ecosystem Modeling Approach in the Coniferous Forest Biome, in B. C. Patten (Ed.) *Systems Analysis and Simulation in Ecology, Vol. III,* pp. 117-138, Academic Press, Inc., New York.

———, 1977, A Strategy of Model Construction, in C. A. S. Hall and J. W. Day Jr. (Eds.), *Ecosystem Modeling in Theory and Practice: An Introduction with Case Histories,* pp. 50-73, John Wiley & Sons, Inc., New York.

Peterson, R. C., and K. W. Cummins, 1974, Leaf Processing in a Woodland Stream, *Freshwater Biol.,* 4: 343-368.

Ricker, W. E., 1958, *Handbook of Computations for Biological Statistics of Fish Populations,* Bulletin 119, Fisheries Research Board of Canada.

Sedell, J. R., F. J. Triska, and N. S. Triska, 1975, The Processing of Conifer and Hardwood Leaves in Two Coniferous Forest Streams. I. Weight Loss and Associated Invertebrates, *Verh. Internat. Verein. Limnol.* 19: 1617-1627.

Warren, C. E., J. H. Wales, G. E. Davis, and P. Douderoff, 1964, Trout Production in an Experimental Stream Enriched with Sucrose, *J. Wildl. Mgt.,* 28: 617-660.

Zeigler, B. P., 1976, The Aggregation Problem, in B. C. Patten (Ed.), *Systems Analysis and Simulation in Ecology,* Vol. IV, pp. 299-311, Academic Press, Inc., New York.

4. PREDICTABILITY OF STREAM ECOSYSTEM MODELS OF VARIOUS LEVELS OF RESOLUTION

Joseph H. Wlosinski

Army Corps of Engineers
Waterways Experiment Station
Vicksburg, Mississippi

G. Wayne Minshall

Department of Biology
Idaho State University
Pocatello, Idaho

ABSTRACT

Data collected in a small desert stream (Deep Creek, Idaho–Utah) were used to create four data sets, each at a different level of resolution. Each data set included four plant groups, two detrital categories, decomposers, nutrients, fish, and macroinvertebrates. The invertebrates were sampled in a manner that allowed simulation of the ecosystem at four levels of resolution: (1) For the finest level of resolution, invertebrates were examined by size class within 15 taxonomic groups for a total of 37 variables. (2) Size classes within taxonomic groups were combined to produce 15 variables. (3) Similar taxonomic groups were combined to produce eight functional groups. (4) The eight functional groups were combined and were simulated as one group (coarsest level of resolution). Validation of the model consisted of comparing model predictions to the 95% confidence intervals of over 1000 measured values. Totals calculated for a number of state variables representing invertebrates at higher levels of resolution were compared with values predicted by models of lower resolution. Of 235 comparisons common to all four levels of resolution, 127, 142, 154, and 142 predicted values were within the 95% confidence intervals of empirical data for the finest through coarsest levels of resolution, respectively. Therefore we conclude that modeling the invertebrates at the functial level gave the best predictions of the Deep Creek ecosystem by the Desert Biome Aquatic Model.

INTRODUCTION

One of the major decisions that must be made when developing ecosystem models concerns the level of resolution at which the ecosystem is to be simulated. Although the literature concerning this point is not void of opinions (see e.g., Botkin, 1975; Caswell et al. 1972; Goodall, 1974; Innis, 1975; Lane et al., 1975; Overton, 1977; Root, 1975; Van Dyne, 1972; and Zeigler, 1972; 1976), few experiments have been made where measured values were compared with predicted values. Here we report on an experiment whose objective was to show the relationship between model resolution and predictability for the Aquatic Model developed by the Desert Biome, United States International Biological Program (US/IBP).

MATERIALS AND METHODS

To meet the objective we had to have: (1) a reasonably complete data set collected on an ecosystem scale that would allow for simulation at different levels of resolution, (2) a model framework general enough to handle the data, (3) information for setting parameters in the models, and (4) a workable means of evaluating and comparing models.

Data Set

From the fall of 1970 to the fall of 1972 and from the spring of 1975 to the fall of 1976, data were collected from Deep Creek as part of the US/IBP Desert Biome ecosystem studies. Deep Creek lies in Curlew Valley on the Utah–Idaho border. All data for this modeling effort were collected in a 300-m reach of stream with an average depth of 0.57 m, an average width of 3.7 m, a slope of 0.000918, and an average yearly discharge of 0.26 m^3/s.

Data normally collected were estimates of the mass of: (1) aquatic macrophytes (by species), (2) benthic diatoms, (3) benthic invertebrates (population numbers as well as biomass), measured at 1-mm size intervals to the lowest practicable taxonomic category, (4) fish (by species), (5) particulate matter, both suspended and benthic and organic and inorganic, (6) dissolved materials, both organic and inorganic, and (7) all materials drifting into and out of the reach. Also measured regularly were solar radiation, temperature, discharge, water depth, and velocity.

Because the invertebrates were measured to size class, usually at the species level, they could be manipulated in a hierarchical scheme. This hierarchy was the basis for a set of models simulating the same ecosystem at different levels of resolution. At the finest level of resolution (level 4), 15 taxonomic groups of invertebrates were used, with the most important groups subdivided into size classes. In all, the mass of invertebrates was represented by 37 state variables. At the next lower level of resolution (level 3), the size classes were combined, representing the invertebrates as 15 state variables. Similar taxonomic groups were then joined, representing invertebrates as eight functional or aggregated taxonomic groups (level 2). At the coarsest level of resolution (level 1), the eight functional groups were combined into one "super" invertebrate. Table 1 shows the relationship of the invertebrates at the four different levels of resolution. The invertebrates modeled included most of those measured. In the first year of sampling (1970–1971), the modeled invertebrates constistuted 99.8% of the total invertebrate biomass collected. This figure was 97.2 and 99.5% in 1975 and 1976, respectively.

Besides invertebrates, each of the four models included four plant groups, one species of fish, two decomposer categories, two detrital categories, and dissolved organic and inorganic matter. Three of the four plant groups were modeled at the species level. These were *Cladophora glomerata*, *Potomogeton pectinatus*, and *Chara vulgaris*. The fourth group included all diatoms. The fish (*Rhinichthys osculus*) was represented by compartments for eggs, young, and adults. Detritus was subdivided into fine and coarse particulates, and decomposers were divided into those found in the benthos and those found in the water column.

Model Framework

The model framework was developed by the Aquatic Section of the Desert Biome, following many of the modeling principles of Goodall (1974). The model is a general, nonlinear, deterministic computer program used for simulating stream ecosystems. It is general in the sense that specific components are not explicitly specified by the model, but are specified by the user at execution time, along with a series of switches and parameters that control and describe each component. This generality

Table 1. Relationship of Invertebrate State Variables
at Four Levels of Resolution

Level 4 (37 variables)	Size*	Level 3 (15 variables)	Level 2 (8 variables)	Level 1 (1 variable)
Hyalella azteca	Eggs 1 2 3 4 5 6 7–8			
Gammarus lacustris	E–6 7–12 13–17	. . . . . .		
Fluminicola rutsaliana	E–5 6–10	. . . . .		
Physa	E–12	. . . .	. . . . .	
Lymnaea	E–16	. . . .		
Gyraulus	E–4	. . . .		
Pisidium	E–4	. . . .		
Chironomidae	E–5 6–11	. . . .		
Dubiraphia giulianii	E–4 5–8 Adults	. . . . . .	. . . .	
Optioservus divergens	E–6	. . . .		
Hydropsyche occidentalis	E–6 9–16	. . . . . .	. . . .	
Cheumatopsyche anna	E–11	. . . .		
Erpobdella punctata	E–4 5–8 9–13 14–20 21–28	. . . . .	. . . .	
Helobdella elongata	E–4 5–8 9–11	. . . . .		
Enallagma anna	E–4 5–12 13–20	. . . .		

*Abbreviation E is eggs.

allowed for the simulation of the Deep Creek ecosystem at different levels of resolution. The computer program is written in FORTRAN IV, with difference equations with a time step of 1 day. If the approximation by difference equations over this time step leads to a negative value of an essentially nonnegative variable, the program reduces the time step as required.

Figure 1 is a simplified diagram of the major state variables and fluxes of energy considered by the model. It is simplified because the five variables shown may actually represent a much larger number of simulated variables, each one with as many flows as is represented in Figure 1. In addition to these variables, physical characteristics, such as depth, width, and water velocity, were simulated. Thus Deep Creek was modeled as a total ecosystem composed of living organisms and their abiotic environment.

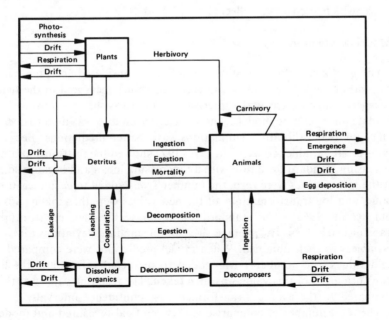

Figure 1. Simplified diagram of the energy flow within the Desert Biome Aquatic Model.

Model Parameters

The number of parameters in the four models from the coarsest through finest levels of resolution was 399, 1149, 1766, and 4212, respectively. Initial parameter estimates were taken directly from the literature or from experiments conducted within the Desert Biome. Most information obtained in this manner was at the species level. When making initial estimates for size classes of invertebrate species, the species value was usually used for all size classes. For the aggregated groups of invertebrates at lower levels of resolution, weighted mean parameters were calculated for the species involved. When a complete set of parameters was estimated, a calibration run would be made, with model predictions being compared to field-measured values. If it was thought that an appropriate change in some of the parameters would improve predictions, the change was made, and another simulation was run. If the prediction improved, the change was accepted.

The data decks used for the calibration runs used information collected in 1971 and 1972. Four decks were calibrated, each at a different level of resolution. Each simulation was for a period of 519 days, commencing April 1, 1971. The same coefficients were used in the validation simulations, which used data collected in 1975 and 1976.

Model Evaluation

The predicted value of a state variable was considered acceptable if it was within the 95% confidence intervals of values measured in the field. The percentage of acceptable predicted values was the criterion used to validate and compare models. The same criterion also was used to decide which of two calibration simulations gave better predictions. Because data for all variables collected in 1971–1972 showed the relationship of an increasing standard deviation with an increasing mean and because some of the mean values were zero, the number 1 was added to each measured value, and log transformations of the new values were then taken (Sokal and Rohlf, 1969). The standard deviation was calculated from the transformed values. The average value of all standard deviations for each variable was then obtained. When model predictions were compared to field values, the number 1 was added to predicted values, and the log transformation of this value was then taken. The resulting value was then compared to the average transformed 95% confidence intervals.

The total number of comparisons between field-measured and model-predicted values for the calibration simulations for levels 1 through 4 were 108, 364, 844, and 1996, respectively. The same figures for the validation years of 1975–1976 were 235, 619, 1339, and 3067, respectively.

Validation data came from the same section of stream that was sampled in 1971–1972. The stream was subdivided into three sections as part of an incident light manipulation experiment. Each section was approximately 100 m long and one section acted as a control. Another section was covered with white plastic from June 21, 1975, to November of that year and in 1976 from May 12 to September 20. The white plastic covering removed 70% of incident solar radiation. A third section was completely shaded with black plastic from June 21, 1976, until November of that year. Since the first sampling of both plants and animals in 1975 occurred on June 17, this date was used for the beginning of the 1975–1976 data sets. Simulation was for 496 days, ending on November 3, 1976, which corresponded to the last sampling date. Since each section was sampled independently, data sets were made for each of the three sections, each at four levels of resolution. In addition, since sampling occurred before the stream was shaded, a second group of data sets was constructed. These data, which were used for simulating the control section, contained values averaged over all three sections. The latter average was considered acceptable since the data in 1971 and 1972 were collected within the entire station and in 1975 were collected before the stream was artificially shaded. Thus a total of 16 validation simulations were made.

The variables used for validation and comparison at various levels of resolution were the four plant groups, coarse detritus, and invertebrate numbers and masses. For invertebrates this was facilitated by having the model internally subtotal or total values for the finer levels of resolution. For example, at the lowest level of resolution, biomass and populations for all invertebrates combined were the state variables. At the next level of resolution, the variables of prediction were the biomass and populations of each of eight functional groups. For comparison with the lowest level of resolution, the biomass and population of these groups were totaled, and the totals then became the variables for comparison. Similar totals were calculated for variables predicted by models of the third and fourth level of resolution.

RESULTS

Calibration Simulations

Final results of the calibration simulations from data collected in 1971–1972 are presented in Table 2. The results are reported as the percentage of model-predicted values that were within the 95% confidence intervals of field-measured values. Although the four plant groups were

Table 2. Percentage of Predicted Values Within the 95% Confidence
Interval of Field-Measured Values for the Data from 1971–1972

Variables	Number of comparisons	Level 4	Level 3	Level 2	Level 1
4 Plant groups	60	80.0	81.7	81.7	81.7
Coarse ditritus	16	87.5	87.5	87.5	87.5
Invertebrates					
Total					
Biomass	16	62.5	68.8	87.5	87.5
Populations	16	50.0	81.3	81.3	75.0
8 Functional groups					
Biomass	128	59.4	70.3	66.4	
Populations	128	62.5	63.3	70.3	
15 Taxonomic groups					
Biomass	240	55.4	70.8		
Populations	240	62.5	68.7		
36 Size groups					
Biomass	576	62.5			
Populations	576	59.2			

simulated as separate variables, their results were combined for reporting
purposes. The results for the biomass, as well as the population numbers
of invertebrates, at the three higher levels of resolution also were
combined.

Predictions for the four plant groups were very similar among levels of
resolution, being accepted between 80 and 81.7% of the time. All levels of
resolution predicted detritus acceptably 87.5% of the time. The results for
invertebrates were much more variable. When compared at the lowest
level of resolution, accepted predictions ranged between 62.5 and 87.5%
for biomass and 50.0 and 81.3% for population numbers. For the
invertebrates at higher levels of resolution, accepted biomass prediction
ranged between 55.4 and 70.8%. For population numbers, predicted
values were accepted between 59.2 and 70.3% of the time.

Validation Simulations

Eight simulations were made of the control section for the period from
June 17, 1975, to Nov. 3, 1976, two for each resolution level. One of the
simulations used initial values that were measured only in the control
section; whereas the other used initial values measured in all three
experimental sections before shading. In general, the results were similar
for the two sets of initial values within each level of resolution; thus their

Table 3. Percentage of Predicted Values Within the 95% Confidence
Interval of Measured Values for the Control Section in 1975–1976

Variables	Number of comparisons	Level 4	Level 3	Level 2	Level 1
4 Plant Groups	96	64.5	65.6	72.9	72.9
Coarse detritus	18	100.0	100.0	100.0	100.0
Invertebrates					
Totals					
Biomass	18	0	50.0	50.0	22.2
Populations	18	11.1	27.8	50.0	11.1
8 Functional groups					
Biomass	144	46.5	57.0	59.0	
Populations	144	34.7	27.1	31.9	
15 Taxonomic groups					
Biomass	270	50.0	63.3		
Populations	270	40.0	39.3		
36 Size groups					
Biomass	648	61.5			
Populations	648	49.7			

results have been combined (Table 3). Across levels of resolution, plants
were predicted within the 95% confidence intervals between 64.5 and
72.9% of the time. For all simulations, coarse detritus was predicted
correctly 100% of the time. Invertebrates, on the other hand, were
predicted very poorly. At the super invertebrate level, for both sets of
initial values, accepted biomass predictions ranged between 0 and 55.6%.
The range for population numbers was between 11.1 and 55.6%. For both
biomass and populations of invertebrates, levels 2 and 3 models were
better predictors than levels 1 and 4. For invertebrate variables at finer
resolution levels, predictions were acceptable between 22.2 and 64.4% of
the time.

Although not as many comparisons were made in the experimental
sections covered with plastic, the results were similar (Tables 4 and 5).
Plants were predicted correctly between 56.3 and 65.7% of the time for
both treatments. Two out of three values for detritus were within the 95%
confidence intervals for all levels simulating the section covered with
black plastic. For the section covered with white plastic, detritus was
predicted acceptably all three times. For the super invertebrate only
biomass for levels 2 and 3, under the black plastic, showed nonzero
predictability. Acceptable predictions for higher level invertebrate variables ranged between 22.2 and 80.0% for biomass and 0 and 43.5% for
populations.

Table 4. Percentage of Predicted Values Within the 95% Confidence Interval
of Measured Values for the Section Covered with White Plastic

Variables	Number of comparisons	Level 4	Level 3	Level 2	Level 1
4 Plant groups	35	62.9	65.7	65.7	65.7
Coarse detritus	3	100.0	100.0	100.0	100.0
Invertebrates					
Totals					
Biomass	3	0	0	0	0
Populations	3	0	0	0	0
8 Functional groups					
Biomass	24	25.0	29.2	62.5	
Populations	24	25.0	8.3	20.8	
15 Taxonomic groups					
Biomass	45	22.2	44.4		
Populations	45	20.0	8.8		
36 Size groups					
Biomass	108	53.7			
Populations	108	43.5			

Table 5. Percentage of Predicted Values Within the 95% Confidence
Interval of Measured Values for the Section Covered with Black Plastic

Variables	Number of comparisons	Level 4	Level 3	Level 2	Level 1
4 Plant groups	32	56.3	56.3	59.3	62.2
Coarse detritus	3	66.7	66.7	66.7	66.7
Invertebrates					
Totals					
Biomass	3	0	33.3	33.3	0
Populations	3	0	0	0	0
8 Functional groups					
Biomass	24	54.2	62.5	62.5	
Populations	24	4.2	4.2	0	
15 Taxonomic groups					
Biomass	45	51.1	80.0		
Populations	45	8.9	13.3		
36 Size groups					
Biomass	108	65.7			
Populations	108	40.7			

DISCUSSION

Predictability of Variables

The most striking difference of model predictability in comparing the results of the 1971–1972 simulations (Table 2) to those of 1975–1976 simulations (Table 6) was for total biomass and populations of invertebrates. In 1971–1972 plants were predicted correctly an average of 81.3% of the time. For the validation simulations of 1975–1976, this figure was 66.1%, a drop of approximately 15%. The predictions for coarse detritus were actually better in 1975 and 1976; correct predictions were 95.8% as compared with 87.5% for 1971–1972. For total invertebrates, however, the combined figures for biomass and populations dropped from a modest 74.2% for the 1971–1972 simulation to a very poor 21.9% for 1975–1976 simulations.

The measured values vs. model predictions for the two time periods were compared to explain the poor performance of model predictability. The models (except for level 4) predicted lower values in 1975–1976 than in 1971–1972 but not nearly as low as the values that were measured. The 1975–1976 figures are from the control section only, although similar or lower figures were measured under the sections covered with plastic. All 1975–1976 measured values were lower than the values measured during the same time of year in 1971–1972 (Figures 2 and 3). Many of the values for the later period were only one-fifth as high as the earlier ones, with some of the differences being nearly an order of magnitude.

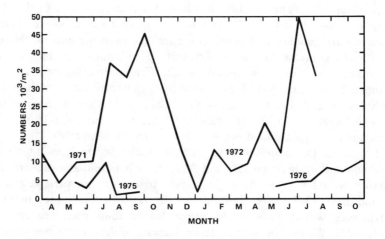

Figure 2. Observed numbers of invertebrates at Deep Creek.

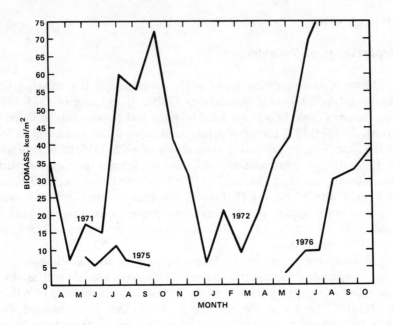

Figure 3. Observed biomass of 15 groups of invertebrates at Deep Creek.

Divergence of predicted and measured values occurred near beginning of the simulations in 1975. Increasing or high values were predicted by all models for both invertebrate numbers and biomass from July through September 1975. This prediction followed a pattern similar to that for data collected in 1971–1972 and to a diagrammatic model of insect numbers and biomass and multivoltine snail biomass presented by Hynes (1970). In contrast to this, invertebrate populations and biomass as measured in the control section showed low and decreasing numbers and biomass in 1975. In 1971 biomass rose from 14.9 to 59.1 kcal/m² between July and August. For the same two months in 1972, the figures were 70.6 and 86.2 kcal/m². In 1975 the July figure was 11.8 kcal/m², falling to 7.3 kcal/m² in August and down to 5.5 kcal/m² in September.

For any of the models to have predicted such a severe drop in biomass and populations between July and August, we believe that a catastrophic change would have to occur in respect to the driving variables; food would have to become limiting, or the ratio of carnivores to non-carnivores would have to be much higher than was measured in 1971–1972. We do not have evidence that any of these three mechanisms limited field populations.

The most notable change in 1975 conditions was that the spring and summer were cool and cloudy. For a 4-month period, May through August, solar radiation averaged 5210 kcal m^{-2} d^{-1} in 1971 and 5290 kcal m^{-2} d^{-1} in 1972. In contrast, in 1975 this figure was 4458 kcal m^{-2} d^{-1}. This reduction caused a concomitant drop in water temperature of approximately 2°C throughout the summer and a postponement of the runoff from the usual period in March or April to the last week of May and the first two weeks of June. Because of the late spring and the high water in May and early June, reproductive cycles may have been disrupted or a greater number of eggs may have been washed out of the system. Without replacement, even a low percentage of carnivores could cause a drop in population numbers and biomass. These scenarios would not be reflected by the models because eggs were not included as part of the population. Since eggs were not counted in the field, they were purposely left out of the models, so population increases were simulated at the time of egg hatching, rather than at egg laying.

Predictability at Various Levels of Resolution

Seven predicted variables were common to all four levels of resolution. These were the four plant groups, coarse detritus, and total invertebrate numbers and weights. Replicate samples of these seven variables, along with the four simulations at each level of resolution, allowed for 235 comparisons between measured and predicted values (Table 6). Of these, 142, 154, 142, and 127 were predicted within the 95% confidence intervals for levels 1 through 4, respectively. The plants were predicted better when invertebrates were modeled at lower resolution (Table 6). Part of this may be caused by the fact that most of the work on setting plant parameters dealt with the level 1 model, although there was very little difference in plant predictability for all four levels from the 1971–1972 data (Table 2). The predictability of detritus was the same for all four levels of resolution. Invertebrate weights and numbers combined were predicted most accurately at level 2, followed by levels 3, 1, and 4.

The results of all seven variables were expressed as a function of the number of parameters contained in models of different resolution (Figure 4). Even though the predictability is lower for the 1975–1976 simulations, the shape of the two curves is very similar. It is possible that the results of the model applied to the data from 1975–1976 show level 2 to be the best predictor and level 4 the worst predictor because of the effectiveness of setting parameters from the 1971–1972 data. If the same ratio that was found for the 1971–1972 data (80 : 87 : 90 : 89) is applied to the 565 correct

Table 6. Percentage of Predicted Values Within the 95% Confidence
Interval of Measured Values for the Two Simulations Concerning the Control
Section Plus the Simulation of the Two Light Experiments*

Variables	Number of comparisons	Level 4	Level 3	Level 2	Level 1
4 Plant groups	163	62.6	63.8	68.7	69.3
Coarse detritus	24	95.8	95.8	95.8	95.8
Invertebrates					
Totals					
Biomass	24	0	41.7	41.7	16.7
Populations	24	8.3	20.8	37.5	8.3
8 Functional groups					
Biomass	192	44.8	54.2	59.9	
Populations	192	29.7	21.9	26.6	
15 Taxonomic groups					
Biomass	360	46.7	63.1		
Populations	360	33.6	32.2		
36 Size groups					
Biomass	864	61.1			
Populations	864	47.8			

*All results are from 1975–1976.

Table 7. Chi-Square Test of Expected Results from 1971–1972 Predictions
Compared to Observed Results of the 1975–1976 Validation Tests

Level	Expected	Observed	Deviation	Deviation^2	X^2
4	131	127	4	16	0.122
3	142	142	0	0	0
2	147	154	7	49	0.333
1	145	142	3	9	0.062
	565	565		$\sum X^2 =$	0.517

predictions for the four levels during validation, the expected frequencies
would be as shown in Table 7. A chi-square analysis of expected vs
observed frequencies was made with a null hypothesis that the frequencies
were equal. Calculations (Table 7) show the chi-square value to be 0.517.
Since the value given in the table for $X^2_{0.05[3]}$ is 7.815 (Rohlf and Sokal
1969), the null hypothesis cannot be rejected.

The large differences for the predictability of invertebrates between levels of resolution (Figure 4) have been mediated by the more constant predictions for plants and detritus. The difference between levels caused by invertebrates alone can be seen in Figure 5. Each line represents the prediction of invertebrate weights and numbers at a particular level of resolution. The predictability of a particular model in relation to different hierarchical levels is striking. For example, nearly 55% of the time the model with 4212 parameters predicted invertebrate variables at the level of size classes correctly. When the size class variables were subtotaled and predictions checked at the species level, predictability fell to 41%. For predicting invertebrate variables at the functional level, the same model dropped to 37% accuracy and to just 4% accuracy in predicting the super invertebrate. Although the differences are not as great, the same type of change occurred for level 3 (1766 parameters) and 2. This same type of hierarchical difference was not found for the results of the 1971–1972 simulations.

State variables of ecosystem models often represent "system components" that are at a lower level of resolution than the variable of interest.

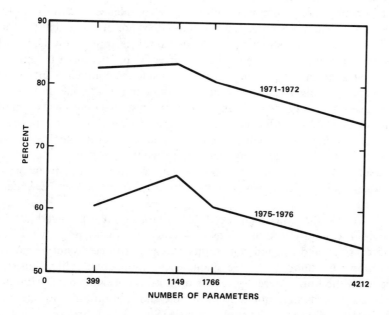

Figure 4. Predictability of seven variables by models of four levels of resolution.

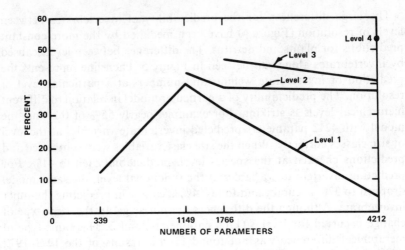

Figure 5. Percentage of invertebrates predicted correctly by models of four levels of resolution in 1975–1976.

For example, a model to predict populations of a particular fish species may have state variables that represent a number of size classes for the species. The prediction from level 4 (where nearly 55% of the system components were predicted within the 95% confidence intervals but only 4% of the system was predicted correctly) does not disprove that this can be done, but it does show a possible pitfall for its application. Unless all the system components are predicted correctly, prediction at the highest levels of aggregation may be incorrect (outside the confidence intervals). A modeler, being unsatisfied with the prediction of a model in which a state variable consisted of aggregated ecosystem components, might choose to subdivide the one state variable. Using this choice he must predict a number of variables accurately instead of only one. Even if the majority of subdivided components were within confidence intervals of measured data, the remaining components could give rise to incorrect totals. This is probably what happened during validation of the Aquatic Model.

We realize the experiment would be of more value if predictability for the four calibration simulations was equal. Unfortunately, within resource, time, and method constraints, this objective could not be met. In addition, the amount of time spent or the number of calibration runs made on the four different levels of resolution were not equal. Although accurate records were not kept, most time and calibration runs were used on level 1, followed by levels 3, 2, and 4. Since all calibration runs were made at one level of resolution before working on another level, experience gained when working at other levels should have benefited level 4 the most, followed by levels 2, 3, and 1. We also realize that running the model in a stochastic fashion, where coefficients could have been varied, may have changed the results. Again, the resources for evaluating these effects were not available. Nevertheless, on the basis of the experiment reported here, modeling the invertebrates at the functional level gave the best predictions of the Deep Creek ecosystem by the Desert Biome Aquatic Model.

ACKNOWLEDGMENTS

We gratefully acknowledge the contributions of an interdisciplinary team within the Desert Biome who helped develop the Aquatic Model. Within this group we especially thank J. Deacon, C. W. Fowler, D. W. Goodall, W. Grenney, A. Holman, R. Jeppson, D. Koob, R. Kramer, D. Porcella, F. Post, and C. Stalnaker.

REFERENCES

Botkin, D. B., 1975, Function Groups of Organisms in Model Ecosystems, in S. A. Levin (Ed.), pp. 98–102, *Ecosystem Analysis and Prediction*, Society for Industrial and Applied Mathematics, Philadelphia.

Caswell, H., H. E. Koenig, J. A. Resh, and Q. E. Ross, 1972, An Introduction to System Science for Ecologists, in B. C. Patten (Ed.), *Systems Analysis and Simulation In Ecology*, Vol. II, pp. 3–78, Academic Press, Inc., New York.

Goodall, D. W., 1974. Problems of Scale and Detail in Ecological Modelling, *J. Environ. Manag.*, 2: 149–157.

Hynes, H. B. N., 1970, *The Ecology of Running Waters*, University of Toronto Press, Toronto.

Innis, G., 1975, One Direction for Improving Ecosystem Modeling, *Behav. Sci.*, 20: 68–74.

Lane, P. A., G. H. Lauff, and R. Levins, 1975, The Feasibility of Using a Holistic Approach in Ecosystem Analysis, in S. A. Levin (Ed.), *Ecosystem Analysis and Prediction*, pp. 111–130, Society for Industrial and Applied Mathematics, Philadelphia.

Overton, W. S., 1977, A Strategy of Model Construction, in C. A. S. Hall and J. W. Day, Jr. (Eds.), *Ecosystem Modeling in Theory and Practice: An Introduction with Case Histories*, pp. 49–73, John Wiley & Sons, Inc., New York.

Rohlf, F. J., and R. R. Sokal, 1969, *Statistical Tables*, W. H. Freeman and Company, San Francisco.

Root, R. B., 1975, Some Consequences of Ecosystem Texture, in S. A. Levin (Ed.), *Ecosystem Analysis and Prediction*, pp. 83–97, Society for Industrial and Applied Mathematics, Philadelphia.

Sokal, R. R., and F. J. Rohlf, 1969, *Biometry*, W. H. Freeman and Company, San Francisco.

Van Dyne, G. M., 1972, Organization and Management of an Integrated Ecological Research Program—with Special Emphasis on Systems Analysis, Universities, and Scientific Cooperation, in J. N. R. Jeffers (Ed.), *Mathematical Models in Ecology*, pp. 111–172, Blackwell Scientific Publications, Oxford, England.

Zeigler, B. P., 1972, The Base Model Concept, in R. R. Mahler and R. A. Ruberti (Ed.), *Theory and Applications of Variable Structure Systems*, pp. 69–93, Academic Press, Inc., New York.

———, 1976, The Aggregation Problem, in B. C. Patten (Ed.), *Systems Analysis and Simulation in Ecology*, Vol. VI, pp. 299–311, Academic Press, Inc., New York.

5. AN ALTERNATIVE FOR
CHARACTERIZING STREAM SIZE

R. M. Hughes and J. M. Omernik
Environmental Protection Agency
Corvallis, Oregon

ABSTRACT

Stream order is often useful in expressing relative stream and watershed size within a physiographically and climatically homogeneous basin. However, there are disadvantages when comparing stream and watershed size on a regional or national scale because of, among other things, the lack of uniform map specifications, the lack of agreement on the definition of a first-order stream, and the problem of deciding the appropriate map scale to determine stream order.

We examined published studies on 71 watershed/stream ecosystems in 31 states within major physiographic and climatic regions of the conterminous United States. Our objective was to demonstrate the value of using discharge characteristics and watershed area instead of stream order to provide a rough but useful characterization of watershed and stream sizes throughout the nation. We found that streams of a given order show vast ranges in discharge and watershed area, greatly overlapping the ranges for higher and lower order streams. Therefore we suggest using mean annual discharge per unit area and watershed area instead of stream order to quantify stream and watershed size.

INTRODUCTION

Quantification of stream characteristics is necessary to study and manage the nation's streams and to facilitate communication among a diverse group of scientists and managers throughout the United States. Currently, stream order (Strahler, 1957) is used by scientists and managers throughout the nation to relate stream characteristics. The term is commonly used to convey an understanding of stream size, watershed size, and, in some instances, even quantity of water. Although stream order has been and will probably continue to be a useful means of expressing relative size within a physiographically and climatically homogeneous basin, the term is often used beyond its capacity.

Several problems arise when stream order is used to represent stream size (Hughes and Omernik, 1981). (1) There is little agreement on how to include perennial, intermittent, and ephemeral streams in determining stream order. Are they considered as equals regardless of flow frequency? If so, note that some hydrologists use all map crenulations in a watershed although some channels only have flows during major storm or snowmelt periods. If not, how permanent must a stream be, given the short history of some stream gauging? Are Alaskan streams that freeze solid during the winter considered permanent or temporary? (2) There is little agreement as to which scale to use in determining order. For instance, depending on the map scale selected, a stream such as Oak Creek at Corvallis, OR, can be categorized as unordered, or first- third- or fourth-order. (3) All regions are not mapped to the same scale, under the same specifications, or during similar weather periods. Differences in stream density (and hence stream order) can be a function of different map compilation or field annotation processes. These differences often can be seen along neat lines between adjoining maps that have been compiled at different times under different specifications. Hence the small streams used to derive stream order are not all mapped in a uniform manner from one region to another in the United States, much less from one country to another.

Aside from the problem of determining stream order, the term provides little quantifiable information about streams and their watersheds. Stream order was developed to describe the linear geomorphic characteristics of small stream networks within a homogeneous physiographic area. It does not, nor was it intended to, address area, relief, or discharge. Smart (1972) felt that stream order was a mediocre approach even for the primary classification of stream networks, adding that watershed area may be preferable. Stream order provides no information about climate in the vicinity of a stream or annual and seasonal variations in discharge. Yet this information is useful for understanding the human uses and the community structure and function of all streams. Moreover, stream order has little or no meaning when considering distributaries, channelized or ditched streams, influent or disappearing streams, or streams arising from or flowing through alluvium, large springs, lakes, wetlands, snowfields, or glaciers. In karst and glaciated regions, streams may have discharges an order of magnitude greater than higher order streams in the same basin. Also, as pointed out by Hynes (1970), the stream order resulting from the junction of two equal-order tributaries can be increased whether a tributary is only a few hundred feet long or several miles long. Finally, the continuous addition of small tributaries of order n–1 to a stream of order n can greatly change the discharge and watershed area of a stream without changing its order. Shreve's (1966) link analysis and Scheidegger's (1965) consistent scheme of stream ordering classify each stream segment by the number of first-order streams flowing into it. This alleviates the last problem but not the others.

We suggest that using watershed area and mean annual discharge per unit area (i.e., unit discharge in cubic meters per second per square kilometer or preferably centimeters per year) rather than stream order will lead to a more accurate understanding of stream size, watershed size, and quantity of water. We believe this use will alleviate many of the difficulties described.

MATERIALS AND METHODS

We examined data on 71 streams in 31 states within most of the major physiographic regions (Fenneman, 1946) and ecoregions (Bailey, 1976) of the conterminous United States (Figure 1, Table 1). Ecoregions are large regional ecosystems with similar climate, landform, soils, vegetation, and fauna. We selected small streams that have been studied rather intensively and were covered by 1 : 24,000 scale U. S. Geological Survey topographic maps. We used these maps to determine watershed areas (by planimeter) and stream orders (from solid and broken blue lines). Unit discharges for the stream sites were determined directly from U. S. Geological Survey data, when possible, or from unit discharge isolines constructed from U. S. Geological Survey data on nearby streams. Unit discharge isolines were used to show regional patterns in runoff by the U. S. Geological Survey (1970) and by Muckleston (1979). Extrapolations from unit discharge isolines are also useful to estimate discharge in regions where watershed boundaries are difficult or impossible to define from topographic maps.

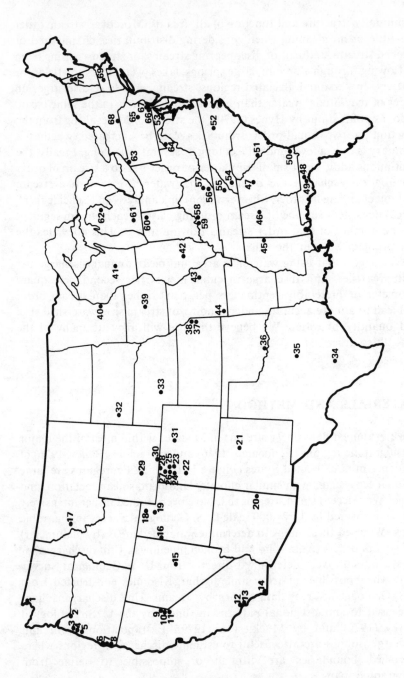

Figure 1. Locations of study sites.

Table 1. Stream Order, Watershed Area, Mean Annual Discharge, and Mean Annual Discharge per Unit Area of Selected Study Streams

Number	Stream	Investigator	Stream order	Watershed area, km^2	Mean annual discharge, m^3/sec	Mean annual discharge per unit area, cm/yr
1	Whatcum Creek	Orrell (1980)	3	145.3	9.85	213.87
2	Bull Run	Fredriksen et al. (1974)	3	277.1	24.09	274.32
3	Siletz River at Siletz, OR	Hughes and Omernik (1981)	5	523.2	45.02	271.53
4	Willamina River at Willamina, OR		5	168.4	7.00	131.06
5	Oak Creek	Kerst and Anderson (1974)	3	7.5	0.34	141.8
6	Winchester Creek	Oregon Department Fish and Wildlife (1969)	3	30.0	1.52	158.75
7	Big Creek		3	8.8	0.44	158.75
8	Two Mile Creek		3	24.1	1.21	158.75
9	Ward Creek	Leonard et al. (1979)	4	25.1	0.63	79.65
10	Blackwood Creek		4	29.0	1.89	110.49
11	Prosser Creek	Needham and Usinger (1956)	3	137.8	2.41	55.12
12	Sespe Creek	Swift et al. (1975)	4	650.1	1.42	6.86
13	Santa Ana river		4	2097.9	4.59	6.86
14	Temecula River		3	339.3	0.74	6.86
15	Hot Creek	Hubbs et al. (1974)	4	33.4	0.13	12.19
16	Thoms Creek	Winget and Reichert (1976)	3	21.2	0.25	36.83
17	Owl Creek	Oswood (1979)	3	362.6	2.97	25.91
18	Temple Fork	Pearson and Kramer (1972)	3	37.3	0.25	20.83
19	Red Butte Creek	Bond (1979)	3	18.9	0.12	20.83
20	Ord Creek	Rinne (1978)	3	21.5	0.14	20.83
21	Santa Fe River at Santa Fe, NM	Molles and Gosz (1980)	3	32.6	0.09	8.64

Table 1, continued

Number	Stream	Investigator	Stream order	Watershed area, km²	Mean annual discharge, m³/sec	Mean annual discharge per unit area, cm/yr
22	Cement Creek	Allan (1975)	4	69.7	0.91	41.4
23	Service Creek	Shirazi et al. (1980 draft)	3	100.5	1.26	39.62
24	Fish Creek		3	89.4	0.39	13.72
25	Grassy Creek		1	66.8	0.04	2.03
26	Yampa River at Steamboat Springs, CO	Hughes and Omernik (1981)	5	1,564.4	13.00	26.16
27	Little Snake River at Lily, CO		5	9660.7	15.85	5.08
28	Little Snake River at Slater, CO		5	738.2	5.97	25.65
29	Little Popo Agie at Lander, WY	Binns and Eiserman (1979)	5	323.8	2.3	22.35
30	Deadman Creek		2	2.3	0.02	24.13
31	North St. Vrain Creek	Pennak and Van Gerpen (1947)	3	274.5	2.25	25.91
32	Rapid Creek	Stewart and Thilenius (1964)	4	1559.2	1.70	3.56
33	Otter Creek	Van Velson (1979)	2	9.1	0.01	3.56
34	San Antonio River at San Antonio, TX	Hubbs et al. (1978)	4	2641.8	5.78	6.86
35	Bosque River at Waco, TX	Lind (1971)	4	4410.8	12.06	8.64
36	Rush Creek	Barclay (1979)	4	60.6	0.13	6.86
37	Mill Creek	Hazel et al. (1979)	4	100.8	0.72	22.35
38	Cedar Creek		4	128.5	0.91	22.35
39	Four Mile Creek	Johnson (1978)	3	50.5	0.30	19.05
40	Valley Creek	Waters (1964)	2	6.7	0.03	13.72
41	Lawrence Creek	Hunt (1969)	1	136.8	1.05	24.13
42	Kaskaskia River at Arcola, IL	Larimore and Smith (1963)	3	9137.5	84.93	29.21
43	Courtois Creek	Ryck (1974)	4	595.7	5.54	29.21
44	James River at Galena, MO	Dieffenbach and Ryck (1976)	5	2556.3	25.16	30.99

#	Stream	Reference				
45	Luxapalila River at Columbia, MS	Arner et al. (1976)	5	2095.3	33.22	50.04
46	White Oak Creek	Lawrence and Webber (1979 draft)	4	22.0	0.36	51.82
47	Rooty Creek	North et al. (1974)	3	105.7	1.21	36.32
48	Ford's Arm	Turner et al. (1977)	3	4.4	0.07	48.26
49	Meginniss Arm		2	8.0	0.12	48.26
50	Satilla River at Brunswick, GA	Benke et al. (1979)	4	9142.7	74.98	25.91
51	Upper Three Runs	McFarlane (1976)	4	490.0	6.70	43.18
52	New Hope Creek	Hall (1972)	4	57.0	0.62	34.54
53	Rhode River	Correll (1977)	4	9.8	0.12	39.62
54	Coweeta Creek	Monk et al. (1977)	3	16.3	0.51	98.3
55	Walker Branch	Harris (1977)	3	1.04	0.02	63.75
56	Buckhorn Creek	Kuehne (1962)	4	113.4	1.61	44.96
57	Clemons Fork	Lotrich (1973)	3	5.7	0.08	44.96
58	Morgan's Creek	Minshall (1967)	1	0.5	0.01	41.40
59	Doe Run	Minckley (1963)	3	182.1	2.38	41.40
60	Black Creek	Gorman and Karr (1978)	2	54.9	0.51	29.21
61	Augusta Creek	Mahan and Cummins (1978)	3	71.5	0.71	30.99
62	Au Sable River at Mio, MI	Richards (1976)	4	4677.5	40.9	27.69
63	Linesville Creek	Coffman et al. (1971)	2	23.1	0.35	48.26
64	Fernow	Kochenderfer and Aubertin (1975)	3	14.8	0.32	69.09
65	Mahantango Creek	Pionke and Weaver (1977)	4	420.1	6.2	46.48
66	Conowingo Creek	Stauffer and Hocutt (1980)	3	278.4	3.65	41.40
67	White Clay Creek	Moeller et al. (1979)	4	122.0	1.60	41.40
68	Owego Creek	Sheldon (1968)	4	479.2	8.38	55.12
69	Roaring Brook	McDowell and Fisher (1976)	2	1.3	0.02	53.59
70	Hubbard Brook	Vitousek (1977)	4	30.8	0.67	69.09
71	Bear Brook	Fisher and Likens (1973)	2	1.3	0.03	69.09

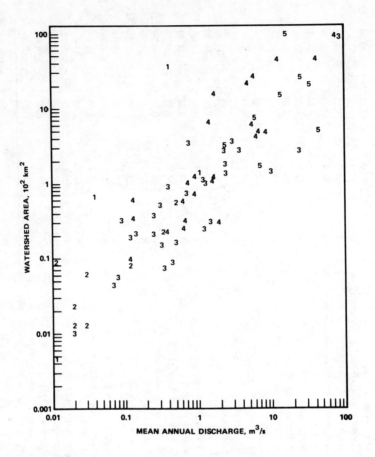

Figure 2. Mean annual discharges and watershed area relative to stream order. Numbers refer to stream order.

RESULTS

A plot of the log of watershed area against the log of mean annual discharge for first- to fifth-order streams is shown in Figure 2. Both watershed area and mean annual discharge vary over an order of magnitude within all stream orders represented. Consequently, streams of a given order may have watershed areas and mean annual discharges that are considerably greater than higher order streams. Similar variability exists even if streams within the same ecoregion (Bailey, 1976) are examined. For example, among the following pairs of streams, 22 and 23, 48 and 49, and 66 and 67, the lower order streams have greater watershed areas and mean annual discharges than the higher order streams although the unit discharges for each pair are similar or identical.

DISCUSSION

The three major advantages of using watershed area and unit discharge instead of stream order to quantify stream and watershed size are: (1) They provide a quick and fairly accurate estimate of evapotranspiration relative to precipitation. (2) They relate watershed and stream characteristics that have considerable biological significance. (3) Uniform understanding of stream size and watershed size is provided regardless of the available scale of topographic maps, the permanence of streams, or the presence of other bodies of water in the channels. This allows more meaningful comparisons of stream and watershed size.

The use of watershed size and unit discharge also leads to the use of other stream-watershed relationships. Unit discharges can be related to precipitation, evapotranspiration, and groundwater recharge when these components of the hydrologic cycle are expressed on the basis of unit area. This allows much more adequate modeling of the fate of precipitation, a very important consideration in watershed studies.

Mean annual values of low, average, and high discharges and their standard deviations, which are important determinants of habitat stability, can be estimated from the same data. Two-year flood flows, which are generally considered the major channel-forming events, can be estimated by plotting peak discharges against their recurrence intervals (Morisawa, 1968). The recurrence interval equals the number of years of record plus 1, divided by the rank of the peak discharge (the highest discharge is ranked as 1, the second highest as 2, etc.).

Minimum discharges and flow-duration curves (plots of discharge against time) can be used to classify watersheds by their water-storage capacities (Orsborn, 1976; Morisawa, 1968). Steep flow-duration curves and low minimum discharges indicate considerable direct runoff and wide fluctuations in flows. Flat flow-duration curves and relatively high minimum discharges indicate substantial storage and more equalized flows. Watershed area can be related to discharge, mean velocity, depth, and width of streams in a downstream direction (Stall and Yang, 1970). This requires gauge data. It is done by using flow frequency and the logarithm of watershed area as predictor variables and discharge, mean velocity, width, and mean depth as dependent variables. The hydraulic geometry equations are then produced by linear regression.

Discharge, mean velocity, width, and mean depth are more meaningful than stream order for predicting changes in production, respiration, particulate organic matter, and community structure along the stream continuum (Vannote et al., 1980). For example, for a total of four rivers in Oregon, Idaho, Michigan, and Pennsylvania, Moeller et al. (1979) stated that mean annual discharge, watershed area, stream links, and stream order have correlation coefficients of 0.96, 0.89, 0.80, and 0.79, respectively, with dissolved organic carbon (DOC) transport. There is considerable intercorrelation among all these parameters. When mean annual discharge was omitted from the stepwise multiple regression analysis, watershed area explained 80% of the variance. Correlations with the first canonical variable (which accounted for 83.5% of the among-group variability in a discriminant analysis) indicated that mean annual precipitation ($r = -0.88$) and watershed area ($r = 0.78$) were the two most important variables out of 15 for explaining the classification of 27 streams in Europe and North America (Cushing et al., 1980). The other variables considered were phosphate, total dissolved solids, langleys per year, maximum diurnal water temperature fluctuation, annual degree days, summer and winter base flows, gradient, nitrate, annual number of storms 5 and 10 times greater than base flow, terrestrial litter input, and stream length/watershed area. Also, where stream order is meaningless, such as in distributaries or disappearing streams or where surface and subsurface watersheds differ, at least discharge can be measured and the data compared with that from more typical streams.

On the other hand, there are three important disadvantages of using watershed area and unit discharge rather than stream order to characterize the size of streams: (1) Watershed area and unit discharge estimates may include considerable error in small, arid, poorly defined, and topographically complex watersheds or where surface and subsurface watersheds differ. (2) Estimates of average discharge may include considerable bias when developed from short-duration gauge data. (3) Watershed area and unit discharge take more effort to determine.

We caution stream ecologists to use watershed area and unit discharge together to characterize watershed and stream size; these parameters have less meaning alone than combined. To better understand the distribution, abundance, and functions of stream biota, we also encourage stream ecologists to incorporate other discharge characteristics into their studies (mean annual values of low, average, and high discharges and their standard deviations and mean velocity, depth, and width and their standard deviations). We are not advocating the use of unit discharge and watershed area in all hydrological or stream ecology models or as descriptors of channel networks. Like stream order, these terms should not be extended into areas for which they were not designed. We only emphasize that unit discharge and watershed area provide a simple, universally useful, and relatively accurate general characterization of stream and watershed size.

REFERENCES

Allan, J. D., 1975, The Distributional Ecology and Diversity of Benthic Insects in Cement Creek, Colorado, *Ecology*, 56(5): 1040–1053.

Arner, D. H., H. R. Robinette, J. E. Frasier, and M. H. Gray, 1976, *Effects of Channelization of the Luxapalila River on Fish, Aquatic Invertebrates, Water Quality and Fur Bearers*, Report FWS/OBS-76-08, U. S. Fish and Wildlife Service, Department of the Interior, Washington, D.C., NTIS, Springfield, VA.

Bailey, R. G., 1976, *Ecoregions of the United States* (map), Forest Service, U. S. Department of Agriculture, Intermountain Region, Ogden, UT.

Barclay, J. S., 1979, The Effects of Channelization on Riparian Vegetation and Wildlife in South Central Oklahoma, in *Strategies for Protection and Management of Floodplain Wetlands and Other Riparian Ecosystems*, Proceedings of a Symposium, Dec. 11–13, 1978. Callaway Garden, GA, Report #GTR-WO-12, pp. 129–138. Forest Service, U. S. Department of Agriculture, Washington, DC.

Benke, A. C., D. M. Gillespie, F. K. Parrish, T. C. Van Arsdall, Jr., R. J. Hunter, and R. L. Henry III, 1979, *Biological Basis for Assessing Impacts of Channel Modification: Invertebrate Production, Drift, and Fish Feeding in a Southeastern Blackwater River*, Environmental Resources Center, Report # ERC 06-79, Georgia Institute of Technology, Atlanta.

Binns, N. A., and F. M. Eiserman, 1979, Quantification of Fluvial Trout Habitat in Wyoming, *Trans. Am. Fish. Soc.*, 108(3): 215–228.

Bond, H. W., 1979, Nutrient Concentration Patterns in a Stream Draining a Montane Ecosystem in Utah, *Ecology*, 60(6): 1184–1196.

Coffman, W. P., K. W. Cummins, and J. C. Wuycheck, 1971, Energy Flow in a Woodland Stream Ecosystem. I. Tissue Support Trophic Structure of the Autumnal Community, *Arch. Hydrobiol.*, 68(2): 232–276.

Correll, D. L., 1977, An Overview of the Rhode River Watershed Program, in D. L. Correll (Ed.), *Watershed Research in Eastern North America, A Workshop to Compare Results*, Feb. 28–Mar. 3, 1977, Edgewater, MD, Vol. 1, pp. 105–123, Smithsonian Institution, Washington, D.C.

Cushing, C. E., C. D. McIntire, J. R. Sedell, K. W. Cummins, G. W. Minshall, R. C. Peterson, and R. L. Vannote, 1980, Comparative Study of Physical-Chemical Variables of Streams Using Multivariate Analyses, *Arch. Hydrobiol.*, 89: 343–352.

Dieffenbach, W., and F. Ryck, Jr., 1976, *Water Quality Survey of the Elk, James and Spring River Basins of Missouri, 1964–1965*, Aquatic Series No. 15, Missouri Department of Conservation, Jefferson City.

Fenneman, N. M., 1946, *Physical Divisions of the United States* (map), U. S. Geological Survey, Reston, VA.

Fisher, S. G., and G. E. Likens, 1973, Energy Flow in Bear Brook, New Hampshire: An Integrative Approach to Stream Ecosystem Metabolism, *Ecol. Monogr.*, 43(4): 421–439.

Fredriksen, R. L., D. G. Moore, and L. A. Norris, 1974, The Impact of Timber Harvest, Fertilization and Herbicide Treatment on Streamwater Quality in Western Oregon and Washington, in B. Bernier and C. H. Winget (Eds.), *Forest Soils and Forest Land Management*, Proceedings of the Fourth North American Forest Soils Conference, Aug 1973, pp. 283–313, Les Presses de L'Universite, Laval, Quebec.

Gorman, O. T., and J. R. Karr, 1978, Habitat Structure and Stream Fish Communities, *Ecology*, 59(3): 507–515.

Hall, C. A., 1972, Migration and Metabolism in a Temperate Stream Ecosystem, *Ecology*, 53: 585–604.

Harris, W. F., 1977, Walker Branch Watershed: Site Description and Research Scope, in D. L. Correll (Ed.), *Watershed Research in Eastern North America, A Workshop to Compare Results*, Feb. 28–Mar. 3, 1977, Edgewater, MD, Vol. I, pp. 5–17, Smithsonian Institution, Washington, DC.

Hazel, R. H., C. E. Burkhead, and D. G. Huggins, 1979, *The Development of Water Quality Criteria for Ammonia and Total Residual Chlorine for the Protection of Aquatic Life in Two Johnson County, Kansas Streams*, Kansas Water Resources Research Institute, University of Kansas, Lawrence.

Hubbs, C. L., R. R. Miller, and L. C. Hubbs, 1974, Hydrographic History and Relict Fishes of the North-Central Great Basin, *Mem. Calif. Acad. Sci.*, No. 7.

Hubbs, C., T. Lucier, G. P. Garrett, R. J. Edwards, S. M. Dean, E. Marsh, and D. Belk, 1978, Survival and Abundance of Introduced Fishes near San Antonio, Texas, *Texas J. Sci.*, 30(4): 369–376.

Hughes, R. M., and J. M. Omernik, 1981, Use and Misuse of the Terms, Watershed and Stream Order, in L. Krumholz (Ed.), *Warmwater Streams Symposium*, Mar. 9–11, 1980, Knoxville, TN, pp. 320–326, Southern Division, American Fisheries Society, Bethesda, MD.

Hunt, R. L., 1969, Effects of Habitat Alteration on Production, Standing Crops and Yield of Brook Trout in Lawrence Creek, Wisconsin, in T. G. Northcote (Ed.), *Proceedings of a Symposium on Salmon and Trout in Streams*, H. R. MacMillan Lectures, 1968, pp. 281–312, University of British Columbia, Vancouver.

Hynes, H. B. N., 1970, *The Ecology of Running Waters*. University of Toronto Press, Toronto.

Johnson, H. P., 1978, *Development and Testing of Mathematical Models as Management Tools for Agricultural Nonpoint Pollution Control*, Annual Report 1977–1978, EPA Grant No. R-804102, Department of Agricultural Engineering, Iowa State University, Ames.

Kerst, C. D., and N. H. Anderson, 1974, Emergence Patterns of Plecoptera in a Stream in Oregon, USA, *Freshwater Biol.*, 4: 205–212.

Kochenderfer, J. N., and G. M. Aubertin, 1975, Fernow Experimental Forest, West Virginia, in *Municipal Watershed Management*, Symposium proceedings, General Technical Report NE-13, pp. 14–24, Forest Service, U. S. Department of Agriculture, Upper Darby, PA.

Kuehne, R. A., 1962, A Classification of Streams, Illustrated by Fish Distribution in an Eastern Kentucky Creek, *Ecology*, 43(4): 608–614.

Larimore, R. W., and P. W. Smith, 1963, The Fishes of Champaign County, Illinois, as Affected by 60 Years of Stream Changes, *Ill. Nat. Hist. Survey Bull.*, 28(2): 299–382.

Lawrence, J. M., and C. Webber, 1979 draft, Annual Report on Evaluation of Selected Chemical, Physical, and Biological Characteristics on White Oak and Wesobulga Creek Prior to Construction of Site No. 17A, Crooked Creek, Watershed, Department of Fisheries and Allied Agricultures, Auburn University, Auburn, AL.

Leonard, R. L., L. A. Kaplan, J. F. Elder, R. N. Coats, and C. R. Goldman, 1979, Nutrient Transport in Surface Runoff from a Subalpine Watershed, Lake Tahoe Basin, California, *Ecol. Monogr.*, 49(3): 281–310.

Lind, O. T., 1971. The Organic Matter Budget of a Central Texas Reservoir, *Reserv. Fish. Limnol. Spec. Publ.*, 8: 193–202.

Lotrich, V. A., 1973, Growth, Production, and Community Composition of Fishes Inhabiting a First-, Second-, and Third-Order Stream of Eastern Kentucky, *Ecol. Monogr.*, 43: 377–397.

McDowell, W. H., and S. G. Fisher, 1976, Autumnal Processing of Dissolved Organic Matter in a Small Woodland Stream Ecosystem, *Ecology*, 57: 561–569.

McFarlane, R. W., 1976, Fish Diversity in Adjacent Ambient, Thermal, and Post-Thermal Freshwater Streams, in G. W. Esch and R. W. McFarlane (Eds.), *Thermal Ecology II*, ERDA Symposium Series, CONF-750425, pp. 268–271, NTIS, Springfield, VA.

Mahan, D. C., and K. W. Cummins, 1978, *A Profile of Augusta Creek in Kalamazoo and Barry Counties, Michigan*, Technical Report No. 3, W. K. Kellogg Biological Station, Hickory Corners, MI.

Minckley, W. L., 1963, *The Ecology of a Spring Stream Doe Run, Meade County, Kentucky*, Wildlife Monograph No. 11.

Minshall, G. W., 1967, Role of Allochthonous Detritus in the Trophic Structure of a Woodland Springbrook Community, *Ecology*, 48(1): 139–149.

Moeller, J. R., G. W. Minshall, K. W. Cummins, R. C. Petersen, C. E. Cushing, J. R. Sedell, R. A. Larson, and R. L. Vannote, 1979, Transport of Dissolved Organic Carbon in Streams of Differing Physiographic Characteristics, *Organ. Geochem.*, 1: 139–150.

Molles, M. C., Jr., and J. R. Gosz, 1980, Effects of a Ski Area Development on the Water Quality and Invertebrates of a Mountain Stream, *Water, Air, Soil Pollut.*, 14: 187–205.

Monk, C. D., D. A. Crossley, Jr., R. L. Todd, W. T. Swank, J. B. Waide, and J. R. Webster, 1977, An Overview of Nutrient Cycling Research at Coweeta Hydrologic Laboratory, in D. L. Correll (Ed.), *Watershed Research in Eastern North America, A Workshop to Compare Results*, Feb. 28–Mar. 3, 1977, Edgewater, MD, Vol. I, pp. 35–50, Smithsonian Institution, Washington, DC.

Morisawa, M., 1968, *Streams—Their Dynamics and Morphology*, McGraw-Hill Book Company, New York.

Muckleston, K. W., 1979, Water, in R. M. Highsmith, Jr., and A. J. Kimerling (Eds.), *Atlas of the Pacific Northwest*, pp. 67–75, Oregon State University Press, Corvallis.

Needham, P. R., and C. L. Usinger, 1956, Variability in the Macrofauna of a Single Riffle in Prosser Creek, California, as Indicated by the Surber Sampler, *Hilgardia*, 24(14): 383–409.

North, R. M., A. S. Johnson, H. O. Hillestad, P. A. Maxwell, and R. C. Parker, 1974, *Survey of Economic–Ecologic Impacts of Small Watershed Development*, Technical Completion Report ERC-0974, Institute of Natural Resources, University of Georgia, Athens.

Oregon Department of Fish and Wildlife, 1969, Stream Surveys of Winchester, Big and Two Mile Creeks, unpublished reports, Research Division, Charleston, OR.

Orrell, R., 1980, *Report to Maritime Heritage Center Technical Committee Members*, Department of Fisheries, Burlington, WA.

Orsborn, J. F., 1976, Drainage Basin Characteristics Applied to Hydraulic Design and Water-Resources Management. in *Proceedings of the Geomorphology and Engineering Symposium*, Sept. 24–25, 1976, State University of New York, Binghamton, pp. 141–171, Dowdere, Hutchinson, and Ross, Inc., Stroudsburg, PA.

Oswood, M. W., 1979, Abundance Patterns of Filter-Feeding Caddisflies (Trichoptera: Hydropsychidae) and Seston in a Montana (U. S.) Lake Outlet, *Hydrobiologia*, 63(2): 177–183.

Pearson, W. D., and R. H. Kramer, 1972, Drift and Production of Two Aquatic Insects in a Mountain Stream, *Ecol. Monogr.*, 42(3): 365–385.

Pennak, R. W., and E. D. Van Gerpen, 1947, Bottom Fauna Production and Physical Nature of the Substrate in a Northern Colorado Trout Stream, *Ecology*, 28(1): 42–48.

Pionke, H. B., and R. N. Weaver, 1977, The Mahantango Creek Watershed—An Interdisciplinary Watershed Research Program in Pennsylvania, in D. L. Correll (Ed.), *Watershed Research in Eastern North America, A Workshop to Compare Results*, Feb. 28–Mar. 3, 1977, Edgewater, MD, Vol. I, pp. 83–103, Smithsonian Institution, Washington, DC.

Richards, J. S., 1976, Changes in Fish Species Composition in the Au Sable River, Michigan, from the 1920's to 1972, *Trans. Am. Fish. Soc.*, 105(1): 32–40.

Rinne, J. N., 1978, Development of Methods of Population Estimation and Habitat Evaluation for Management of the Arizona and Gila Trouts, in J. R. Moring (Ed.), *Proceedings of the Wild Trout—Catchable Trout Symposium*, Feb. 15–17, 1978, pp. 113–125, Oregon Department of Fish and Wildlife, Eugene, OR.

Ryck, F. M., Jr., 1974, *Water Quality Survey of the Southeast Ozark Mining Area, 1965–1971*, Aquatic Series No. 10, Missouri Department of Conservation, Jefferson City.

Scheidegger, A. E., 1965, *The Algebra of Stream Order Numbers*, Professional Paper 525B: 187–189, U. S. Geological Survey, Washington, DC.

Sheldon, A. L., 1968, Species Diversity and Longitudinal Succession in Stream Fishes, *Ecology*, 49(2): 193–198.

Shirazi, M. A., R. M. Hughes, and J. M. Omernik, 1980 draft, Land and Water Quality Interrelated by Broad Geographic Characteristics—A Feasibility

Study, Corvallis Environmental Research Laboratory, U. S. Environmental Protection Agency, Corvallis, OR.

Shreve, R. L., 1966, Statistical Law of Stream Numbers, *J. Geol.*, 74: 17–37.

Smart, J. S., 1972, Channel Networks, in V. T. Chow (Ed.), *Advances in Hydroscience*, Vol. 8, pp. 305–346, Academic Press, Inc., New York.

Stall, J. B., and C. T. Yang, 1970, *Hydraulic Geometry of 12 Selected Stream Systems of the United States*, WRC Research Report No. 32, Water Resources Center, University of Illinois, Urbana.

Stauffer, J. R., Jr., and C. H. Hocutt, 1980, Inertia and Recovery: An Approach to Stream Classification and Stress Evaluation, *Water Resour. Bull.*, 16(1): 72–78.

Stewart, R. K., and C. A. Thilenius, 1964, *Stream and Lake Inventory and Classification in the Black Hills of South Dakota, 1964*, Job Nos. 14 and 15, D-J Project F-1-R-13, Department of Game, Fish and Parks, Pierre, SD.

Strahler, A. N., 1957, Quantitative Analysis of Watershed Geomorphology, *Trans. Am. Geophys. Union*, 38: 913–920.

Swift, C. C., A. W. Wells, and J. S. Diana, 1975, *Survey of the Freshwater Fishes and Their Habitats in the Coastal Drainages of Southern California*, Inland Fisheries Branch, California Department of Fish and Game, Sacramento.

Turner, R. R., T. M. Burton, and R. C. Harriss, 1977, Lake Jackson Watershed Study: Description of Sites, Methodology and Scope of Research, in D. L. Correll (Ed.), *Watershed Research in Eastern North America, A Workshop to Compare Results*, Feb. 28–Mar. 3, 1977, Edgewater, MD, Vol. I, pp. 19–33, Smithsonian Institution, Washington, DC.

U. S. Geological Survey, 1970, *The National Atlas of the United States of America*, Washington, DC.

Vannote, R. L., G. W. Minshall, K. W. Cummins, J. R. Sedell, and C. E. Cushing, 1980, The River Continuum Concept, *Can. J. Fish. Aquat. Sci.*, 37: 130–137.

Van Velson, R., 1979, Effects of Livestock Grazing upon Rainbow Trout in Otter Creek, Nebraska, in O. B. Cope (Ed.), *Proceedings of the Forum—Grazing and Riparian/Stream Ecosystems*, Nov. 3–4, 1978, pp. 53–55, Trout Unlimited, Denver, CO.

Vitousek, P. M., 1977, The Regulation of Element Concentrations in Mountain Streams in the Northeastern United States, *Ecol. Monogr.*, 47: 65–87.

Waters, T. F., 1964, Recolonization of Denuded Stream Bottom Areas by Drift, *Trans. Am. Fish. Soc.*, 93: 311–315.

Winget, R. N., and M. K. Reichert, 1976, Aquatic Survey of Selected Streams with Critical Habitats on National Resource Lands Affected by Livestock and Recreation, unpublished final report, U. S. Bureau of Land Management, Salt Lake City.

6. RIVER WASTE HEAT ASSIMILATION VS ORGANIC WASTE DISCHARGE

Erhard F. Joeres

Department of Civil and Environmental Engineering
University of Wisconsin
Madison, Wisconsin

Nathaniel Tetrick

Environmental Division
Proctor & Gamble Corporation
Cincinnati, Ohio

ABSTRACT

The issue of water as a constraint to power plant siting was investigated at the Wisconsin River between Wisconsin Dells and Lake Wisconsin. This stretch of river has been the subject of much interest since the construction of the 1000-MW coal-fired Columbia power plant. An attempt was made to determine future trade-offs between using the river for waste heat assimilation or organic waste discharge. Simulations of water quality were carried out to evaluate the effect of potential heat discharges from projected power plants at three possible sites on the levels of dissolved oxygen, biochemical oxygen demand, and algae growth during periods of extremely low flow. Model parameters, boundary conditions, and assumptions made to specify the hydraulic parameters of the system to model the river are discussed. Results indicate that future development along the river will adversely affect dissolved oxygen levels in Lake Wisconsin but that nutrient loadings are of greater consequence than waste heat discharges. Neither waste heat nor nutrient discharges were found to limit projected biochemical oxygen demand levels caused by organic waste discharges.

INTRODUCTION

Investigation of any kind of ecological process presupposes that the process under scrutiny contributes in some way to the interactions of the elements of a broader system. The purpose of this paper is to illustrate the extrapolation of river ecosystem modeling to the much broader public sphere of public policy and environmental decision making. This is an area of very complex considerations intersecting in a general hierarchy:

1. Articulation and quantification of social and political objectives
2. Detailing of economic implications relating to the measurement of benefits and costs, as well as their spatial and temporal distribution
3. Description of the physical translation and transport of pollutants; an understanding of transport processes is necessary (transport processes for water are the hydraulic and hydrodynamic behavior characteristics)
4. Understanding of the chemical and biological impacts of waste on the local ecosystem

The particular problem chosen to illustrate these aspects is the question of water as a constraint to power plant siting and operation. Water is needed for cooling power plant condensers; whatever technological variations exist in the generator design and cooling process used, the aquatic environment is affected either through volume losses caused by cooling tower or pond evaporation or through heat addition from return flow. Evaporation-depleted return flows will further cause a net increase in constituent concentrations. If tolerable lower bounds of water quality can be established as a minimum standard beyond which the aquatic ecosystem cannot be disturbed, the availability of water may be the limiting factor in planning and operating generating facilities. This study addresses the potential future use of a 37-mile portion of the Wisconsin River under projected hypothetical scenarios of electric power development, so that the trade-offs involved in using the river for waste heat assimilation rather than organic waste disposal from communities, industries, and non-point sources can be better understood.

PROBLEM SETTING

In the broadest sense, selecting power plant sites is referred to as a "facility location" problem. This problem assumes knowledge of demand for power and the potential locations where the needed supply could be generated. The analysis then becomes one of choosing the "best" sites for plant construction from many possible locations and specifying plant size and the demand area served. The social and political objectives will largely dictate the types of plants to be considered [in Wisconsin there is currently a de-facto moratorium on additional nuclear power development (Wisconsin Public Service Commission, 1978)] and, to some extent, which sites could exist under a multi-interest consensus.

All too often one item of information that has been lacking in this process is site capacity limitation dictated by local environmental impact. Before this limit can be properly set, the complex mechanisms of what actually does happen to the environment must be examined.

One presumes that regulatory agencies operate to achieve and maintain general water-quality standards set as the minimum lower bounds to protect the ecosystem. The problem then becomes assessing the collective impact upon water quality of waste discharge and cooling-water discharge and, once the limiting water quality is reached, describing how much of one use will have to be sacrificed to gain more of the other.

The Columbia generating station is located on the Wisconsin River approximately 30 miles north of Madison (Figure 1). It consists of two separate 527-MW thermal generators using a 480-acre cooling lake in concert with two parallel cooling towers to provide condenser cooling. Short of unforeseen circumstances, there are no plans to discharge hot water from this station to the Wisconsin River. The configuration of the cooling lake provides a 1700-ft flow path, allowing the condenser water to cool from 100.2 to 84.7°F; the Wisconsin River serves mainly as a source of makeup water to this cooling cycle to replace water lost through evaporation and seepage into the ground (total losses are approximately 7370 gal/min).

To create conditions where river quality would indeed be unacceptable to current regulations, we assumed that the Columbia plant would discharge cooling water, that two additional power plant sites existed upstream (arbitrarily chosen), and that various wastewater discharge configurations might exist on the river. Table 1 summarizes the conditions simulated for the river. Wastewater flows were projections for the year 2000 for the communities presently discharging. These projections are based on the currently scheduled construction of a joint treatment plant to serve the cities of Wisconsin Dells and Lake Delton and on changes planned for the city of Portage (Wisconsin Department of Natural Resources, 1972). No evidence was found that any significant increase would occur during the next 20 years in all other major sources of wastewater discharge into the Baraboo River, Duck Creek, and Rowan Creek. The expansion of the Portage Wastewater Treatment Plant involves a controversial planned relocation of its effluent point, from the nearby Fox River into the Wisconsin River.

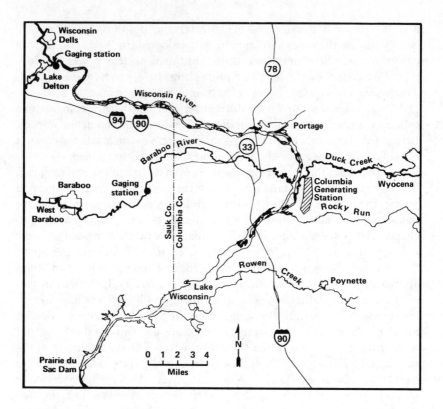

Figure 1. Wisconsin River from Wisconsin Dells to the Prairie du Sac dam.

WATER QUALITY MODELING

It was necessary to combine three separate models to simulate the behavior of the Wisconsin River under the various scenarios outlined. The most important is the QUAL-3 Water Quality Model, which used hydraulic data from the Corps of Engineers HEC-2 Hydraulic Model (1976) and superimposed heat routing from a model developed by Paily and Macagno (1976). A brief description of the modeling follows.

The QUAL-3 model used in this study was developed by Patterson and Rogers (1979) as an improved version of QUAL-2 and QUAL-1, which were originally developed by Norton et al. (1974) and by the Texas Water Development Board (1971, 1976). This model was chosen because of its extensive use in water quality monitoring and modeling of rivers in

Table 1. Scenario Options Considered for Model Simulations

Scenario class	Code	Description
Temperature	T1	Natural river temperature; no heat discharge
	T2	Heat discharge equivalent to 550 MW of electricity
	T3	Heat discharge equivalent to 1086 MW of electricity
Discharge from Portage	P1	Discharge into Fox River
Wastewater Treatment	P2	Discharge into Wisconsin River, no phosphorus/ nitrogen removal
Plant	P3	Discharge into Wisconsin River, with same phosphorus/nitrogen treatment as required for discharge into Fox River
Time of day	D	Daytime simulation
	N	Nighttime simulation
Location of additional	C1	Present Columbia site (river mile 108.5)
power plant	C2	River mile 119.0
	C3	River mile 130.0
Background nutrients	A1	Present-day levels of nitrogen and phosphorus
in river	A2	One-half present-day levels of nitrogen and phosphorus
	A3	Simulated levels of nutrients containing nitrogen concentrations limiting to growth of algae
	A4	One-half simulated levels of nutrients in scenario A3

Wisconsin, including much of the upper half of the Wisconsin River above Wisconsin Dells. The model was used to simulate, or route, levels or chlorophyll-a, nitrates, nitrites, ammonia, organic nitrogen, sediment oxygen demand, dissolved oxygen, and biochemical oxygen demand (BOD).

The model is based on the assumption that concentrations of these constituents in a river can be expressed by the convective-dispersive transport equation, where the equation considers the effects of convection, dispersion, individual constituent changes, and all sources or sinks for each constituent. The equation is (Patterson and Rogers, 1979):

$$A \frac{\partial C}{\partial t} = \frac{\partial}{\partial x} AE \frac{\partial C}{\partial Ex} - \frac{\partial}{\partial x} (A\bar{U}C) + AR_s \qquad (1)$$

where

A = Cross-sectional area of flow in the river (L^2)
C = Concentration of the constituent being routed (mg/l)
E = Longitudinal dispersion coefficient (L^2/T)
t = Time (T)
x = Distance along the longitudinal direction (L)
U = Mean velocity in stream, with respect to cross section (L/T)
R_s = Sources and sinks of the constituent being routed (mg l^{-1} T^{-1}).

Applying the convective-dispersive transport relation to the QUAL-3 model for simulations of river conditions requires that Eq. 1 be modified to consider the river in discrete elements to allow solution by finite differences. For an extensive bibliography covering water-quality modeling and model applications containing specific references to the constituents and reactions modeled with QUAL-3, the reader is referred to Patterson (1980). Patterson describes the use of QUAL-3 for regulatory purposes by the Wisconsin Department of Natural Resources to establish wasteload permits for industries and cities along the Lower Fox River in Wisconsin. For general references on differential equations and numerical solution methods, the reader is referred to Berg and McGregor (1966) and Greenspan (1974). It is important to recognize that all of the constituent reactions and interactions are contained in the last term of Eq. 1, R_s. Any quantity routed through the model may follow one of four pathways:

1. Continue into the next stream reach with no change
2. Be lost to the water system by a removal mechanism, e.g., settling, withdrawal, or decay
3. Enter the system from the atmosphere or any waste input or tributary
4. Be transformed into another substance by biological or chemical reactions

Figure 2 summarizes all the constituent pathways in the source/sink terms of Eq. 1. For a detailed discussion of individual models that simulate temporal changes in concentrations of chlorophyll-a, algae, organic nitrogen, ammonia nitrogen, nitrite nitrogen, nitrate nitrogen, phosphorus, carbonaceous BOD, particulate BOD and benthic oxygen demand, coliforms, and dissolved oxygen and the temperature dependence of reaction rates, the reader is referred to Patterson and Rogers (1979).

Extensive information is required to specify reaction coefficients for QUAL-3. Values of reaction coefficients used in this study, summarized in Table 2, are based on values used by the Wisconsin Department of Natural Resources (1976) in simulations of a portion of the upper Wisconsin River (river miles 210 to 235—miles are numbered from the confluence of the Wisconsin and Mississippi rivers). The model includes provisions for variation in many of these values. On the basis of conversations with personnel of the Wisconsin Department of Natural Resources and of field measurements (Tetrick and Joeres, 1980), the values were not changed for the entire length of the Wisconsin River modeled in this study.

To show the range of model flexibility, we will consider how the reaeration coefficient was selected. Two methods of computing the reaeration coefficient (K_2) were used. Where the river was wide enough

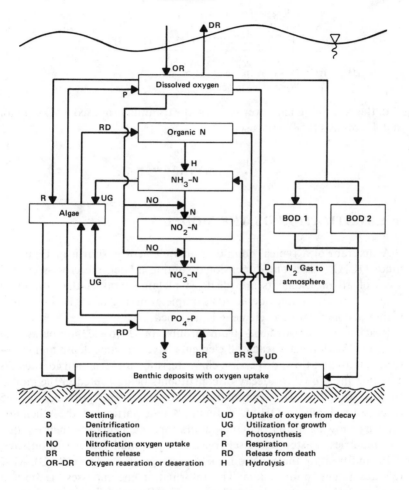

Figure 2. Possible pathways of interaction and feedback in the QUAL-3 water quality model.

for wind to be a factor (in reaches 8 through 11, 19 through 21, and 23 through 25), K_2 was computed as a function of wind speed (Wisconsin Department of Natural Resources, unpublished, c). This relation is expressed as:

$$K_2 = \frac{1 + 86,400 \ [\cosh(B)]}{2D \ \sinh(B)} \ (1.806 \times 10^{-9} \ S)^{0.5}$$

where

$$D = 0.3048 \text{ (DEPTH)}$$
$$S = 0.04 \text{ W} / \text{D}$$
$$B = \frac{10^{-6} [(-0.57835501\text{W} + 15.7735859)^2 \text{ S}]^{1/2}}{4.24971 \times 10^{-5}}$$
$$\text{(DEPTH)} = \text{Depth of the river (ft)}$$
$$\text{W} = \text{Wind velocity (m/s)}$$

In the remaining narrower reaches, the formula proposed by O'Connor and Dobbins (1958) was used:

$$K_2 = \frac{2.25 \times 10^{-8} \text{ U}}{(\text{DEPTH})^{129,600}}$$

SCHEMATIZATION OF THE MODEL

A diagram of a typical computational element is shown in Figure 3. Since QUAL-3 is a one-dimensional routing model, the cross-sectional areas of all the elements are idealized rectangles rather than the more realistic channel shapes depicted. Complete mixing of each constituent within a computational element is assumed.

Reaches are constructed from groups of 2 to 20 computational elements. Within each reach all elements have the same depth and cross-sectional area of flow, dispersion characteristics, and reaction coefficients affecting the growth or decay of constituents being routed in the model. In this study the 370 computational elements were grouped into 25 reaches for input to QUAL-3. Figure 4 is a portion of the schematic drawing showing the particular computational elements and their relation to tributaries, wastewater discharges, and potential power station sites.

In addition to the reaction coefficients given in Table 2, QUAL-3 requires extensive input data. This information encompasses the amount and strength of wastewaters and nutrients influent to the river, upstream and tributary water quality, water flows, and temperatures. Although the model may be run to give time-varying river conditions in response to time-varying inputs, our analysis focused on a simpler steady-state investigation for a selected critical period, namely, the lowest expected river flow that will persist for 7 days with a recurrence interval of 10 years (called the 7Q10). The designation of the 7Q10 as the critical design condition is a matter of widely accepted judgment. Even with protective environmental legislation and compliance with such rules by waste dischargers, the stochastic behavior of streamflow implies that there is still a measurable risk of environmental damage. The choice of a low flow-index

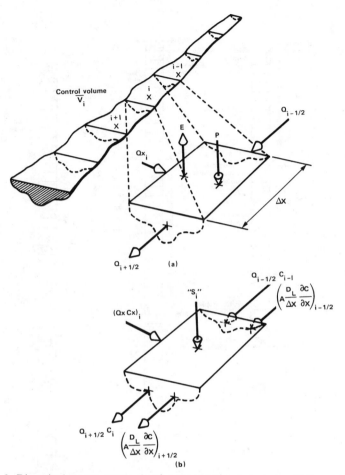

Figure 3. Discretized stream system showing computational elements with transport relations. (a) Hydrologic balance. (b) Material balance. (Data from Texas Water Development Board, 1971.)

value and its probable recurrence from the known probability distribution of flow makes the environmental risk explicit. The hydraulic parameters needed were the average cross-sectional area in each river element, the average depth, and a measure of roughness (Manning's n).

Table 3 summarizes sources of data available. Some of the information could be applied directly. Direct measurements of hydraulic parameters or river velocity were not available, however, and had to be generated in a separate model. Given data for the elevation of the riverbed at known locations, the model computed flows for comparison with known values measured at some locations.

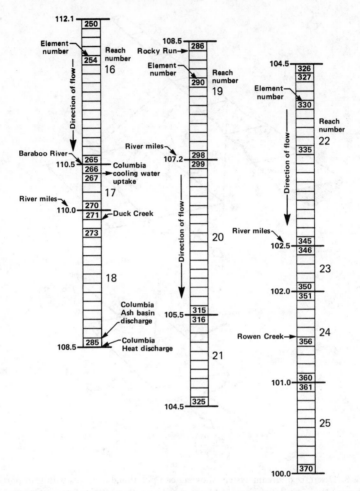

Figure 4. Schematic of QUAL-3 elements for reaches 16 through 25 on the Wisconsin River.

To be able to achieve a realistic flow and elevation condition, we had to assume that flow in some parts of the cross section was "shut off" so that higher velocities and greater elevation changes could be achieved in the remaining, flowing portion of the river. Since the Wisconsin River is a wide, shallow river, this procedure is not unrealistic. An example cross section is shown in Figure 5. For the analysis of the downstream portion reaching into Lake Wisconsin, water elevations and lake depths yielded water volumes; dividing these by the 7Q10 flow gave detention time. With given reach length and average width, we could determine velocity, cross-sectional area of flow, and depth as inputs for the QUAL-3 model.

Table 2. Reaction Coefficients Used in QUAL-3 Simulations of Water Quality of Wisconsin River in Vicinity of Portage

Reaction coefficients in QUAL-3 model	Description	Units	Suggested range of values	Reliability of suggested values*	Temperature dependent	Values used in this study
ALPLAϕ	Ratio of chlorophyll-a (Chl-a) to algae biomass (A)	μg Chl-a/mg A	50-100 (2-50)	Fair	No	5
CKORGN	Rate of hydrolysis of organic Nitrogen per unit of algae	d^{-1} mg A^{-1}	0.0005-0.005	Fair	Yes	0.000
ALPHA1	Nitrogen fraction of algae biomass	mg n/mg A	0.04-0.10	Good	No	0.06
ALPHA2	Fraction of algae	mg P/mg A	0.01-0.015	Good	No	0.01
ALPHA3	O_2 production per unit of algae respired	mg O/mg A	1.4-2.5	Fair	No	2.00
ALPHA4	O_2 uptake per unit of algae respired	mg O/mg N	1.5-2.3	Fair	No	1.50
ALPHA5	O_2 uptake per unit of NH_3 oxidation	mg O/mg N	3.23-3.43	Excellent	No	3.4
ALPHA6	O_2 uptake per unit of NO_2 oxidation	mg O/mg N	1.11-1.14	Excellent	No	1.4
GROMXX	Maximum specific growth rate of algae	d^{-1}	1.0-3.0	Good	Yes	1.60
RESPTT	Algae respiration rate	d^{-1}	0.05-0.5	Fair	Yes	0.15
CKNH3	Rate constant for biological oxidation of NH_3–NO_2	d^{-1}	0.05-1.5	Good	Yes	0.80
EXCOEF	Light extinction coefficient	ft^{-1}	0-20	Fair	No	0.38
CKNϕ2	Rate constant for biological oxidation of NO_2–NO_3	d^{-1}	0.5-2.5	Good	Yes	2.50
DNKK	Denitrification rate	d^{-1}	0.1-0.8	Fair	No	0.4
ALGSET	Local settling rate for algae	ft/d	0.0-6.0	Fair	No	0.4

Table 2, continued

Reaction coefficients in QUAL-3 model	Description	Units	Suggested range of values	Reliability of suggested values*	Temperature dependent	Values used in this study
SNH3	Benthos source rate for NH$_3$	mg N/(d-ft^2)	+	Poor	No	0
SPHϕS	Benthos source rate for NH$_3$	mg P/(d-ft^2)	+	Poor	No	0
CK4	Organic nitrogen settling rate	ft/d	+	Poor	No	0.05
CK1	Carbonaceous BOD decay rate	d^{-1}	0.01–2.0	Good	Yes	0.30
CK2	Reaeration rate	d^{-1}	0.0–100	Good	Yes	‡
CK3	Term-2 carbonaceous BOD decay rate	d^{-1}	0.01–2.0	Fair	No	0.08
CK5	Particulate BOD sink rate	ft/d	0–100	Fair	No	2.0
CKN	Nitrogen half-saturation constant for algae growth	mg/l	0.015–0.2	Fair	No	0.02
CKP	Phosphorus half-saturation constant for algae growth	mg/l	0.001–0.5	Fair	No	0.01
CKL	Light saturation constant for algae growth	langleys/min	0.21	Good	No	0.21
EXPOQV	Velocity correction factor		1.00–1.2	Fair	No	1.11
SONET	Daily solar radiation	langleys		Good	No	530

*Data from Patterson and Rogers, 1979.
†Determined during model calibration.
‡See text for computational procedure.

Table 3. Sources and Types of Raw Data Collected for Analysis of Conditions in the Wisconsin River

Number	Title	Parameter	Location	Source
1	Pollution investigation survey	Quantity and strength of waste-water discharges; chemical sampling and temperature data spot checks	Wisconsin River and Trib-utaries below confluence of Duck Creek	Wisconsin Department of Natural Resources (WDNR)
2	Pollution investigation survey	Quantity and strength of waste-water discharges; chemical sampling and temperature data	Wisconsin River and tributaries between Lemonweir and Baraboo Rivers	WDNR
3	Flood plain information (FPI)	Location of cross sections of Wisconsin River channels	Wisconsin River in Sauk and Columbia Counties	U. S. Army Corps of Engineers
4	Flood plain information (FPI)	Location of cross sections of Wisconsin River channels	Wisconsin River in Columbia County	U. S. Army Corps of Engineers
5	Cross-section data (HEC-2-0877)	Computer-card coded data containing cross sections (de-scribed in numbers 3 and 4) for the HEC-2 water surface profile program)	Wisconsin River from Wisconsin Dells to the I-90/94 bridge	WDNR
6	Water quality sampling data	Complete chemical analysis of water; temperature	1. Wisconsin River at Prairie du Sac 2. Baraboo River: County Trunk Highway X near Baraboo 3. Hydroelectric plant at Wisconsin Dells	WDNR
7	Permit files of National Pollut-ant Discharge Elimination System (NPDES)	Detailed data regarding chemi-cal nature and quantity of wastewater from regulated dischargers; anticipated future discharges	1. Wisconsin Dells publicly owned treatment plant (POW1P) 2. Lake Delton POWTP 3. Portage POWTP 4. Columbia Generating Station	WDNR (on file)

Table 3, continued

Number	Title	Parameter	Location	Source
8	River sediment deposition studies	Depth of water upstream, alongside and downstream of Wisconsin state highway bridges; elevation of water surface	1. I-90/94 bridge 2. State Highway 33 bridge (Portage) 3. State Highway 78 bridge	Wisconsin Department of Transportation
9	Surface waters of the United States	Daily discharge records	1. Wisconsin River near Lake Delton 2. Dell Creek near Lake Delton 3. Baraboo River, County Trunk Highway X near Baraboo	U. S. Geological Survey (USGS)
10	Water resources investigation (45-74)	Low-flow frequency of Wisconsin streams at sewage treatment plants (7Q10 flows)	1. Wisconsin River at Lake Delton 2. Baraboo River 3. Duck Creek 4. Rocky Run 5. Rowan Creek	USGS
11	Hydrologic investigation (HA-390)	Low-flow frequency of Wisconsin Streams	(same as number 10)	USGS
12		Water depth contours in Lake Wisconsin (map)		WDNR
13	Field study	Level of dissolved oxygen, level of 5-day and ultimate BOD, depth of water, levels of nitrate, nitrite, ammonia, phosphorus, total nitrogen, and temperature	At selected locations between Columbia Generating Plant and Lake Wisconsin	Project measurements (Appendix F, Table F-1)

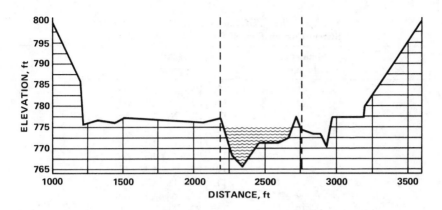

Figure 5. Plot of HEC-2 cross section at river mile 107.45, showing elevation or river bottom. Shaded region is the area where water is assumed to be standing but not flowing. Vertical dashed lines indicate boundary of water flow assumed in HEC-2 model. The diagram assumes the viewer is looking upstream.

Examining flow records for the river allowed the occurrence of the 7Q10 to be placed in July and August; thus air and water temperatures for those periods could be used.

Finally, since the version of QUAL-3 used could not route river temperatures, it became necessary to carry out separate simulations of heat discharges from the postulated power plants. A one-dimensional model developed by Paily and Macagno (1976) was modified for this purpose. The model was originally applied to the winter response to power plant discharge on the Mississippi and, thus, had to be adapted to summer conditions. The model solves the convective-dispersive equation for fully mixed river temperature, T:

$$\frac{\partial T}{\partial t} + \frac{1}{A} + \frac{\partial}{\partial x}(QT) = \frac{1}{A}\frac{\partial}{\partial x} AE \frac{\partial T}{\partial x} + R_s \qquad (3)$$

where

  t = Independent time variable
  A = Cross-sectional area of flow
  Q = Flow rate
  E = Dispersion
  R_s = Temperature changes caused by the river-atmosphere heat flux

RESULTS

Simulations show that given today's conditions of waste loading and the 7Q10 of 1800 ft^3/s discharge for the Wisconsin River, cooling water

Table 4. Simulations Performed on the QUAL-3 Model with
Different Combinations of Possible Future Conditions*

Simulation or run	Scenario option				
1	T1,	P1,	D,	C1,	A1
2	T1,	P2,	D,	C1,	A1
3	T1,	P2,	N,	C1,	A1
4	T2,	P2,	D,	C1,	A1
5	T2,	P2,	N,	C1,	A1
6	T2,	P3,	D,	C1,	A2
7	T2,	P3,	N,	C1,	A2
8	T1,	P3,	D,	C1,	A2
9	T1,	P3,	N,	C1,	A2
10	T1,	P3,	D,	C1,	A1
11	T3,	P1,	D,	C1,	A1
12	T3,	P2,	D,	C1,	A2
13	T2,	P2,	D,	C2,	A1
14	T2,	P3,	D,	C2,	A2
15	T2,	P2,	D,	C3,	A1
16	T2,	P3,	D,	C3,	A2
17	T2,	P2,	D,	C1,	A3
18	T2,	P3,	D,	C1,	A4

from each 100 MW of generated power will raise river temperature by 0.84°F. If we consequently set, for example, a limiting summer rise in temperature of 5°F, the river could absorb the waste heat from a 593-MW plant.

More important, however, is the behavior of the system under potential future conditions of stress. Table 4 summarizes the simulations actually carried out using the notation of future scenarios given in Table 1. Each of these allowed derivation of associated river profiles of constituent concentrations vs. distance downstream. Figure 6 illustrates the dissolved oxygen (DO) and BOD profiles under present-day conditions, i.e., without heat discharge to the river. Note that during low-flow/high-temperature summer conditions, the DO profile dips almost to 4 mg/l in Lake Wisconsin at mile point 100. Figure 7 shows the nutrient profiles for three of the runs.

If heat is now introduced to the river 8.5, 19, or 30 miles above the critical river reach at mile point 100 (where dissolved oxygen is at a minimum), very little change is observed in the DO level at the constraining point (Figure 8).

A very different behavior is observed when each simulation run has in common the waste heat from a 550-MW power plant at the current

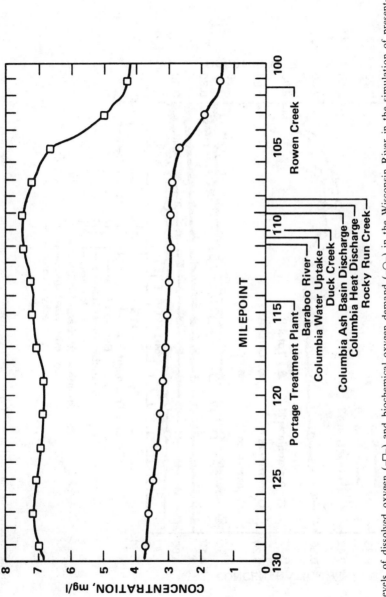

Figure 6. Levels of dissolved oxygen (–□–) and biochemical oxygen demand (–○–) in the Wisconsin River in the simulation of present-day conditions (run 1).

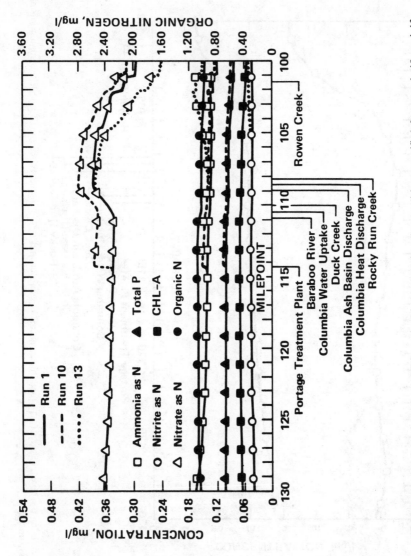

Figure 7. Levels of nutrients in the Wisconsin River, assuming the conditions specified for runs 1, 10, and 13.

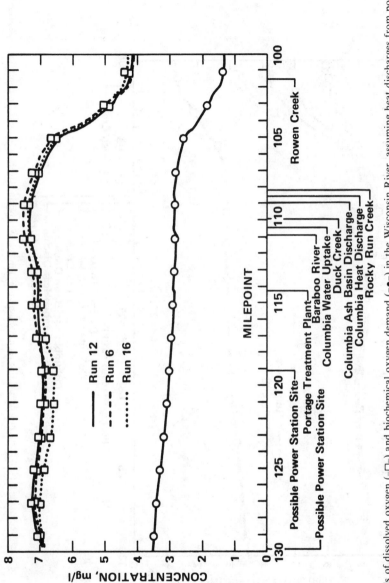

Figure 8. Levels of dissolved oxygen (–□–) and biochemical oxygen demand (–●–) in the Wisconsin River, assuming heat discharges from power plants at miles 108.5 (run 6), 119.0 (run 12), and 130 (run 16).

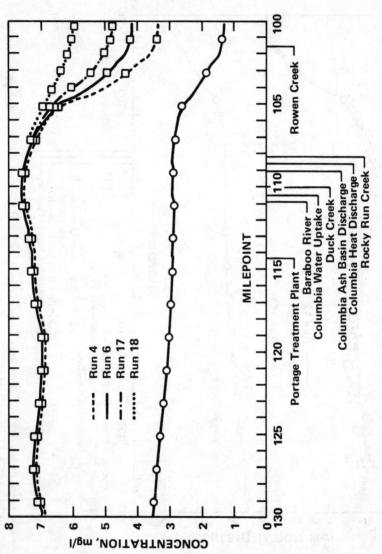

Figure 9. Levels of dissolved oxygen (–□–) and biochemical oxygen demand (–●–) in the Wisconsin River, assuming the conditions specified for runs 4, 6, 17, and 18.

Columbia station site; each curve at the critical point in ascending observed DO concentration assumes progressively more stringent control of background nutrients in the river (Figure 9). Whether or not the state of Wisconsin could actually implement policies to achieve the nutrient reductions stipulated is an open question. Certainly the model illustrates that future water quality in the Wisconsin River will be far less affected by simultaneous use of the river for organic waste and cooling water discharge than by state policies governing the control of nutrient discharges from all possible point and non-point sources.

CONCLUSION

A 37-mile stretch of the Wisconsin River from Wisconsin Dells to Lake Wisconsin was modeled to illustrate water-quality response to organic waste and thermal loadings. The study illustrated that it is possible to model interactions in a complex river system and to show the change in state variables as the system is stressed. Even though the many assumptions in the input data may cause specific levels of predicted parameter values to be uncertain, knowledge of system sensitivity (as illustrated for nutrient and dissolved oxygen changes) is of great value and will be accurate so long as the fundamental mechanisms of the system are structured properly.

ACKNOWLEDGMENTS

This study example and results were part of research carried out by Nathaniel Tetrick under the direction of E. Joeres (Tetrick and Joeres, 1980). The investigation was a 1-year subproject of a major and comprehensive effort sponsored by the Environmental Protection Agency (EPA), the Wisconsin Power and Light Company, the Madison Gas and Electric Company, the Wisconsin Public Service Corporation, the Wisconsin Public Service Commission, and the Wisconsin Department of Natural Resources from 1971 to 1978 to document the impacts of a coal-fired power plant on the environment. The overall study was titled *The Impacts of Coal-Fired Power Plants on the Environment*. Approximately 29 separate studies dealing with chemical constituents, chemical transport mechanisms, biological effects, social and economic effects, and integration and synthesis are appearing in the EPA Ecological Research Series.

Data sets covering aquatic chemistry, trace elements, organic contaminants, air pollution modeling, meteorology, hydrogeology, plant damage, aquatic invertebrates, fish, cooling lake ecosystem assessment, wetlands birds, wetlands plants, remote sensing, citizen concerns and attitudes, and power company air effluent measurements are available from the Data Center, Columbia Impact Study, Water Resources Center, University of Wisconsin, Madison.

REFERENCES

Berg, Paul W., and James L. MacGregor, 1966, *Elementary Partial Differential Equations*, Holden-Day, San Francisco.

Greenspaw, Donald, 1974, *Discrete Numerical Methods in Physics and Engineering*, Academic Press, Inc., New York.

Norton, W. R., L. A. Roesner, D. E. Evenson, and J. R. Monser, 1974, Unpublished, Computer Program Documentation for the Stream Quality Model QUAL-II, Interim Technical Report Prepared for the U. S. Environmental Protection Agency by Water Resources Engineers, Inc., Walnut Creek, Cal.

O'Connor, D. J., and W. E. Dobbins, 1958, Mechanism of Reaeration in Natural Streams, *Trans. Am. Soc. Civil Eng.*, 123: 644–666.

Paily, P. P., and E. O. Macagno, 1976, Numerical Prediction of Thermal Regime of Rivers, *J. Hydraulics Div., Am. Soc. Civil Eng.*, 102: 255–274.

Patterson, Dale, 1980, *Water Quality Modeling of the Lower Fox River for Wasteload Allocation Development*, Wisconsin Department of Natural Resources, Water Quality Evaluation Section, Madison.

———, and J. Rogers, 1979, QUAL-III Water Quality Model Documentation, Wisconsin Department of Natural Resources, Water Quality Evaluation Section, Madison.

Tetrick, N., and E. Joeres, 1980, *Water Constraints in Power Plant Siting*, EPA-600/3-80-077, U. S. Environmental Protection Agency Ecological Research Series.

Texas Water Development Board, 1971, *Simulation of Water Quality in Streams and Canals: Theory and Description of the QUAL-I Mathematical Modeling System*, Report No. 128, Austin.

———, 1976, QUAL-I: Program Documentation and Users Manual, Systems and Engineering Division, Austin.

———, 1976, *HEC-2 Water Surface Profiles: User's Manual With Supplement*, Hydrologic Engineering Center, Computer Program 723-X6-L202A, Davis, CA.

Wisconsin Department of Natural Resources, 1972, *Pollution Investigation Survey: Lower Wisconsin River*, Division of Environmental Protection, Madison.

———, (unpublished), c, Listing of input data for QUAL-3 Water Quality Model: FORTRAN, Available from the Wisconsin Department of Natural Resources, Madison.

———, 1976, Tabulations of Water Quality Chemical Data for the Wisconsin River at Prairie due Sac and Wisconsin Dells, and for the Baraboo River at County Trunk Highway X East of Baraboo, Wisconsin, Data available from Wisconsin Department of Natural Resources, Madison.

Wisconsin Public Service Commission, 1978, Order 05-EP-1, Aug. 17.

SECTION 2

SYNTHESIS STUDIES

Papers in the previous section outlined and discussed several frameworks for interpreting and predicting dynamics in lotic systems. Yet all models are ultimately empirical. At some point a measurement must be made; data must be collected. Papers in this section highlight measurement of physical, chemical and biological processes that emphasize feedback among various system components. Current ecological research tends to argue the importance of single processes (eg. predation, competition, nutrient limitation) as regulators of ecosystem structure. Clearly, the relative importance of these individual processes changes through time and space. We will only come to usefully understand the dynamics of ecosystems when we have described the complex, many dimensional response surfaces that delineate shifts in the controlling processes. Lotic systems, with their complex hydrological cycles and varied watershed characteristics, seem to be ideal systems to begin description of the response surfaces.

The relative importance of light, nitrogen and phosphorus in regulating production of periphyton is measured by Triska et al. In this study, the role of historic, episodic events and their influence on shaping community structure is indicated. Knight offers some intriguing experimental evidence for regulation of lotic system structure by consumer organisms in the Silver Spring system. He hypothesizes that consumer populations will be maintained at a given level only if they stimulate, either directly or indirectly, gross productivity to a degree equal to that which they remove it.

Burton and King offer an example of much needed long-term survey studies for evaluating model predictions and testing theory relevant to lotic systems. In an applied vein, their data suggest that while significant effort was mounted to abate nutrient loading problems from point sources, increased urbanization effectively changed the nature of the

127

problem to one of more diffuse loading, the net result being no real change in nutrient dynamics of the Red Cedar River.

The relative importance of allochthonous vs autochthonous carbon inputs to lotic production dynamics remains an unbounded problem. Matson and Klotz present data relevant to this issue for the Shetucket River, Connecticut.

Boundaries of ecosystems remain difficult to define, both for purposes of characterizing individual systems and for comparison of systems. Lotic systems are a perfect example. Brinson et al. and Yarbro describe stream channel–floodplain interactions with regard to nitrogen and phosphorus retention and cycling. The relevance of seasonal timing of flooding events and temperature dependent rates of metabolic processes are especially evident in these papers.

7. EFFECT OF SIMULATED CANOPY COVER ON REGULATION OF NITRATE UPTAKE AND PRIMARY PRODUCTION BY NATURAL PERIPHYTON ASSEMBLAGES

F. J. Triska, V. C. Kennedy, and R. J. Avanzino
U. S. Geological Survey
Menlo Park, California

Barbara N. Reilly
San Francisco State University
San Francisco, California

ABSTRACT

Within periods approximating periphyton succession in forested drainages, day length and temperature fluctuate slowly while nutrient regime and community structure are reset irregularly by meteorological events. To test canopy impact on biological response to a simulated reset, or nutrient input event, an experiment was conducted in five Plexiglas channels set in Little Lost Man Creek in California. An unshaded control channel (40 μg NO_3–N/l and 10 μg PO_4–P/l background) received no nutrient addition. Treatment channels (0, 30, 66, and 92% shaded), amended with approximately 100 μg NO_3–N/l and 25 μg PO_4–P/l, simulated onset of autumnal storms.

Periphyton biomass increased at least fivefold except in the 92% shaded channel. Maximum biomass was essentially similar in both the control and the fully lighted treatment channel, but chlorophyll was half as great in the control. Biomass-to-chlorophyll ratio in the control stream increased throughout the experiment (indicating nitrogen limitation) but decreased by half or more in treatment flumes within 4 days. Diel fluctuations in nitrate concentrations were observed, but their magnitude did not increase directly with biomass accumulation because of gradual periphyton senescence. Uptake was related to net community primary production. Results suggest that the cycling distance of biologically important nutrients in a stream reach depends on input location on the watershed (stream order and canopy cover) and time elapsed since the previous resetting event.

INTRODUCTION

Periphyton communities in streams draining forested watersheds develop in response to a myriad of environmental factors, e.g., day length, temperature, nutrient regime, and biological interactions, which regulate periphyton succession.

Within periods approximating periphyton succession, day length and temperature change slowly, whereas nutrient regime and community structure are reset irregularly by meteorological events. Resetting results in a period of renewed vigorous growth because of removal of mature periphyton community and/or increase in nutrient concentration. The extent of both scouring and nutrient input depends on storm intensity and season. High inland ranges of California and the Pacific Northwest (Cascades and Sierra Nevada) have their highest discharge in late spring as snow melts. In coastal areas periphyton removal occurs with both winter and spring runoff and just after the onset of the fall storms. In addition to scouring, storms can also cause a large increase in nutrient concentration. Kennedy and Malcolm (1977) reported a 35-fold increase in nitrate concentration during fall storm runoff in the Mattole River, California, and an increase from 40 to 150 $\mu g/l$ in Little Lost Man Creek, California. Both the inland and coastal ranges are subject to long dry summers, which provide excellent opportunities to observe periphyton communities for long periods with minimum disturbance.

The purpose of this study was to test canopy impact on biological responses to a simulated reset (partially colonized surfaces with minor nitrate and phosphate enhancement). The experimental site was Little Lost Man Creek, Humboldt County, California (Figure 1).

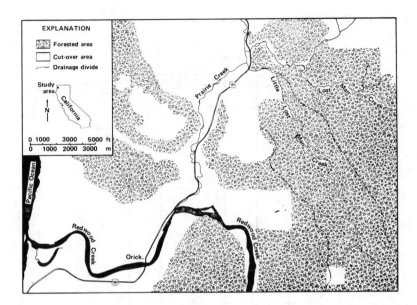

Figure 1. Map of experimental drainage area and its approximate location in northern California.

MATERIALS AND METHODS

Materials

The experiment was conducted in a series of five Plexiglas channels between Aug. 17 and Sept. 20, 1979 (Figure 2). A header box provided with spearate mixing chambers for each channel and separate "V" notched weirs regulated flow to each channel at 10 l/min. Water was supplied through PVC pipe by gravity from an upstream station and filtered (300 μm) at the header box to remove coarse particulate organic matter. Nutrient solutions, when added, were pumped from a common source using a separate peristaltic pump for each channel. Nutrient enhancement was targeted at 100 μg/l NO_3–N and 25 μg/l PO_4–P, approximately double night background concentrations and comparable to fall storm concentrations. Diel fluctuations of background nitrate concentration was approximately 25%. Chloride was added to serve as a

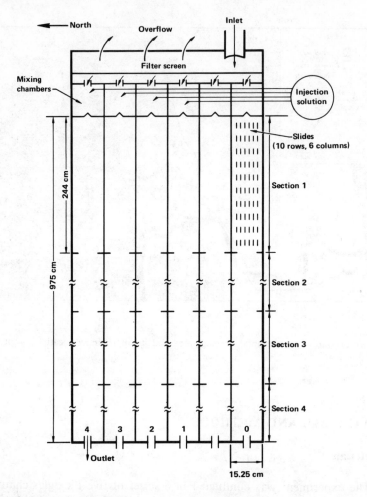

Figure 2. Schematic diagram of experimental channels. Each section of each channel contained 10 rows and 6 columns of acrylic plastic slides. Flumes were shaded and nutrient amended as follows: Channel 0, background nutrients, 100% light; channel 1, NO_3 + PO_4 amendment, 100% light; channel 2, NO_3 + PO_4 amendment, 30% shaded; channel 3, NO_3 + PO_4 amendment, 66% shaded; and channel 4, NO_3 + PO_4, 92% shaded. Each channel had a surface area of 12.4 m^2 available for colonization.

conservative tracer for determining nitrate loss. Nutrient addition started Aug. 24 and ended Sept. 11, 1979. Water travel time was approximately 20 min.

Each channel was 15.25 cm wide and 9.75 m long and consisted of four successive longitudinal sections (see Figure 2), each of which contained 10 rows and 6 columns of Plexiglas slides (10.16 by 15.24 cm) mounted perpendicular to the bottom. Each row consisted of six slides spaced 2.5 cm apart. Slides were roughened by sandblasting to better simulate mineral surfaces. The standardized surface texture and area, constant flow, and uniform channel geometry made access to nutrients by periphyton identical in each channel. Scouring of the periphyton community was simulated using partially colonized substrates. Slides were placed in a variety of stream habitats 5 days before placement in the Plexiglas channels. The developing populations were further acclimated to simulated canopy cover in the channel for 4 days before nutrient addition. Combinations of nutrient addition and canopy simulation were:

Channel 0	Channel 3
Background nutrient, full sunlight	Nutrient amendment, 66% shade
Channel 1	Channel 4
Nutrient amendment, full sunlight	Nutrient amendment, 92% shade
Channel 2	
Nutrient amendment, 30% shade	

Shading was done with commercially available greenhouse shading of woven nylon. The 30% shading was available only in black color and simulated a natural patchy pattern of light intercepted by the canopy. The 66% and 92% shading was green and translucent, further simulating changes in light quality and intensity.

Sampling

Visual observation of the flumes during the acclimatization period indicated a general longitudinal decrease in colonization from upstream to downstream and a lateral decrease from center to the sides of each channel. Each channel was sampled by section, and only the first three sections were sampled. The fourth section was assumed to have similar biomass to the third. To compensate for longitudinal variation, we estimated total flume biomass and chlorophyll totals by summing separate estimates from each section and reported totals on a per channel basis. Each channel had a surface area for algal colonization of 12.4 m^2. To compensate for differences between the center and sides of each channel, we set each sectional sample at six slides, one from each column, all randomly selected. Thus on each sampling date three samples (18 slides) were taken from each channel. No location was sampled more than once during the experiment. Slides were placed in individual plastic bags and returned to our field laboratory. Each sample yielded four subsamples. Since the slides were mounted perpendicular to the bottom and parallel to flow, each slide had two sample surfaces. Thus each subsample consisted of one side from columns 1 through 3 or one side from columns 4 through 6 to compensate for horizontal variation. After scraping, periphyton was placed in plastic bags and frozen immediately. Scraped slides were returned to the channels to keep the surface area constant. Approximately 6% of the area was sampled on each date. No correction was made for partial colonization of slides returned to the channel. Further processing was done after the experiment was over. Chlorophyll estimates on one subsample from each section of each treatment on each sampling date were determined immediately after return from the field to minimize chlorophyll deterioration in storage. Since biomass was estimated from slides, only attached biomass was counted. Sloughed tissue deposited on the bottom or transported out of the system was not included in biomass estimates.

Methods

Light input was estimated by a LiCor 500 integrating light meter equipped with an LI 190S quantum sensor, which measures photosynthetically active radiation. The sensor was mounted above the water, downstream of the experimental channels, and near the center of the stream. Results are reported as mean microeinsteins per square meter per second ($\mu E\ m^{-2}\ s^{-1}$) based on a full 24-h day. For comparison to other literature data, we converted light to lux, using conversion factors supplied by LiCor, Inc.

Water temperature was measured continuously. The temperature sensor was placed in a shaded riffle upstream of the experimental channels. Maximum and minimum temperature and the light data are reported for days when primary production was estimated. Chlorophyll determinations were made by extraction in 90% acetone shaken with magnesium carbonate. Samples were extracted approximately 20 h in darkness, under refrigeration before reading. Absorbance was read on a Spectronic 710 spectrophotometer at 665 mμ for chlorophyll a and at 750 mμ to correct for turbidity. Readings were made before and after acidification to correct for phaeopigments (Wetzel and Westlake, 1974). Extracted solids were saved to estimate biomass.

Biomass was determined by oven-drying a second nonextracted sample at 50°C. Ash content was estimated by ignition at 500°C for 4 h. When biomass was large enough to permit ashing a duplicate subsample, ash content was generally within 5%. To estimate biomass from the extracted sample, we first determined ash amount from the extracted sample by ignition. From the amount of ash and the known percent ash from the unextracted sample, biomass was estimated for the extracted sample. Biomass estimates were also obtained on a third sample following removal of invertebrates.

Estimates of primary production were made on five dates in each flume by 24-h upstream–downstream oxygen determinations. Oxygen concentration was determined using YSI Model 57 temperature-compensated oxygen meters with YSI Series 5700 oxygen probes. Membranes were changed and probes were calibrated before each run. Output from the oxygen meters was fed to a switching device that read each station consecutively at 200-s intervals. Output from the switching device was then fed into a battery-operated chart recorder. Calibration points (time and O_2 concentrations) were marked on the chart at least 10 times daily to assist in calculating actual O_2 readings. Oxygen concentrations were calculated at hourly intervals, and production was calculated by the method of Owens (1974). We assumed 1 mole of carbon fixed for each mole of oxygen produced. Carbon content was assumed to be 50% ash-free dry weight for conversion of oxygen data to estimates of biomass. Carbon content of ash-free biomass at the middle of each flume, subsequently determined on Aug. 24, 1979, was $46 \pm 3\%$ for all channels.

Water was sampled five times daily for background NO_3, PO_4 and Cl concentrations and for nutrient output at the base of each flume. Samples were collected before sunrise (approximately 0600 h) and at 1000, 1400, 1800, and 2200 h. Previous studies at Little Lost Man Creek indicated this scheme was sufficient to obtain maximum diel fluctuations of NO_3 and PO_4 (M. J. Sebetich, personal communication). After collection, samples were filtered immediately at streamside through 0.45 μm Millipore filters into 250-ml acid-washed bottles (HCl) and stored frozen ($-20°C$) until analysis. Analyses were made on a Technicon AutoAnalyzer II with a precision for NO_3-N and PO_4-P of ± 1 μg/l below 100 μg/l and $\pm 1\%$ above, and for Cl $\pm 1\%$ at 5 mg/l and above. When this procedure is used, water samples can be frozen for extended periods with no significant loss of nutrients in comparison with samples processed immediately (R. J. Avanzino, personal communication).

Coarse filtering of input water to remove large detritus presumably formed a barrier to colonization by large grazing invertebrates. Other than this, no effort was made either to encourage or to discourage invertebrate colonization. Although most large invertebrates were effectively screened out, smaller invertebrates were able to invade the experimental channels. As a result, one set of samples from each sampling date was carefully picked for invertebrates. Each sample was examined by both eye and dissecting microscopes. Drifting invertebrates presumably had an equal probability of entering any channel. Once within the channel, invertebrates could either exit or be recruited into the community but could not migrate to other channels.

Invertebrates picked from the samples were either preserved in 50% ethyl alcohol for identification by L. J. Tilley (U. S. Geological Survey, Menlo Park) or dried at 50°C and weighed.

RESULTS

Light and Temperature

Light input on days when primary production was measured varied from 468 μE m^{-2} s^{-1} on the longest day to 294 μE m^{-2} s^{-1} on the shortest overcast day, a 38% difference (Table 1). Local conditions, such as cloud cover and morning fog, which are characteristic of the region, influence light input. As a result light input appeared to increase between Aug. 28 and Sept. 12, although actual day length decreased. Temperature throughout the experiment ranged from 14.5 to 20°C. Diel fluctuations were generally less than 2°C

Biomass

At the time nutrient addition commenced, periphyton biomass was 4.5 ± 2.5 g/channel in all channels despite colonization under a range of shading (Figure 3). After 4 days of nutrient enhancement, the most heavily shaded channel maintained essentially constant biomass and from this point had significantly lower biomass than unshaded channel 1 for the remainder of the study (Student's t-test $p < 0.01$). Two weeks after the injection began, channel 1, with nutrient amendment but no shading, was significantly higher in biomass (40.5 ± 2.94 g SD) than either the 30% (23.25 ± 1.43 g SD, $p < 0.05$) or the 66% (21.33 ± 2.80 g SD, $p < 0.05$)

Table 1. Ambient Light Input and Water Temperature
on Dates when Primary Production was Estimated

| Date | Temperature, °C | | Light, μE m^{-2} s^{-1}, |
	Maximum	Minimum	24-h average
Aug. 24–25	16.7	15.5	468
Aug. 28–29	16.9	15.6	317
Sept. 6–7	17.2	15.6	363
Sept. 12–13	17.2	16.7	372
Sept. 18–19	15.6	14.4	294

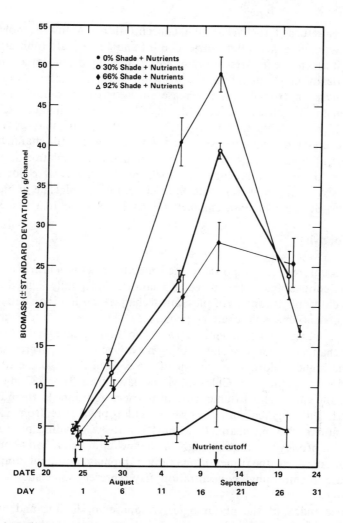

Figure 3. Periphyton biomass (grams of ash-free dry weight) in each experimental channel on each sampling date. Error bars are one standard deviation. Nutrient injection was initiated (I) Aug. 24 and concluded (C) Sept. 12. Channel designations are: Channel 1, ●; channel 2, ○; channel 3, ◆; and channel 4, △.

shaded channels. On Sept. 11, 19 days after nutrient addition commenced, all shaded channels remained significantly lower than the unshaded channel 1 (49.0 ± 2.3 g SD; channel 2, 39.6 ± 1.0 g SD, $p < 0.05$; and channel 3, 28.2 ± 4.7 g SD, $p < 0.01$), and biomass was maximum. On Sept. 12, nutrient addition ended and nutrient concentration in all treated channels returned to background concentrations. A final sample was taken Sept. 20, 8 days after nutrient cutoff and 4 weeks after initiation of nutrient amendment. After nutrient cutoff, periphyton biomass declined dramatically in all channels. The decrease in biomass after cutoff was generally proportional to the increase during nutrient addition. Despite differences in nutrient concentrations, the two fully lighted channels eventually supported approximately equal amounts of biomass: 52.2 $\pm$ 6.9 g/channel in the control vs. 49.0 $\pm$ 2.3 g/channel in the channels with nutrient addition (Figure 4). The rate of biomass accumulation appeared faster in the nutrient-amended channel; the differences were not statistically significant, however. After nutrient cutoff, the decline in biomass was almost equal in both channels (17.63 $\pm$ 2.7 g vs. 16.3 $\pm$ 0.31 g).

Chlorophyll

Although biomass response to the simulated reset was nearly identical for the control and nutrient-treated channels under fully lighted conditions, chlorophyll content of the control channel was less than half that of the nutrient-amended channel (Figure 5). The fact that the channel without nutrient amendment received somewhat greater light than the other channels because of side lighting may have been in part responsible for the higher chlorophyll content of the control flume at the time the experiment commenced. Chlorophyll increased from 31 to 182 mg/channel in the fully lighted flume without nutrient addition. In the nutrient-treated, fully lighted flume, however, chlorophyll rose from 22 to a maximum of 403 mg/channel. Channel 2, with 30% shading, produced as much chlorophyll as the fully lighted channel although it had about 20% less biomass at the time of maximum biomass accumulation. This increase in chlorophyll concentration may represent some adaptation to shading.

Some index of the physiological response to shading and nutrient addition can be obtained by considering the biomass-to-chlorophyll ratio for the first section of each channel during nutrient addition (Figure 6). The first channel section provided the most valid index of a response since it was closest to the nutrient source; downstream channel sections represented a response to stream chemistry already altered by upstream biological activity. In the control channel, biomass-to-chlorophyll ratio rose from 102 at the start of the experiment to a high of 295 at the conclusion of nutrient input 20 days later. On the other hand, all nutrient-treated channels had lower biomass-to-chlorophyll ratios within 4 days after nutrient amendment. In fact, the biomass-to-chlorophyll ratio decreased by approximately one-half in the nutrient-treated channels.

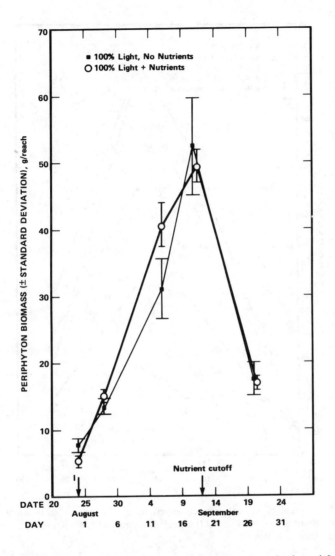

Figure 4. Comparison of periphyton biomass between the control, channel 0 (■), and channel 1 (○) which was nutrient-amended and unshaded. Estimates are in grams of ash-free dry weight ± standard deviation.

Channel 1 exhibited the least percentage decrease in biomass-to-chlorophyll ratio (46%); channels 2, 3, and 4 had decreases of 65, 55, and 51%, respectively. Channel 4 consistently had the lowest biomass-to-chlorophyll ratio, presumably because of, in part, shade adaptation.

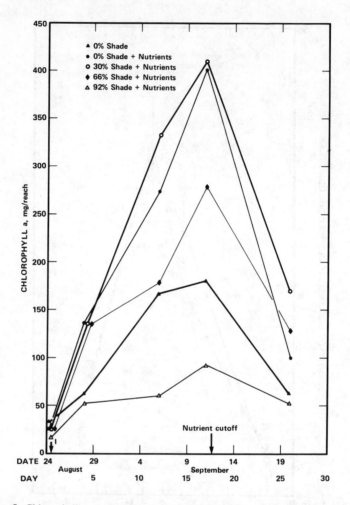

Figure 5. Chlorophyll a (mg) in each experimental channel on each sampling date. Channel designations are: Control (channel 0), ▲; channel 1, ●; channel 2, ○; channel 3, ◆; and channel 4, △.

Biomass-to-chlorophyll ratio generally increased longitudinally in the channels, and, in equivalent locations, nutrient-amended channels always had a lower ratio than the unamended control.

Primary Production

Net community primary production was highest in the full-light and nutrient-amended flume (Figure 7). Both the 30%-shaded and the 66%-

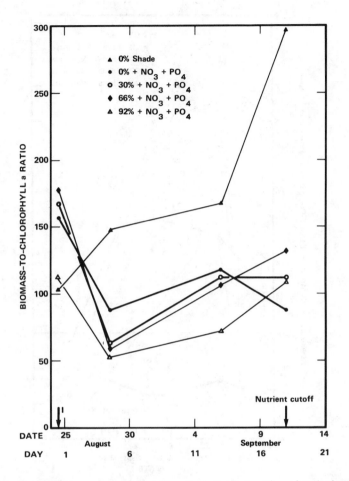

Figure 6. Ratio of biomass to chlorophyll a, from the first section of each channel on each sampling date. Channel designations are: Channel 0, ▲; channel 1, ●; channel 2, ○; channel 3, ◆; and channel 4, △. Nutrient injection was initiated (I) Aug. 24 and concluded (C) Sept. 12, 1979.

shaded channels had higher rates of net community primary production than did the control (channel 0) 4 days after the start of nutrient injection. On other sampling dates, channel 0 had higher rates of net community primary production than any of the shaded, nutrient-amended channels. Large declines in net community primary production observed on the last sampling date were presumably caused by the oxygen demand exerted by decomposing algae that had entered the detritus pool. Biotic activity in the 92%-shaded channel was heterotrophic throughout the experiment.

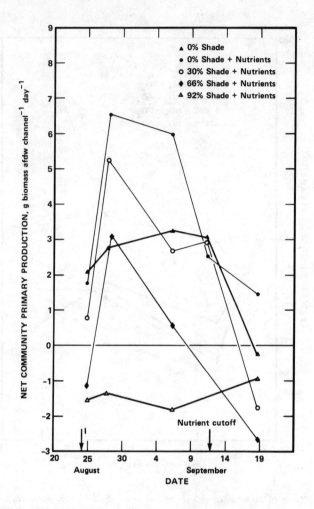

Figure 7. Net community primary production in each channel for each sampling date. Estimates are in grams of ash-free dry weight/day. Channel designations are: Channel 0, ▲; channel 1, ●; channel 2, ○; channel 3, ◆; and channel 4, △.

This does not indicate a total absence of primary production since a net increase in oxygen was observed during midafternoon on most sampling dates. It indicates only that heterotrophic processes dominated overall. As expected, the highest rates of net community primary production occurred when the rate of biomass increase was fastest. In nutrient-amended channels 2 and 3, net community primary production declined sharply after August 28 although biomass continued to accumulate until September 12.

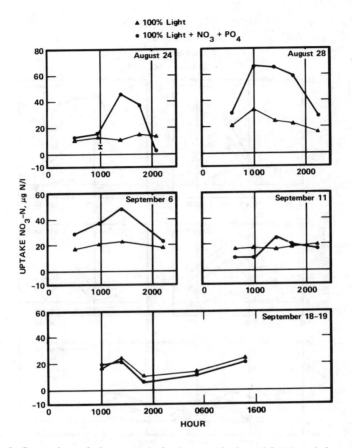

Figure 8. Comparison of nitrate uptake in the control, channel 0 (▲), and the nutrient-amended treatment channel, channel 1 (●). Uptake is defined as input concentration minus output concentration. Nutrient injection (I) started 1000 hrs on Aug. 24.

Nitrate Uptake

Nitrate uptake, the difference between input and output concentrations, exhibited a definite diel pattern in all channels (Figures 8 and 9). Uptake was greatest at midafternoon and least after dark. The magnitude of diel fluctuation depended on nutrient and light regimes and on the age of the community.

Comparison of nitrate uptake through time in fully lighted channels with and without nutrient amendment (Figure 8) indicates that, before the injection (I) was begun on August 24, uptake was similar in both flumes.

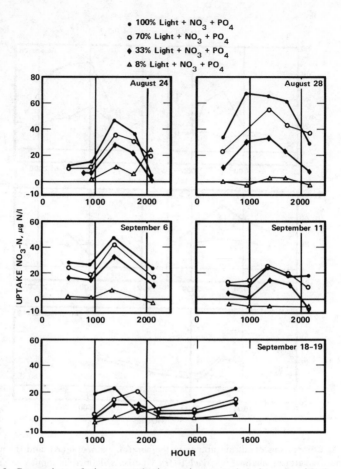

Figure 9. Comparison of nitrate uptake in nutrient-amended channels with three levels of shade treatment: Channel 1, ●; channel 2, ○; channel 3, ◆; and channel 4, △. Uptake is defined as input concentration minus output concentration.

Once nutrient injection commenced, the observed response was immediate, with much larger uptake in the nutrient-amended flume. This high rate of uptake continued through August 28 as the community grew exponentially. In the control flume the amount of uptake also increased, but not as dramatically as in channel 1. Differences in uptake became less between control and treated channels (September 6 and 11) as the community matured. On September 19, a week after the cessation of nutrient amendment, diel uptake patterns were similar in both channels.

Nitrate uptake was influenced by shading as well as by nutrient concentration (Figure 9). Even the 92%-shaded channel showed some degree of diel fluctuation. Except for channel 1, at 1600 h on September 18, nutrient uptake was proportional to light input. Occasionally water samples taken after the community matured had higher output than input nitrogren concentrations, possibly because of remineralization of organic detritus.

The capacity of periphyton to influence channel nitrate chemistry was related to the successional stage of the community. In the lighted, nonamended control and in all the nutrient-amended treatments, maximum diel uptake decreased as the community aged. The maximum nitrate uptake occurred when biomass accumulation rate was greatest. On September 11, when periphyton biomass was maximum, senescence of the community was indicated by a decline in nitrate uptake and in net community primary production.

Four hours after the start of nutrient addition, maximum uptake was linearly related to the amount of shading (Figure 10). Within 4 days this relationship had broken down, and midafternoon uptake declined at all levels of shading (Figure 11); this provides evidence of gradual senescence in the periphyton community.

Invertebrates

Changes in invertebrate numbers through time in each channel (Figure 12) was a general response to differences in periphyton biomass related to shading. Until nutrient cutoff, channel 1, the open channel with nutrient

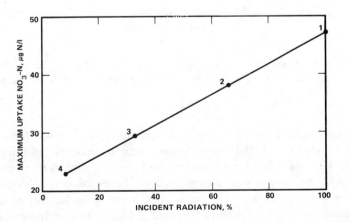

Figure 10. Nitrate uptake in channels 1 to 4, 4 hr after the start of nutrient addition. Nitrate uptake (input concentration minus output concentration) was maximum at 1400 hr on all sampling dates.

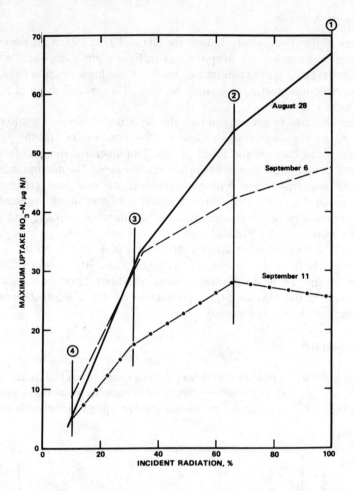

Figure 11. Maximum nitrate uptake (1400 hr) as a function of shading on three different sampling dates as the periphyton community gradually underwent senescence. The channel number is circled.

addition, had the largest numbers of invertebrates. This was followed in turn by channel 0, the open channel with no nutrient addition, and channel 3 (66% shade + nutrient) recorded the third highest numbers. This apparent exception was inexplicable in terms of either biomass or chorophyll and presumably represented some other habitat advantage provided by denser shading. Channel 2 (30% shade + nutrients) was higher than channel 4 (92% shade + nutrients).

A subsample of 65 invertebrates was preserved in 50% ethyl alcohol and indentified. Since identified invertebrates were selected by observed

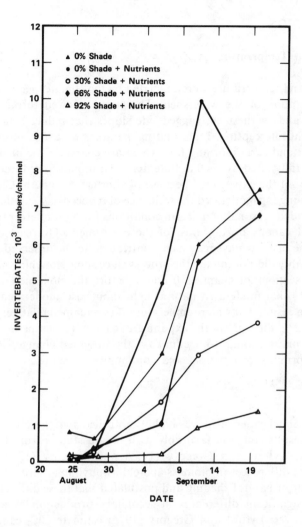

Figure 12. Numbers of invertebrates in each channel on each sampling date.

morphological differences, taxonomic examination was not an estimate of invertebrate community composition. In all, 22 taxa were identified. Almost all were Chironomidae larva, particularly Orthocladinae. Since virtually all larvae were Chironomidae, grazer biomass was small (<0.1 g dry weight/m^2). Growth of invertebrate larvae was observed during the experiment as the mean individual weight increased from 0.6×10^{-4} g on September 6 to 1.1×10^{-4} g on September 11, and finally to 1.9×10^{-4} g on September 20.

DISCUSSION

Light and Temperature

Light and temperature were regulated by the climate, geomorphology, and vegetation of the watershed. The stream flowed through a long, narrow valley with steep, rugged side slopes that reduced the hours of direct light interception. Light and temperature were also moderated by late night and early morning fog, which are characteristic of the region. Fog generally lifted by 1000 h. Potential solar input was reduced from the beginning of the experiment because of shorter day length. Cloud cover and morning fog moderated the influence of reduced day length, however, and provided reasonably uniform conditions for light interception by the simulated canopy during most of the experiment. Overcast conditions during the last few days may have contributed to the observed sloughing but probably did not initiate the process since some sloughing occurred in the first sections of channels 0 and 1 before the cloudiness. The same conditions that moderated natural light input also moderated temperature. As a result, diel water temperature flux remained relatively constant at about 2°C throughout the experiment. Changes in temperature did not seem dramatic enough to account for the observed changes in nutrient uptake, primary production, and community structure.

Biomass

The channel experiment was highly controlled and, therefore, artificial, but the results compare favorably with in situ studies from the Oregon Cascades. Lyford and Gregory (1975) and Gregory (1980) reported significant differences in biomass and chlorophyll in a third-order stream as a result of natural shading and postulated that these differences could lead to significant differences in secondary production as a result of canopy cover. Lyford and Gregory (1975) estimate 80% canopy cover (shading) for their forested section and 15% shading for the clear-cut area. This 65% difference in light interception resulted in an ash-free dry weight of 2.0 and 3.5 g/m^2 of periphyton biomass for shaded and open areas, respectively, in a mid-September sampling (Gregory, 1980). In our channel experiment with light interception, we measured biomass levels of 3.92 (0% shading) and 2.25 (66% shading) g periphyton AFDW/m^2. In WS 10, a first-order stream with an estimated 99% shading, Lyford and Gregory (1975) reported approximately 0.3 g/m^2 of periphyton during mid-September in comparison with a maximum of 0.6 g/m^2 at 92% shading in this study. Maximum ash-free dry weights in the control channel (4.2 g/m^2) and unshaded, nutrient-amended channel (3.9 g/m^2) also compare favorably with the 4.0 g/m^2 estimated from identical slides exposed simultaneously at five stations in Little Lost Man Creek (unpublished data). Thus the levels of biomass obtained in the flumes were representative of some actual lotic systems with similar nutrient and light regimes.

Our biomass levels were substantially less than those reported by Stockner and Shortreed (1978) for a flume experiment at Carnation Creek on Vancouver Island, Canada. Similar control and treatment concentrations of NO$_3$ and PO$_4$ produced a maximum ash-free dry weight biomass of approximately 10 g/m^2 for a control and 37.5 g/m^2 for a NO$_3$ + PO$_4$ treated channel. A major difference between the experiments was that the days were substantially longer in the Canadian study. In our experiment, slower current velocity, longer channel length, and channels with a higher ratio of colonization surface area to water surface area produced a set of physical-chemical conditions in which biomass was apparently the maximum sustainable in each channel. On the basis of water surface area, periphyton biomass in this experiment was 37.25 g/m^2 in the control channel and 35.01 g/m^2 in the unshaded nutrient-amended channel.

Although biomass was lower in this study than in that of Stockner and Shortreed (1978), the overall pattern of community development was quite similar. The periphyton community grew rapidly in the latter study, producing maximum ash-free dry weight in enriched (NO$_3$ + PO$_4$) channels after 38 days. From that point the biomass declined to approximately 3 g/m^2 between days 38 and 59. Within an additional 40 days, neither control nor treatment biomass exceeded 7 g/m^2. In our experiment, in both control and treatment flumes, maximum periphyton was attained within 18 days from a previously seeded base (9 days); then a similar decline in biomass followed. Treatments that generated the highest biomass had the largest decline.

Although both nutrient and light regimes influenced maximum periphyton biomass, neither seemed to control its eventual resetting. Timing of the community decline was not related to nutrient treatment since both the control and treatment communities (channels 0 and 1), with different regimes, declined similarly in the 8-day interval between nutrient cutoff and the final sample. If light was the regulating agent, then the gradual decrease in day length in the shaded flumes would presumably accelerate sloughing in comparison with those which had more light. This did not occur in the periphyton communities in our experiment; communities in channels 1 through 3 all declined at the same time. A third alternative, the gradual senescence and physiological death of organisms closest to the substrate surface provides a logical hypothesis to explore in the process of community resetting.

Chlorophyll

As in the case of periphyton biomass, chlorophyll increased with biomass and then fell as the community declined. However, the unshaded channel without nutrient addition had only about 40% as much chlorophyll as the unshaded channel with nitrate and phosphate amendment, but biomass was essentially equal. Furthermore, the chlorophyll differences were maintained throughout the experiment. Protein is the major reservoir of cellular nitrogen (Rhee, 1978), but some difference in chlorophyll might also be expected since the chlorophyll molecule contains significant quantities of nitrogen. White and Payne (1977) found a 38% mean increase in chlorophyll production after 5 days of nitrogen addition to batch cultures of phytoplankton from Lake Taupo, New Zealand. This was between July and September, when nitrogen was considered limiting. In a similar study by the same researchers (1978) at Lake Rotorya, New Zealand, 5-day assays after nitrogen amendment with and without phosphorus and micronutrients indicated a minimum of 86% stimulation in chlorophyll production. The degree to which the increase was a function of actual growth rather than a decrease in biomass-to-chlorophyll ratio was not specified. It is generally well known that nitrogen starvation results in greatly diminished algal chlorophyll content (Droop, 1974; Fogg, 1959; Hase et al., 1957; Skoglund and Jensen, 1976); however, there is little evidence for rapid decrease in biomass-to-chlorophyll ratio when inorganic nitrogen is added to lotic field communities. In our study the biomass-to-chlorophyll ratio decreased by half within 4 days of NO_3 and PO_4 amendment at all experimental light levels (see Figure 6). After the initial drop, biomass-to-chlorophyll ratios rose gradually as the community aged. Conover (1975) found that the pigment pool of *Thalassiosira fluviatilis*, a marine diatom, was highest in young and mature cells and decreased as cells became senescent. Skoglund and Jensen (1976) noted that chlorophyll levels in certain diatoms are influenced by nitrate. Rhee (1978), however, found that chlorophyll content increased as a function of carbon fixation in nitrogen- and phosphorus-limited cultures with continuous illumination.

Light also influences biomass-to-chlorophyll ratio. Meeks (1974) stated that, within limits, cellular chlorophyll content is inversely proportional to light intensity during growth. In this study, from August 28 to September 6, biomass-to-chlorophyll ratios decreased as shading increased in all the treated channels (see Figure 6). This adaptation to shading was previously reported in lotic ecosystems by Lyford and Gregory (1975). Chlorophyll estimation can serve as a valuable tool in a treatment-response laboratory stiuation for light and nutrients but is less valuable as a field estimate of community structure, particularly when the age of the community is unknown.

Primary Production

Pristine streams of northern California and the Pacific Northwest are often regarded as nitrogen limited because of low nitrate concentration (Fredriksen, 1971; 1972; Gregory, 1980), active conservation of nitrogen by the undisturbed watershed (Fredriksen, 1971; 1972, Scrivener, 1975), and low nitrogen-to-phosphorus (N/P) ratios (Gregory, 1980). Redfield (1958) suggested that aquatic primary production would be nitrogen limited at an N/P atomic ratio below 15. Rhee (1978) later empirically determined the N/P ratio to be 30 for one species of algae. Using a conservative N/P ratio of 10, Thut and Haydu (1971) found that half the surface waters in the state of Washington would be nitrogen limited. In this study the N/P atomic ratio was approximately 9 in both the control and treatment channels. Another indicator of potential nitrogen limitation is the presence of nitrogen-fixing algae, particularly *Nostoc* in streams. *Nostoc* was found occasionally in well-lighted riffles of Little Lost Man Creek. It never constituted an important component of the periphyton, however, possibly because of shading by riparian vegetation.

Under certain conditions, adding nitrate to a presumably nitrogen-limited, first-order stream has little effect on primary production. Gregory (1980) found that nitrate addition had little effect on accumulation of algae, gross primary production, net primary production, community respiration, P/R ratio, or community structure of diatoms at low ambient light levels (500 to 1000 lux) in Watershed 10, H. J. Andrews Experimental Forest, Oregon. The absence of a nitrate effect was observed despite the fact that nitrate levels were so low that they were often undetectable during summer months. In our experiment the 92% shading in channel 4 resulted in extremely low biomass and a stable community, obviously limited by light (see Figure 7). Daily net community primary production was continually negative. McIntire and Phinney (1965) reported small amounts of net O_2 evolution at 861 lux during 8-h light periods. In our study, although 92% shading would produce a light regime similar to Gregory's 500 to 1000 lux, small increases in O_2 concentration were also observed in midafternoon. The community was dominated by heterotrophy, however, and was probably energetically based on dissolved organic carbon, upstream input of fine organic detritus, and cell death within the community. On the other hand, net primary production was limited by nutrients, at least initially, in the control channel (see Figure 7).

Rates of net community primary production in the nutrient-amended flume were almost twice as high as in the control flume until biomass was maximum. Physical access to nutrients, light and sites for attachment may have played a more decisive role as the community matured. Failure to generate significantly higher biomass than the control, despite higher net production, probably indicates greater sloughing in the amended channel. In the two partially shaded channels, net community primary production was higher than in the control during the period of exponential growth. After August 28, net community primary production dropped quickly, possibly because of self shading as the community developed. Thus net community primary production in each channel was a response to a specific set of nutrient and light conditions for each habitat and was dependent on the age of the community.

The results indicate that, in small first-order streams of a forested watershed, net community primary production may actually be light limited despite extremely low concentrations of inorganic nitrogen and phosphorus. In open reaches of higher order streams, nutrients are the more likely factor limiting periphyton growth. In habitats more recently sloughed or scoured, net community primary production may be higher than in a more space-limited or senescent community, even if the nutrient and light regimes are identical.

Nitrate Uptake

A major goal of the project was to determine the potential ability of periphyton to impact nitrogen cycling through a drainage network. Results indicated that biotic uptake by an immature algal community is significant, especially during the day, in well-lighted habitats. The experimental channels provided ideal habitats to measure this potential since nitrate in the water column could be neither renewed nor diluted by groundwater, and the system was free from potential uptake by riparian vegetation.

The definite diel pattern with greatest uptake in midafternoon indicated that the algal component of periphyton was primarily responsible. Dark uptake definitely occurred, and its significance varied by shade treatment and, therefore, by amount of biomass (Figures 8 and 9). Dark uptake was greatest in all channels during the period of most active periphyton growth (August 24–28). In the control channel, dark uptake was almost as high as during the day, presumably because of nitrogen limitation. Sebetich (personal communication) found dark uptake to be a highly significant proportion of the periphyton nitrate budget. Eppley et al. (1971) reported both night and day uptake of nitrate in nitrogen-limited chemostat cultures of two marine phytoplankton. Nitrogen assimilation was much greater during the day than at night, however. Similarly, Grant and Turner (1969) reported light-stimulated uptake up to 23 times greater than in the dark, presumably because nitrate uptake and reduction are energetically linked to photosynthesis (Eppley and Coatsworth, 1968; Eppley et al., 1971; Healy, 1973; Cloern, 1977). Uptake response to the initiation of nitrate addition on Aug. 24 was both large and immediate. A similar initial rapid response was previously observed in pure culture (Cloern, 1977).

Gregory (1980) recently reported the differential capacity of the periphyton to alter the nutrient chemistry of first-, third-, and fifth-order streams. He attributed the large differences in diel uptake patterns to light interception by the canopy, although the nutrient regimes and current velocity of the streams were also different. Data from our study experimentally confirm his findings under controlled conditions of light, current velocity, and nutrient concentration.

The decreased nutrient uptake found as the periphyton community matured was unexpected but illustrates the importance of considering community processes rather than static estimates of community structure, such as biomass. Pryfogle and Lowe (1979) reported that dead cells can constitute a large proportion of the biomass of natural periphyton communities. Estimates of live biomass in their study varied from approximately 80% during early spring to about 50% in late August. Thus a major portion of the periphyton community may be nearly physiologically inert to nutrient uptake or primary production, while still constituting a major component of community structure.

Since the capacity for both primary production and nutrient uptake are biologically coupled (Rhee, 1978) and are related to the history of the community (Healy, 1973), both parameters should be measured simultaneously to describe adequately the ecological role of periphyton. Considering maximum uptake vs. net community primary production indicates a strong correlation between the two processes (Figure 13). Within 4 h after injection started (the circled data point on the figure), maximum nitrate uptake and net community primary production had a direct linear correlation ($r = 0.997$). If all data before nutrient cutoff are considered, maximum diel uptake is still significantly correlated to net community primary production ($r = 0.742$).

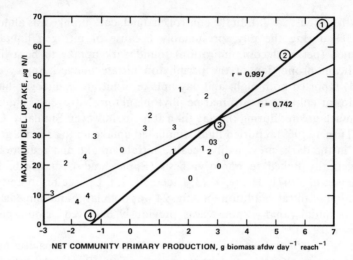

Figure 13. Maximum diel uptake of nitrate as a function of net community primary production, 4 hr after nutrient amendment (circled data points) and for all sample dates. Data points are the channel number.

Invertebrates

The scarcity of macroinvertebrates prevented any significant grazer impact on the periphyton community. When the study ended, invertebrate numbers were large, but their mean weight was extremely small. Even feeding by increased numbers of larger herbivores (i.e., those whose weight had increased) was probably insufficient to impact the periphyton, considering the amount of biomass and net community primary production. Gregory (1980), in a grazing study with the snail *Juga silicula*, found that herbivore biomass equivalent to periphyton biomass was required to prevent accumulation of algal biomass in flume studies. Lower amounts of grazer biomass (2.2 g/m^2) had essentially no effect on accumulation of periphyton.

The appearance of large numbers of invertebrate larvae was significant in interpreting the general physiological state of the periphyton community. Busch (1978) found, after a simulated freshet, that periphyton succession occurs in four distinct stages: (1) A rapid increase in algal biomass and production; (2) an increase in faunal biomass (particularly chironomids) and export; (3) reorganization of the benthic system, with an increase in primary production and a decrease in community respiration; and (4) a hypothesized development of a more stable community.

Our system mimicked the first two phases of Busch's system but did not enter stage 3 for two reasons: (1) Invertebrate biomass was not sufficient to impact periphyton biomass. (2) Sloughed biomass entered the detrital pool rather than being transported from the system. Entry into the detrital pool, in turn, influenced the measurements of primary production since the upstream-downstream method determines production of the total system, whereas the chamber methodology used by Busch (1978) considered only production and respiration of the periphyton film.

Extrapolation to Natural Lotic Systems

Through time it has become increasingly apparent that the biological community of lotic systems is the result of a long history of infrequent, episodic events that shape the channel and build discrete habitats. The community that emerges on these habitats reflects the geology and geomorphology of the region and is influenced energetically by terrestrial vegetation as a contributor of allochthonous materials and an interceptor of light. Between the long-term episodes there are shorter intervals during which local meteorological events make minor modifications to the stream bottom, enchance nitrate concentration, transport fine organic detritus, and reset the periphyton community. The flume experiment indicates how the periphyton community can potentially play a role in regulating nitrogen transport, especially under low-flow conditions. Biological nitrate uptake regulates the distance an ion will travel down the watershed. Travel distance, in turn, varies according to reach order and location, (i.e., canopy cover) and day length. In first-order, heavily forested streams, interception of nitrate is minor (Gregory, 1980), and thus distance is long. In open, high-order streams, the diel fluctuations in NO_3 concentration can be quite large, and thus travel distances are short.

The age of the communities and the time elapsed since the last meteorological resetting may also influence nitrogen uptake kinetics and, consequently, travel distance as demonstrated in the flumes. In young communities with high rates of net primary production, uptake was 45% of available nitrate in less than 10 m of travel distance. In the same channel, uptake by the mature periphyton film was only 20% of available nitrate although biomass increased threefold. The smaller differences in input and output nitrate concentration do not per se indicate a concomitant decline in nitrate uptake since some nitrate may have been supplied by remineralization of sloughed tissue. The simultaneous decrease in the difference between input-output nitrate concentration and net community primary production does, however, tend to indicate an actual physiological decline rather than the asymptotic steady state where nitrogen uptake equals regeneration. A long-term in situ study of diel nitrate uptake patterns under natural conditions has not yet been conducted.

The magnitude of nitrate uptake may not be a unique property of the reach so much as it is a result of the history of the community and canopy density of adjacent terrestrial vegetation. Thus both nitrate uptake and net community primary production should be measured simultaneously whenever possible to assess the abilities of the periphyton community to cycle nitrogen in lotic systems.

Periphyton in the channels behaved similarly to nitrogen-deficient cultures in their ability for rapid nitrate uptake, under both light and dark conditions when nutrient concentration increases. This ability may enable the community to take advantage of episodic events, such as minor storms, which produce nitrate-enhanced runoff. This capacity may also satisfy future nitrogen requirements for the periphyton community after major scouring events since stream water returns to background nitrate concentration. The ability of algae to use small episodic events as an important nitrate source has previously been reported for phytoplankton populations in marine environments (McCarthy and Goldman, 1979).

Just as limiting factors can change in lentic environments (Storch and Dietrich, 1979), they can also change through time and space in lotic environments from first- through higher-order streams. In the smallest, first-order woodland streams, light can be limiting to periphyton production even when nitrate concentration is nearly undetectable (Gregory, 1980). When one considers the myriad of possible light patterns imposed on lotic systems by the adjacent terrestrial vegetation, it is easy to understand why environmental variability is large.

Lane and Levins (1977) commented on the inability to extrapolate from enrichment studies on algae cultures to modeling predictive behavior in field situations. They assert that a major weakness underlying any effort at prediction is the assumption that the system is adequately represented by its conceptual model. In lotic ecology the conceptualizations themselves are still emerging (Cummins, 1974; Webster et al., 1975; Vannote et al., 1980). A major problem in the transfer of laboratory data to lotic systems is that ecosystem process rates are not products of individual species biochemistry, but rather are the result of interactions of organisms with other species and with their physical and chemical environment. Devices like in situ channel systems provide an opportunity both to control interactions and to examine more holistically how community level interactions impact nitrogen cycling through time and space in a drainage network.

DISCLAIMER

Any use of trade names is for descriptive purposes only and does not imply endorsement by the U. S. Geological Survey.

REFERENCES

Busch, D., 1978, Successional Changes Associated with Benthic Assemblages in Experimental Streams, Ph.D. Thesis, Oregon State University, Corvallis.

Cloern, J. E., 1977, Effects of Light Intensity and Temperature on *Cryptomonas ovata* (Cryptophyceae) Growth and Nutrient Uptake Rates, *J. Phycol.*, 13: 389–395.

Conover, S. A. M., 1975, Partitioning of Nitrogen and Carbon in Cultures of the Marine Diatom *Thalassiosira fluviatilis* Supplied with Nitrate, Ammonium, or Urea, *Mar. Biol.*, 32: 231–246.

Cummins, K. W., 1974, Structure and Function of Stream Ecosystems, *BioScience*, 24: 631–641.

Droop, M. R., 1974, Heterotrophy of Carbon, in W. D. P. Stewart (Ed.), *Algal Physiology and Biochemistry*, Botanical Monographs, Vol. 10, Chap. 19, pp. 530–559, University of California Press, Berkeley.

Eppley, R. W., and J. L. Coatsworth, 1968, Uptake of Nitrate and Nitrite by *Ditylum brightwellii*—kinetics and mechanisms, *J. Phycol.*, 4: 151–156.

———, J. N. Rogers, and J. J. McCarthy, 1971, Light/Dark Periodicity in Nitrogen Assimilation of the Marine Phytoplankters *Skeletonema costatum* and *Coccolithus huxleyi* in N-Limited Chemostat Culture, *J. Phycol.*, 7: 150–154.

Fogg, G. E., 1959, Nitrogen Nutrition and Metabolic Patterns in Algae, *Symp. Soc. Expl. Biol.*, 13: 106–125.

Fredriksen, R. L., 1971, Comparative Chemical Water Quality—Natural and Disturbed Streams following Logging and Slash Burning, in J. T. Krygier and J. D. Hall (Eds.), *A Symposium—Forest Land Uses and the Stream Environment*, pp. 125–137, Oregon State University Press, Corvallis.

———, 1972, Nutrient Budget of a Douglas-Fir Forest on an Experimental Watershed in Western Oregon, in J. F. Franklin, L. J. Depster, and R. H. Waring (Eds.), *Proceedings—Research on Coniferous Forest Ecosystems—A Symposium*, pp. 115–131, Pacific Northwest Range and Experiment Station, Portland, OR.

Grant, B. R., and I. M. Turner, 1969, Light Stimulated Nitrate and Nitrite Assimilation in Several Species of Algae, *Comp. Biochem. Physiol.*, 29: 995–1004.

Gregory, S. V., 1980, Effects of Light, Nutrients, and Grazins on Periphyton Communities in Streams, Ph.D.Thesis, Oregon State University, Corvallis.

Hase, E., Y. Morimura, and H. Tamiya, 1957, Some Data on the Growth Physiology of *Chlorella* Studied by the Technique of Synchronous Culture, *Archs. Biochem. Biophysics*, 69: 149–165.

Healey, F. P., 1973, Inorganic Nutrient Uptake and Deficiency in Algae, *CRC Critical Reviews in Microbiology*, pp. 69–113.

Kennedy, V. C., and R. L. Malcolm, 1977, *Geochemistry of the Mattole River of Northern California*, Open-File Report 78–201, U. S. Geological Survey, Menlo Park, CA.

Lane, P., and R. Levins, 1977, The Dynamics of Aquatic Systems. 2. The Effects of Nutrient Enrichment on Model Plankton Communities, *Limnol. Oceanogr.*, 22: 454–471.

Lyford, J. H., Jr., and S. V. Gregory, 1975, The Dynamics and Structure of Periphyton Communities in Three Cascade Mountain Streams, *Verh. Internat. Verein. Limnol.*, 19: 1610–1616.

McCarthy, J. J., and J. C. Goldman, 1979, Nitrogenous Nutrition of Marine Phytoplankton in Nutrient-Depleted Waters, *Science*, 203: 670–672.

McIntire, C. D., and H. K. Phinney, 1965, Laboratory Studies of Periphyton Production and Community Metabolism in Lotic Environments, *Ecol. Monogr.*, 13: 237–258.

Meeks, J. C., 1974, Chlorophylls, in W. D. P. Stewart (Ed.), *Algal Physiology and Biochemistry*, Botanical Monographs, Vol. 10, Chap. 5, pp. 161–175, University of California Press, Berkeley.

Owens, M., 1974, Methods for Measuring Production Rates in Running Waters, in R. A. Vollenweider (Ed.), *A Manual on Methods for Measuring Primary Production in Aquatic Environments*, IBP Handbook No. 12, 2nd ed., Chap. 3.41, Blackwell Scientific Publications, London.

Pryfogle, P. A., and R. L. Lowe, 1979, Sampling and Interpretation of Epilithic Lotic Diatom Communities, in R. L. Weitzel (Ed.), *Methods and Measurements of Periphyton Communities, A Review*, Report ASTM STP 690, pp. 77–89, American Society for Testing and Materials, Philadelphia.

Redfield, A. C., 1958, The Biological Control of Chemical Factors in the Environment, *Am. Sci.*, 46: 205–221.

Rhee, G-Yull., 1978, Effects of N : P Atomic Ratios and Nitrate Limitation on Algal Growth, Cell Composition and Nitrate Uptake, *Limnol. Oceanogr.*, 23: 10–25.

Scrivener, J. C., 1975, *Water, Water Chemistry, and Hydrochemical Balance of Dissolved Ions in Carnation Creek Watershed, Vancouver Island, July 1971–May 1974*, Fisheries and Marine Technical Report No. 564, Department of the Environment, Ottawa, Canada.

Skoglund. L., and A. Jensen, 1976, Studies on N-Limited Growth of Diatoms in Dialysis Culture, *J. Exp. Mar. Biol. Ecol.*, 21: 169–178.

Stockner, J. G., and K. R. S. Shortreed, 1978, Enhancement of Autotrophic Production by Nutrient Addition in a Coastal Rainforest Stream on Vancouver Island, *J. Fish. Res. Board Can.*, 35: 28–34.

Storch, T. A., and G. A. Dietrich, 1979, Seasonal Cycling of Algal Nutrient Limitation in Chautauqua Lake, New York, *J. Phycol.*, 15: 399–405.

Thut, R. N., and E. P. Haydu, 1971, Effects of Forest Chemicals on Aquatic Life, in J. T. Krygier and J. D. Hall (Eds.), *A Symposium—Forest Land Uses and the Stream Environment*, Oregon State University Press, Corvallis.

Vannote, R. L., G. W. Minshall, K. W. Cummins, J. R. Sedell, and C. E. Cushing, 1980, The River Continuum Concept, *Can. J. Fish. Aquat. Sci.*, 37: 130–137.

Webster, J. R., J. B. Wade, and B. C. Patten, 1975, Nutrient Recycling and the Stability of Ecosystems, in F. G. Howell, J. B. Gentry, and M. H. Smith (Eds.), *Mineral Cycling in Southeastern Ecosystems*, ERDA Symposium Series CONF-740513), pp. 1–17, NTIS, Springfield, VA.

Wetzel, R. G., and D. F. Westlake, 1974, Periphyton, in R. A. Vollenweider (Ed.), *A Manual on Methods for Measuring Primary Production in Aquatic Environments*, IBP Handbook No. 12, 2nd ed., Chap. 2.3, Blackwell Scientific Publications, Oxford, England.

White, E., and G. W. Payne, 1977, Chlorophyll Production, in Response to Nutrient Additions, by the Algae in Lake Taupo Water, *N. Z. J. Mar. Freshwater Res.* 11: 501–507.

———, and G. W. Payne, 1978, Chlorophyll Production, in Response to Nutrient Additions, by the Algae in Lake Rotorua Water, *N. Z. J. Mar. Freshwater Res.* 12: 131–138.

8. ENERGY BASIS OF ECOSYSTEM CONTROL AT SILVER SPRINGS, FLORIDA

Robert L. Knight*

Department of Environmental Engineering Sciences
University of Florida
Gainesville, Florida

ABSTRACT

A consumer-control hypothesis is presented that predicts feedback of controlling action equal to energy invested in consumer components of ecosystems. This hypothesis was tested with field measurements of system productivity at Silver Springs, Florida, after major alterations in fish populations were noted, and in replicable flow-through microcosms located in the river channel. Overall system metabolism of Silver Springs was found to be unchanged; this possibly indicates adaptation to loss of dominant consumers by replacement with functionally similar fish species. Microcosm experiments with snails (*Goniobasis floridense*) and mosquito fish (*Gambusia affinis*) demonstrated metabolism enhancement by low levels of these consumers corresponding to natural stream populations.

INTRODUCTION

The controlling influence of consumers (herbivors, 1° carnivores, etc.) in natural ecosystems has received attention in a series of review papers by Chew (1974), Mattson and Addy (1975), Owen and Wiegert (1976), Batzli (1978), and Kitchell et al. (1979). These investigators outline some mechanisms that consumer populations use to regulate system energy

---

*Present address: CH₂M HILL, P.O. Box 1647, Gainesville, FL.

flow, such as nutrient regeneration and/or cropping of producers to optimal growth levels. Efford (1972) and Chew (1974) suggest the need for examining such effects by experimentally regulating consumers over realistic density ranges. A hypothesis predicting consumer effect on system functioning is presented here, along with evidence of consumer control of system metabolism at Silver Springs, Florida.

Lotka (1922) proposed a principle of thermodynamics for open systems which states that selection in the struggle for existence is based on maximum energy flow (power). Later, Odum and Pinkerton (1955) and Odum (1968; 1971; 1979) suggested ways that control actions generate more power and, thus, are selected for in natural ecosystems.

Biological food chains represent concentrations of energy, with each trophic level requiring energy and materials diverted from the structure of the primary producers. A reduction of primary productivity represents a decrease in the main source of free energy input to biological systems and, therefore, a decrease in system power. For energy flow to be maximized in ecosystems, the structure of primary producers diverted to higher trophic levels must be compensated by energies fed back from those consumers to increase the productivity-to-biomass ratio of the remaining producers. Thus a control hypothesis is suggested by the maximum power principle (McKellar and Hobro, 1976; Odum, 1979): In ecosystems where species selection is based on maximum power, consumer components must have controlling actions that are energetically equal to their energy consumption in the system. The hypothesis suggests that controllers, such as consumers, will have an energy cost to the system that may not exceed their value as a stimulant to productivity and that natural selective processes will adjust controller density to an optimal level that results in maximum system productivity. This hypothesis predicts a subsidy-stress curve (E. P. Odum et al., 1979) for all system controllers. Figure 1 illustrates such a curve that has been empirically determined in many experiments studying the effect of a range of perturbation intensities. If we consider consumer density as the perturbation and gross productivity as the affected parameter, the control hypothesis predicts selection of the consumer density that results in maximum gross productivity under fixed conditions of light and nutrient inputs.

Silver Springs, Florida, is an ideal laboratory for testing factors controlling productivity in an aquatic ecosystem (Odum, 1957). Major changes in the fish populations within the last decade provided a natural experiment on the effect of consumers on system productivity. Consumer control of productivity was also studied in a series of flow-through microcosms located in the river channel. Colonization by productive periphyton communities was rapid, and screens allowed experimental

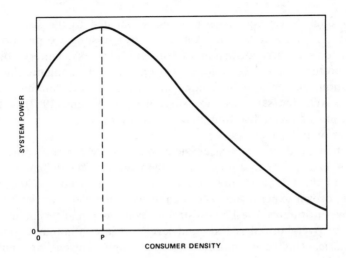

Figure 1. Subsidy-stress curve of consumers on system power (energy flow). The control hypothesis outlined in the text predicts this relationship for consumers with the point P representing optimal consumer density.

control of consumer density. Comparing consumer densities that enhance system energy flow to natural densities allows a test of the consumer-control hypothesis.

METHODS

Study Site

Silver Springs is a natural environmental feature located in north-central Florida in Marion County, east of Ocala. Two billion liters of water flow each day from one major boil and numerous smaller boils forming the Silver River, which flows with only minor dilution for 11 km to its confluence with the Oklawaha River. Chemistry data published by the U.S. Geological Survey (1978) for the main spring boil indicate water of moderate hardness (212 mg/l $CaCO_3$), low dissolved oxygen (DO) (2.0 mg/l), low organic carbon (2.0 mg C/l), and moderate levels of the primary plant nutrients (nitrate–nitrite nitrogen, 0.44 mg N/l), and ortho-phosphorus, 0.05 mg P/l). The chemistry of this spring water has been virtually unchanged since measurements were taken in 1907 (Rosenau et al., 1977).

Although Silver Springs was recently donated to the University of Florida, it is leased by the American Broadcasting Corporation and operated as a tourist attraction, offering guided tours in battery-powered, glass-bottom boats. Restrictions on swimming and fishing prevent direct alterations in the aquatic ecosystem. The biological communities are qualitatively the same as those described by Odum (1957), with the exception of major fish dominance changes, as discussed in the results section of this paper.

System productivity measurements were made for the entire upper section of the river to a point where the river narrows and enters a more shaded run at 1200 m downstream from the main boil (76,000 m^2).

Microcosm experiments were conducted underwater in the center of the river just above the 1200-m station. Water depth at this point is 3 m, and the river bottom is wide and level with a continuous covering of *Sagittaria*. This cover indicates consistent light, current, and nutrient-input characteristics.

System Measurements

Measurements of system metabolism were made using the upstream–downstream DO-change method of Odum (1956; 1957). Dissolved oxygen and temperature were measured with a submersible probe at 2-hr intervals for a 24-hr period at the main boil and at 1200 m downstream on Aug. 31, Oct. 5, and Dec. 13, 1978, and on Mar. 7, Apr. 15, May 16, June 19, and Aug. 15, 1979. Calculations of diffusion were made from percent saturation values of DO in the same manner as Odum (1957). Accrual of oxygen from the side boils was found to be unchanged and was taken as 0.61 mg O_2/l (Odum, 1957). An average depth of 1.8 m was used to convert volume measurement to an area basis. Solar energy input was integrated from a recording solar pyranometer on the days when diel measurements were taken.

Populations of fish > 10 cm in length were estimated on five dates by visual survey. The investigator wore a face mask with snorkel and was towed, holding the bow of a boat, in a criss-cross fashion down the spring run counting all fishes seen and reporting numbers in general groups or specific species to an assistant in the boat. A survey of the 76,000-m^2 area took about 1 hr to complete. Northcote and Wilke (1963) reported good agreement between visual fish counts and rotenone poisoning techniques for larger fish in clear-water environments. At Silver Springs it may be assumed that the visual counts are an underestimate, having greatest accuracy for the larger free-ranging fish (such as shad, bass, mullet, and bluegill sunfish) and less accuracy for the secretive or smaller species (such as spotted gar or small sunfish).

Snail populations were estimated in a shallow cove near the 1200-m station for comparison to microcosm data. A stiff-handled net with a 0.067-m^2 opening was used to sweep the *Sagittaria* beds, which are the habitats of the snails studied. Several passes were made, and the captured snails were sorted and weighed for live weight. These values were converted to a square-meter basis by multiplying the volume of water and plants sampled by the average depth of the sample area.

Microcosms

Microcosms used for measurement of consumer control of productivity were completely submerged, flow-through units (Figure 2). The microcosms consisted of two PVC fittings joined by replaceable, 0.4-mm, clear polyethylene tubing and were 6 m in length. The upstream fitting consisted of a PVC reducer (15 to 7.6 cm) connected to a short piece of PVC pipe (i.d., 7.6 cm). Plastic tubing 9 cm in diameter was connected to the PVC pipe with a hose clamp at the upstream end and to another piece of PVC pipe at the downstream end which was fitted with a standard garden hose connector. Two racks held the eight microcosms, fastened parallel to one another and to the current. The racks were anchored to the stream bottom by ropes and held off the bottom by floats at a depth of about 2 m. The ends of each microcosm were covered by plastic screen with mesh size large enough to allow water flow and colonization but small enough to retain snails and fish placed inside. Because these systems were submerged, they required no corrections for oxygen diffusion and were relatively safe from vandalism by pleasure boaters on the river.

Metabolism of the enclosed microecosystems was measured by upstream–downstream DO changes. For DO measurement, water was pumped from the upstream and from the downstream ends of each microcosm to an oxygen probe using a sealed impeller pump. Upstream DO was measured alternately with downstream DO at 5-min intervals. In this fashion, one production measurement for each of the eight microcosms could be made in a 40-min period.

The Hydrolab Surveyor oxygen probe and meter was readable to about 0.02 mg O_2/l. Because a single oxygen-monitoring system was used to compare upstream and downstream DO values, absolute accuracy of the readings was not a limiting factor. Rather, the stability of the readings was important, and the meter used was very stable. Upstream–downstream DO differences in the microcosms were often as great as 2.0 mg O_2/l on sunny days.

Oxygen rate-of-change curves were integrated for each microcosm to calculate system metabolism. Daylight DO changes measure net production, and nighttime changes represent system respiration. Nighttime

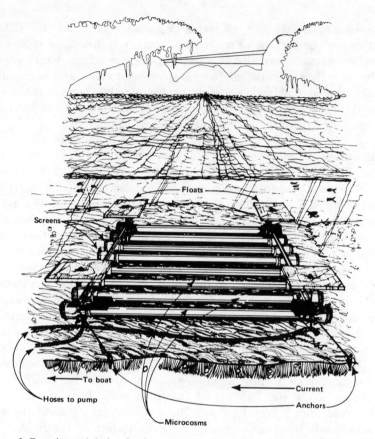

Figure 2. Experimental design for flow-through microcosms tested in Silver River. Eight polyethylene tubes are arranged parallel to the river current, 1 m above the river bottom and 2 m below the water surface. Screens over both ends of the microcosms allow manipulation of consumer density. Hoses to a boat on the surface allow measurement of chemical changes, such as dissolved oxygen over the length of the tubes (6 m).

respiration was found to be quite low (<10% of net production); therefore respiration was not of use in comparing the microcosms. If daytime respiration is assumed to be roughly equal to nighttime respiration, then net production may be nearly equal to gross production.

Photosynthetically active radiation (PAR) was measured concurrently with each daylight oxygen reading by using a submersible sensor at the same depth as the microcosms. Values of PAR were automatically integrated over a 100-sec interval for each reading. Light measurements inside the microcosms were not made. Solar radiation, measured as moles of photons (Einsteins), was reported as total heat energy by applying the conversion factor of 52.27 cal/Einstein calculated for sun and sky radiation by McCree (1972). Microcosm net production is reported as normalized net production (mg O_2/cal) to make data comparable between different days during an experiment.

RESULTS

Fish Populations

In his more comprehensive study, Odum (1957) reported that striped mullet (*Mugil cephalus*) and several species of catfish (*Ictalurus* spp.) were among the dominant consumers of Silver Springs. Since that time, and most probably in the last ten years (James Lowry, personal communication), these two dominants have virtually disappeared from the Silver River wildlife. Observations during the 2 years of this study indicated small scattered groups of mullet and catfish. Fish counts on five dates are reported in Table 1. These data indicate a new important component in the fish populations since Odum's study, i.e., the "blue shad" more commonly known as gizzard shad (*Dorosoma cepedianum*). All of those observed were greater than 30 cm in length. A mixture of sunfish of several species (*Lepomis* spp.) and largemouth bass (*Micropterus salmoides*) were the other dominant consumer fishes at the time of the study. Average live weight of these fishes during a 1-year period over the entire area was estimated as 11.5 g/m². About two-thirds of this mass was made up of herbivorous species, and one-third was carnivorous fish.

Snail Populations

The apple snail, *Pomacea paludosa,* and *Goniobasis floridense* were the only abundant snails in the area sampled. The four samples collected had a mean live snail biomass of 104 g/m², with a 61% coefficient of variation (CV). Biomass of *G. floridense* was measured as 42 g/m², with a CV of 75%.

Community Metabolism

Silver Springs continues to be an extremely productive natural ecosystem (Table 2). Spring and summer values for gross primary production on clear days were nearly 30 g O_2 m⁻² d⁻¹. The low value measured was on

Table 1. Results of Five Fish Counts Over the Entire Spring Area (76,000 m²)*

Fish species	Oct. 20, 1978		Apr. 11, 1979		May 16, 1979		July 17, 1979		Oct. 22, 1979	
	Number	Weight, kg	Number	Weight, kg	Number	Weight, kg	Number	Weight, kg	Number	Weight, kg
Blue shad (*Dorosoma cepedianum*)	689	545.7	308	243.9	931	737.4	575	455.4	677	536.2
Sunfish (*Lepomis*)	1135	130.5	1599	183.9	736	84.6	1333	153.3	339	39.0
Largemouth bass (*Micropterus salmoides*)	114	91.2	229	183.2	155	124.0	277	221.6	111	88.8
Golden shiner (*Notemigonus crysoleucas*)	1027	195.1	105	20.0	0	0	77	14.6	55	10.5
Striped mullet (*Mugil cephalus*)	71	84.6	0	0	9	10.7	0	0	2	2.4
Spotted gar (*Lepisosteus platyrhincus*)	17	9.6	26	14.8	14	8.0	18	10.3	14	8.0
Bowfin (*Amia calva*)	5	10.0	2	4.0	1	2.0	2	4.0	7	2.0
Chain pickerel (*Esox niger*)	8	4.3	71	38.1	3	1.6	6	3.2	7	3.8
Chubsucker (*Erimyzon*)	8	5.9	49	35.8	7	5.1	30	24.0	5	4.0
Needlefish (*Strongylura marina*)	3	0.3	0	0	1	0.1	0	0	1	0.1
Black crappie (*Pomoxis nigromaculatus*)	0	0	0	0	0	0	2	0.8	0	0
Total	3077	1077.2	2389	723.7	1857	973.5	2320	887.2	1212	694.8
Fish biomass, g/m²	14.2		9.5		12.8		11.7		9.1	

*Fish were counted if they were readily visible (>10 cm) by a diver at the water's surface. Fish weights were calculated from measured weights of average-sized fish that I collected and those collected by members of Florida Game and Fresh Water Fish Commission on Jan. 22, 1980.

Table 2. Metabolism of Silver Springs, Florida, Estimated from Oxygen Changes Between the Boil and a Point 1200 m Downstream (76,000 m^2)

Date	Insolation, kcal m^{-2} d^{-1}	Metabolism, g O$_2$ m^{-2} d^{-1}			P/R
		Photosynthesis, gross	Photosynthesis, net	Respiration, night	
Aug. 31, 1978	4463	23.4	7.6	15.8	0.7
Oct. 5, 1978	3331	17.8	6.9	10.9	0.8
Dec. 13, 1978	2270	17.1	6.7	10.4	0.9
Mar. 7, 1979	3540	18.3	15.2	3.0	3.1
Apr. 15, 1979	4888	17.9	3.8	14.1	0.6
May 16, 1979	5119	27.0	18.0	9.0	1.3
June 19, 1979	4030	28.5	15.7	12.8	0.9
July 17, 1979	2228	13.3	1.2	12.1	0.5
Aug. 15, 1979	4109	23.5	15.5	8.0	1.3

July 17, 1979, when low light during a long afternoon thundershower reduced productivity to 13 g O$_2$ m^2 d^{-1}. Calculations of the photosynthesis-to-respiration ratio (P/R) indicated a balanced trophic status for the river, with an average P/R ratio throughout the year of 1.1.

Microcosm Studies

Clean microcosms placed in the Silver River underwent a rapid successional development. Colonization by a diverse periphyton assemblage proceeded from the upstream to the downstream end of each tube. Small consumers, such as chironomid and trichopteran larvae, were abundant in screened tubes, and small fish, such as darters, were observed inside unscreened tubes. Within 3 weeks, growth of filamentous algae and fungi had nearly filled the tubes, lowering the current rate from 9 cm/sec initially to about 2 cm/sec. By this time periphytic growth on the outside of the tubes had begun to lower the light levels reaching the inside, and growth slowed down.

During optimal growth periods, the diverse microcosm communities demonstrated rapid response of net production to light intensity. Upstream–downstream DO change decreased within a few minutes when clouds obscured the sun and increased as quickly when greater sunlight was again available. Net productivity was directly related to light intensity, with greatest efficiency at low-light intensity and reduced efficiency at higher light levels. Typical curves of net production response to light input showed only slight indication of leveling, even at the highest light intensity values recorded.

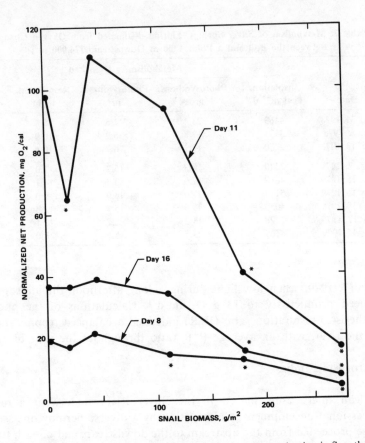

Figure 3. Effect of a range of snail densities on normalized net production in flow-through microcosms at Silver Springs, Florida, on 3 days in December 1979. Asterisks (*) indicate that the value is more than two standard errors from control mean.

Net primary production was slightly enhanced with respect to controls in microcosms receiving low densities of the herbivorous snail *Goniobasis floridense* (Figures 3 and 4 and Tables 3 and 4). This apparent stimulation was not statistically significant, however, because of high variability in the control microcosms. The characteristic subsidy-stress curve was observed on several of the measurement days, with apparent stimulation of +0.22 g O_2 m^{-2} hr^{-1} at a snail density of 44 g/m^2 in the first experiment and +1.87 g O_2 m^{-2} hr^{-1} at a density of 22 g/m^2 in the second snail experiment. Higher snail densities caused reduction in microcosm net production in comparison with controls. In the first experiment the highest snail density (270 g/m^2) lowered net production by –2.38 g O_2 m^{-2} hr^{-1}. In the second

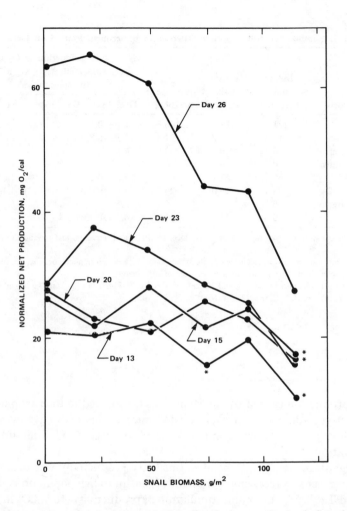

Figure 4. Effect of a range of snail densities on normalized net production in flow-through microcosms at Silver Springs, Florida, on 5 days in February and March 1980. Asterisks (*) indicate that the value is more than two standard errors from control mean.

experiment the highest snail density lowered net production by -3.18 g O_2 m^{-2} hr^{-1}. These high-density microcosms had thinner periphytic algal growth on the walls and were visibly different from the controls.

Most snails were recovered when the microcosms were harvested. In the first experiment 41% of the snails originally put into the high-density microcosm were lost or had died by the time they were recounted. In the low-density microcosms the snails not only survived the experiment but actually gained weight. It appears, therefore, that some of the increased

Table 3. Summary of Silver Springs Microcosm Experiment Started on Dec. 5, 1979*

Microcosm	Initial snail, weight, g	Final snail, weight, g	Average weight, g	Net production, g O_2/m^2 Consumer effect,‡ g O_2 m^{-2} hr^{-1} (in parentheses)		
				Dec. 13‡	Dec. 16‡	Dec. 21‡
1	171.8	121.8	146.8	1.10(–0.73)	0.12(–0.85)	1.03(–2.38)
2	113.8	81.2	97.5	3.30(–0.85)	0.29(–0.60)	2.02(–1.75)
3 (Control)	0	0	0	4.8	0.66	3.95
4 (Control)	0	0	0	4.57	0.58	4.57
5	60.1	62.1	61.1	3.77(–0.23)	0.66(–0.04)	4.61(–0.11)
6	28.4	19.0	23.7	5.30(+0.06)	0.78(+0.13)	5.14(+0.22)
7 (Control)	0	0	0	5.54	0.82	5.84
8	10.4	11.5	10.9	4.21(–0.15)	0.45(–0.36)	4.86(+0.04)

*Snails were released in microcosms on Dec. 10, 1979, and harvested and reweighed on Jan. 10, 1980.

‡Consumer effect was calculated as algebraic change between given microcosm net production and average control net production divided by time of measurement.

‡Dec. 13, 5.33 hr, 5.08 E/m^2, and 265 cal/m^2; Dec. 16, 0.67 hr, 0.14 E/m^2, and 7.3 cal/m^2; and Dec. 21, 1.58 hr, 2.63 E/m^2, and 137.5 cal/m^2.

net production caused by the snails was transferred to their trophic level as net growth. Snails in overcrowded microcosms may have eventually reached a stimulatory population level through food limitation and starvation.

Significant enhancement of net primary productivity was observed in the microcosms receiving a gradient of mosquito fish populations (Figure 5 and Table 5). Maximum stimulation of productivity (+0.9 g O_2 m^{-2} hr^{-1}) was found at a fish density of 3.2 g/m^{-2} (live weight). Of equal significance was the observation that these predaceous fish lowered net productivity of the microcosms at high fish densities. Thus a fish biomass between 15 and 31 g/m^2 lowered productivity by –0.8 g O_2 m^{-2} hr^{-1} in comparison with control microcosms without fish (Table 5).

Some mosquito fish mortality was observed, but this was caused by the initial shock of seining and introduction to the microcosms. Since microcosm 8 became detached shortly after the fish were added and most of them escaped, the data from this microcosm were not included in the summaries. Average weights of the fish at the beginning and end of the experiment indicated no consistent gain or loss of weight by surviving fish during the length of the study.

Table 4. Summary of Silver Springs Microcosm Experiment Started on Feb. 20, 1980

Microcosm	Initial snail weight, g	Final snail weight, g	Average weight, g	Net production, g O_2/m^2 Consumer effect,‡ (g O_2 m^{-2} hr^{-1} (in parentheses)				
				Mar. 4‡	Mar. 6‡	Mar. 11‡	Mar. 14‡	Mar. 17‡
1 (Control)	0	0	0	2.27	10.56	17.75	0.084	7.74
2 (Control)	0	0	0	1.98	8.07	15.94	0.069	12.62
3	51.1	50.1	50.6	1.93(−0.16)	8.85(−0.18)	16.65(−0.57)	0.094(−0.55)	10.09(−1.79)
4	25.7	25.6	25.7	2.19(+0.19)	9.71(+0.35)	15.22(−0.84)	0.125(+1.15)	14.13(−0.24)
5 (Control)	0	0	0	1.89	8.28	25.44	0.159	23.89
6	40.1	40.0	40.1	1.56(−0.65)	7.49(−0.70)	18.71(−0.19)	0.105(+0.054)	10.26(−1.73)
7	11.9	12.3	12.1	1.98(−0.09)	7.61(−0.64)	16.44(−0.61)	0.138(+1.87)	15.22(+0.18)
8	66.2	57.7	62.0	1.05(−1.33)	6.06(−1.37)	12.20(−1.40)	0.059(−2.47)	6.48(−3.18)

*Snails were released in microcosms on Feb. 26 and harvested and reweighed on Apr. 2, 1980.

‡Consumer effect was calculated as algebraic change between given microcosm net production and average control net production divided by time of measurement.

‡Mar. 4, 0.75 hr, 1.94 E/m², and 101.4 cal/m²; Mar. 6, 2.12 hr, 6.89 E/m², and 360.1 cal/m²; Mar. 11, 5.37 hr, 14.40 E/m², and 752.7 cal/m²; Mar. 14, 0.02 hr, 0.07 E/m², and 3.7 cal/m²; and Mar. 17, 2.60 hr, 4.65 E/m², and 243.1 cal/m²

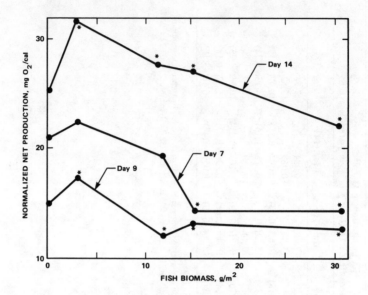

Figure 5. Effect of a range of fish densities on normalized net production in flow-through microcosms at Silver Springs, Florida, on 3 days in April 1980. Asterisks (*) indicate that the value is more than two standard errors from control mean.

Table 5. Summary of Silver Springs Microcosm Experiment Started on Apr. 7, 1980

| | | | | Net production, g O_2/m^2 Consumer effect$\ddagger$ (g O_2 m^{-2} hr^{-1}) (in parentheses) | | |
| | Initial fish | Final fish | Average | | | |
Microcosm	weight, g	weight, g	weight, g	Apr. 14$\ddagger$	Apr. 16$\ddagger$	Apr. 21$\ddagger$
1	8.6	7.7	8.15	2.38(−0.78)	11.83(−0.29)	21.77(+0.29)
2	24.0	9.4	16.7	2.40(−0.77)	11.28(−0.40)	17.78(−0.38)
3 (Control)	0	0	0	3.93	13.64	19.03
4	2.8	0.7	1.75	3.75(+0.17)	15.55(+0.40)	25.41(+0.90)
5 (Control)	0	0	0	3.59	13.50	19.58
6	6.8	6.2	6.5	3.20(−0.21)	10.81(−0.49)	22.03(+0.33)
7 (Control)	0	0	0	2.99	13.06	21.58
8	Microcosm lost					

*Fish were released in microcosms on Apr. 7 and harvested and reweighed on Apr. 30, 1980.
$\ddagger$ Consumer effect was calculated as algebraic change between given microcosm net production and average control net production divided by time of measurement.
$\ddagger$ Apr. 14, 1.43 hr, 3.32 E/m^2, and 173.5 cal/m^2; Apr. 16, 5.33 hr, 17.98 E/m^2, and 939.8 cal/m^2; Apr. 21, 5.97 hr, 15.98 E/m^2, and 835.3 cal/m^2.

DISCUSSION

System Comparison

In his landmark paper on energy flow in aquatic ecosystems, Odum (1957) warned of one danger in working in an apparent steady-state system: "Most terrible and healthy for the poor ecologist is the realization that anyone can check his field work at any later time...." Direct comparisons of gross primary production can be made between this study and Odum's study 25 years before because of the overlap of methods and the correction factors applied. Figure 6 illustrates such a comparison with

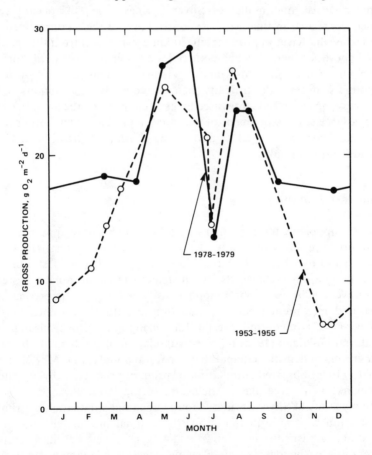

Figure 6. Comparison of gross primary production during the study reported in this paper with the data measured by Odum (1957) at Silver Springs, Florida, for the entire aquatic community to a point 1200 m downstream from the main spring boil.

uncanny similarity of values, including a low summer value resulting from afternoon thunderstorms typical of the area and the time of year. Winter values during the more recent study are substantially higher than those reported by Odum, possibly because of exceptionally clear days encountered in this study or because of an actual increase in productivity. Insolation data between the two studies could not be accurately compared because of the semiquantitative estimations of cloud cover used by Odum.

Consistent productivity changes resulting from changes in consumer populations were not found. Replacement of the absent catfish and mullet species with a similar density of blue shad may be significant, however.

There is much speculation over the reason for the disappearance of the popular catfish and mullet of Silver Springs. Their disappearance is closely correlated in time with the construction of the Rodman Dam on the Oklawaha River approximately 50 km downstream from the springs. One hypothesis suggests a competitive food resource in the newly formed Lake Oklawaha (Sam McKinney, personal communication). Another hypothesis concerns the difficulty encountered by the catadromous mullet in navigating the infrequently used locks circumventing the dam. In either case, a large-scale experiment of consumer replacement may occur if the Rodman Dam is removed as part of the cleanup of the ill-fated Cross Florida Barge Canal.

Consumer Control

Data supporting the consumer-control hypothesis have appeared in the literature long before this study. Besides evidence from terrestrial systems, summarized by Chew (1974) and Mattson and Addy (1975), stimulatory roles of consumers have also been demonstrated in aquatic systems. Hargrave (1970) varied amphipod density in in situ microcosms and measured maximum primary production at natural consumer density. Cooper (1973) found similar results for grazing by a herbivorous minnow. Flint and Goldman (1975) reported stimulation of productivity by crayfish grazers at densities comparable to natural populations. McKellar and Hobro (1976) observed stimulation of primary production by grazing of zooplankton in large plastic enclosures in the Baltic Sea.

In my study at Silver Springs, I found apparent stimulation of primary production with an aquatic herbivore. Optimal productivity was measured at snail densities between 22 and 44 g/m^2. These densities compare favorably to the density of 42 g/m^2 measured in the natural river ecosystem.

Mosquito fish significantly increased net production in the submerged microcosms at a live-weight density of 3.2 g/m^2. This density is similar to the overall carnivorous fish density of 4 g/m^2 estimated by fish counts for the entire river area. One simple mechanism for this stimulation may be that the fish were reducing numbers of herbivores, which, in turn, were reducing algal populations. Another possible mechanism is an increase in nutrient levels in the microcosms because of the waste products of the fish.

Explaining the stimulation of net production in the microcosms as a result of increased nutrient recycling by fish at low densities cannot account for the decrease in primary production measured at high fish densities. This observation may be best explained in terms of high fish predation, which lowers herbivores (such as chironomids) to suboptimal levels and, thus, results in lowered productivity. Thus mosquito fish control system production indirectly through predation on herbivores, and there appears to be an optimal level of mosquito fish that gives maximum primary production. In this case I was the controller who set mosquito fish density to the proper level, but in the river system higher level carnivores normally fill this role.

Thus the consumer-control hypothesis may span all trophic levels; although energy flows up from the lower levels, control may normally flow down from the upper levels. Each controller may be regulated by the next controller up the hierarchy scale. Top consumers of one system may be controlled by more concentrated energies of the next larger system. Catastrophes, such as violent weather, floods, earthquakes, or epidemics, may regulate populations of animals with long generation times (Elton, 1927).

Because of reduced system productivity, energy may not reach consumers that do not optimize the density of their prey. Thus, at unnaturally high densities, snails and fishes in the Silver Springs microcosms lowered the energy available to their own trophic levels. Under this unnatural condition, primary control may have been effected by resource limitation. It appears, however, that maximum stable population levels of the Silver Springs biota are not attained by over exploitation and starvation, but rather by a harmonious system of feedback controls.

ACKNOWLEDGMENTS

I want to thank the numerous friends who were willing to assist with field work and to the management and personnel of the American Broadcasting Corporation. The Department of Environmental Engineering Sciences at the University of Florida provided boat, van, and oxygen-monitoring equipment. I am greatly indebted to Howard T. Odum who served as my major professor during the period of this research.

REFERENCES

Batzli, G. O., 1978, The role of herbivores in mineral cycling, in *Environmental Chemistry and Cycling Processes,* Symposium Proceedings, Augusta, Ga., April 28, 1976, NTIS, 760429, pp. 95-112.

Chew, R. M., 1974, Consumers as regulators of ecosystems: an alternative to energetics, *Ohio J. Sci.* 74:359-370.

Cooper, D. C., 1973, Enhancement of net primary productivity by herbivore grazing in aquatic laboratory microcosms, *Limnol. Oceanogr.* 18:31-37.

Efford, I. E., 1972, An interim review of the Marion Lake Project, in: Kajak, Z. (ed.) *Productivity Problems in Freshwaters,* Symposium Proceedings, Kazimierz Dolny, Poland, May 6, 1970, IBP-Unesco, pp. 89-109.

Elton, C., 1927, *Animal Ecology,* MacMillan Company, New York.

Flint, R. W., and C. R. Goldman, 1975, The effects of a benthic grazer on the primary productivity of the littoral zone of Lake Tahoe, *Limnol. Oceanogr.* 20:935-944.

Hargrave, B. T., 1970, The effect of a deposit-feeding amphipod on the metabolism of benthic microflora, *Limnol. Oceanogr.* 15:21-30.

Kitchell, J. R., R. V. O'Neill, D. Webb, G. W. Gallepp, S. M. Bartell, J. F. Koonce, and B. S. Ausmus, 1979, Consumer regulation of nutrient cycling, *Bioscience* 29:28-34.

Lotka, A. J., 1922, Contribution to the energetics of evolution, *Proc. Natl. Acad. Sci.* 8:147-151.

McCree, K. J., 1972, Test of current definitions of photosynthetically active radiation against leaf photosynthesis data, *Agric. Meteorol.* 10:443-453.

McKellar, H., and R. Hobro, 1976, Phytoplankton-zooplankton relationships in 100 liter plastic bags, *Contrib. Asko Lab., Univ. Stockholm, Sweden, No. 13,* 83 pp.

Mattson, W. J., and N. D. Addy, 1975, Phytophagous insects as regulators of forest primary production, *Science* 190:515-522.

Northcote, T. G., and D. W. Wilke, 1963, Underwater census of stream fish populations, *Trans. Am. Fish. Soc.* 92:146-151.

Odum, E. P., J. T. Finn, and E. H. Franz, 1979, Perturbation theory and the subsidy-stress gradient, *Bioscience* 29:349-352.

Odum, H. T., 1956, Primary production in flowing waters, *Limnol. Oceanogr.* 1: 102-117.

————, 1957, Trophic structure and productivity of Silver Springs, Florida, *Ecol. Monogr.* 27: 55-112.

————, 1968, Work circuits and system stress, in: Young, Y (ed.), *Primary Production and Mineral Cycling in Natural Ecosystems,* Univ. Maine Press, p. 81-138.

————, 1971, *Environment, Power and Society,* John Wiley, New York, 331 pp.

————, 1979, Energy quality control of ecosystem design, in: Dame, R. F. (ed.), *March-Estuarine Systems Simulation,* Symposium Proceedings, Belle W. Baruch Lab., S.C., Univ. S.C. Press, pp. 221-235.

————, and R. C. Pinkerton, 1955, Time's speed regulator; the optimum efficiency for maximum power output in physical and biological systems, *Am. Sci.,* 43: 331-343.

Owen, D. F., and R. G. Wiegert, 1976, Do consumers maximize plant fitness?, *Oikos* 27:488-492.

Rosenau, J. C., G. L. Faulkner, C. W. Hendry, and R. W. Hull, 1977, *Springs of Florida,* Bull. No. 31 (revised), Florida Geological Survey, 461 pp.

U.S. Geological Survey, 1978, *Water Resources Data for Florida,* Vol. I, Report #FL-77-1.

9. ALTERATIONS IN THE BIODYNAMICS OF THE RED CEDAR RIVER ASSOCIATED WITH HUMAN IMPACTS DURING THE PAST 20 YEARS

Thomas M. Burton

Department of Zoology,
Department of Fisheries and Wildlife, and
    Institute of Water Research
Michigan State University
East Lansing, Michigan

Darrell L. King

Department of Fisheries and Wildlife and
    Institute of Water Research
Michigan State University
East Lansing, Michigan

ABSTRACT

Studies conducted on the Red Cedar River, a warmwater, third-order stream in south-central Michigan, in 1958–1962 and in 1978–1979 allow evaluation of alterations in sediment accrual rate, nutrient load, and primary production over a period of about 20 yr. Total phosphorus input to the lower sections of the river basin has remained unchanged; thus urbanization inputs have offset reductions in total phosphorus inputs resulting from upgrading domestic wastewater treatment from primary to tertiary. Total annual phosphorus discharge from the upper portion of this urbanizing basin has increased 2.4-fold over the period. Discharge of phosphorus from the lower basin only increased 1.2-fold over 1959 values, however, despite the increased upstream loading and 1.6 times more annual hydrologic discharge. Inorganic sediment accrual to the river bottom has been significantly reduced. Macrophyte productivity has declined markedly throughout the river. Periphyton production of organic matter has declined to about half of 1961 values for the entire river, with this reduction occurring in the three downstream zones of the river. Nutrient loading is essentially the same for an upstream and a downstream station; this indicates that nutrient input is derived more from diffuse sources at present. Changes observed during the past 20 yr are especially related to human activities within the basin, changes in sewage treatment facilities, and urbanization.

INTRODUCTION

In recent years efforts to improve water quality in streams have intensified. Many of these efforts originated with the passage of the Federal Water Pollution Control Act of 1972 (P.L. 92-500), which mandated best practicable technology for point-source discharges by 1977, with the ultimate goal of zero discharge by 1985. This timetable was relaxed somewhat by 1977 amendments (P.L. 95-217). Even so, most point sources of pollution now receive best practicable wastewater treatment. The effects of this upgrading of treatment of point-source pollution in receiving waters has not been studied in detail. These effects could vary from no noticeable effect on streams with limited point-source inputs to substantial effects on streams with major inputs. It is conceivable that even in streams with major point-source inputs, increases in non-point-source pollution from increased urbanization or more intensive agricultural practices in the watershed may negate gains derived from the adoption of best practicable pollution control technology.

Since adoption of best practicable wastewater treatment has been expensive, it is worth determining whether it has achieved the goal of improved water quality. If streams have indeed recovered, the recovery process needs to be documented and studied in detail. Such detailed studies would form the basis for models to predict the recovery of streams after further alleviation of pollution inputs and would provide a better understanding of stream structure and function. To meet these objectives, a study was initiated in 1978 on the Red Cedar River, a third-order, warmwater stream in the south-central portion of the lower peninsula of Michigan. This paper documents changes in primary production and water quality in the Red Cedar River which may have resulted from control of point-source pollution.

DESCRIPTION OF THE STUDY AREA

The Red Cedar River is a warmwater, third-order stream in the south-central lower peninsula of Michigan. It arises from the outflow of Cedar Lake in Livingston County and flows northwesterly for approximately 80 km through Livingston and Ingham Counties, Michigan, where it joins the Grand River (a tributary of Lake Michigan) in the city of Lansing. Although the total drainage area of the river is 122,000 ha, the actual study area only included 91,900 ha. The 71-km study section of the river extended upstream from the Farm Lane bridge on the Michigan State University campus in East Lansing. The 91,900-ha watershed included in the study area was intensively studied in the 1950s and 1960s (Ball and Bahr, 1975; Ball et al., 1968; 1969a; 1969b; Brehmer, 1958; Brehmer et al., 1968; 1969; Garton, 1968; Grzenda and Ball, 1968; Grzenda et al., 1968; Grzenda, 1960; Kevern, 1961; King, 1964; King and Ball, 1964; 1966; 1967; Linton and Ball, 1965; Vannote, 1961, 1963).

The Red Cedar River varies from 7.6 to 24.4 m in width over the study area. Average stream gradient is 0.45 m/km, and discharge varies seasonally between the record low of 0.08 m^3/s, which occurred in July 1931, and the record high of 168 m^3/s which occurred in April 1975. Maximum discharge normally occurs during spring runoff (March to May), with July and August being the usual time of minimum discharge. Maximum discharge during the course of this study (Oct. 1, 1978, to Sept. 30, 1979) was 55 m^3/s on Mar. 7, 1979. The minimum discharge, 0.51 m^3/s, occurred during September 1979. Mean discharge was 4.2 m^3/s. Discharge during the course of productivity studies (June to September 1979) varied from 0.51 to 7.08 m^3/s.

The Red Cedar River drains an area of gently rolling to level glaciated terrain with many depressions and broad swamps, with the highest and lowest elevations being 331 and 251 m, respectively (Ball and Bahr, 1975). Land use in the entire watershed is 57% cropland, 18% forest, 14% urbanized land, and 11% other uses (Ball and Bahr, 1975). The human population is about 150,000 at present, in contrast with 13,000 in 1900 and 500 during early historical times of occupation by Potawatomi Indians (Ball and Bahr, 1975). The actual study area contains a population of less than 90,000 people, however, with all but about 10,000 people concentrated in the East Lansing–Meridian Township (Okemos) area along the lower 9 km of the study area.

The climate of the study area varies between continental and semi-marine. Freezing temperatures usually occur from October into May. The mean annual temperature is 8.3°C. Precipitation averages 760 mm/yr, and there are twice as many cloudy days as clear days.

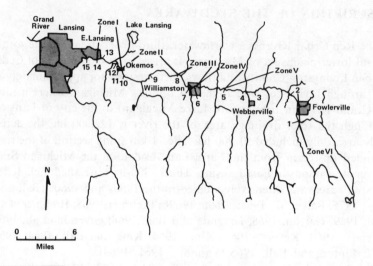

Figure 1. Map of the Red Cedar River showing experimental zones and water quality sampling sites.

The study area of the Red Cedar River was divided into six zones for studies of aufwuchs production (Figure 1). Four of these zones were identical to those used in earlier studies; zone IV used in the earlier studies was not sampled in 1979; and zone VI, a zone upstream of any municipal influence, was added in 1979. Water quality samples were taken at eight main river stations in 1958–1959 and at twelve stations in 1978–1979 (Figure 1). Stations 1, 3, 4, 6, and 14 were in both studies; stations 7, 10, and 13 were sampled only in 1958–1959; and stations 2, 5, 8, 9, 11, 12, and 15 were sampled only in 1978–1979 (Figure 1). Thus a total of 15 stations were sampled during one or both of the study periods.

METHODS

Since the purpose of this study was to compare present data with historical records, the methods used in earlier studies were used whenever possible. Thus aufwuchs production was measured by determining accrual of organic and inorganic material on Plexiglas substrates according to the methods used by King in 1959–1961 (King, 1964; King and Ball,

1966). The substrates used were cut from flat sheets of 0.64-cm Plexiglas, each with a total exposed area, including the sides, of 1.4 dm^2 when attached by metal clamps to the supporting racks in the 1959–1961 studies and a surface area of 1.5 dm^2 in the 1979 studies. The supporting racks were made of a wood crossbar bolted securely to a steel fence post, which was driven into the bottom of the stream. Pairs of vertically and horizontally placed substrates were oriented with the 0.64-cm edge facing into the current. These paired racks of substrates were placed at two preselected random points within each of the study areas of the river (Figure 1). The positions of the paired racks were changed to new preselected random points within the zone every 18 days in 1959–1961 or every 21 days in 1979. One pair of racks of horizontal and vertical substrates was placed at the midpoint of exposure for the other pair of racks to give overlapping exposure periods throughout the growing season from June to September.

The exposure depth of the substrates was maintained at six-tenths of the distance from the water surface to the bottom of the river. This it was possible to compare accrual rates of aufwuchs in different areas of the river.

Six different exposure periods were allowed for the accrual of aufwuchs in 1959–1961—3, 6, 9, 12, 15, and 18 days. In 1979 seven exposure periods were used—3 or 4, 7, 10 or 11, 14, 17 or 18, and 21 days. The change was made so that sampling could be done on the same days each week. Two substrates were placed on the rack for each of the exposure periods.

When the substrates were removed from the river, they were placed in individual plastic freezer bags and frozen. The material that accumulated on the artificial substrates was divided into inorganic sediments, inorganic material produced on site (diatom frustules, etc.), organic sediments, and organic matter produced on site by autotrophic and heterotrophic biota, according to the techniques of King and Ball (1966).

Macrophytes were sampled at seventeen randomly selected transects within each zone in 1979. All aboveground and belowground biomass was removed at ten preselected 0.09-m^2 point samples along a 31-m transect in earlier studies, and all biomass was removed from the entire 10-m-long by 1-m-wide transect in 1979. Macrophyte samples were dried to a constant weight at 60°C for determination of biomass.

Water samples were taken at stations 1, 3, 4, 6, 7, 10, 13, and 14 in 1958–1959 and at stations 1–6, 8, 9, 11, 12, 14, and 15 in 1978–1979 (Figure 1). In the 1958–1959 studies the samples were analyzed by methods outlined by Ball et al. (1968) or by the use of standard, automated EPA-approved techniques in 1978–1979 (Environmental Protection Agency, 1979).

RESULTS AND DISCUSSION

Composition of Algal Communities

In this study, as well as in past studies (Ball et al., 1969a; King and Ball, 1966), the algae that colonized the Plexiglas substrates were almost exclusively diatoms. In a previous study (Peters, 1959), the same diatoms dominant on the artificial substrates were also dominant on natural substrates in the Red Cedar River. The most common genera in the present study, listed in order of abundance, were *Navicula*, *Melosira*, *Cocconeis*, *Cymbella*, *Fragilaria*, and *Cyclotella*. This list differs from past observations (Ball et al., 1969a; King and Ball, 1966) only in the presence of *Melosira* as a dominant genus and in the relative sparseness of *Gomphonema* and *Synedra*. *Diatoma hiemale* was listed as dominant only in winter months previously and was not observed in this study, but *Gomphonema* and *Synedra* tended to be dominant in fall months (Ball et al., 1969a). Thus the composition of the algal community has been dominated by the same algal genera for more than 20 yr, with *Navicula* and *Cocconeis* being the dominant forms in summer months. *Melosira* was the only new addition to the list of dominant algae observed.

Organic Production of Aufwuchs

Estimates of accrual rates of organic and inorganic matter, sedimentation rates, and production on artificial substrates for the entire Red Cedar River, based on 1961 data from King and Ball (1966) and from the 1979 study, are summarized in Table 1. Net production of organic matter was reduced to less than half the 1961 levels, but organic matter sedimentation rates more than doubled on the vertical and increased by more than 40% on the horizontal substrates (Table 1). The lowered on-site production resulted in a decrease of total organic matter accrual of 33% on the vertical and 31% on the horizontal substrates even though organic matter sedimentation rates had increased substantially (Table 1). Inorganic matter sedimentation rates also dropped 44% for the vertical and 65% for the horizontal substrates. Thus, although organic sedimentation rates were increased markedly for the entire river, total organic matter accrual rate, total inorganic matter accrual rates, inorganic sedimentation rates, and organic and inorganic matter production rates were all substantially decreased for the entire river, with 1979 values being 30 to 65% lower than 1961 values.

Table 1. Organic and Inorganic Matter Accrual Rates, Sedimentation Rates, and Production on Artificial Substrates for the Red Cedar River

	1961* mg m^{-2} day^{-1}	1979† mg m^{-2} day^{-1}	Change, %
Organic matter accrual rate, vertical	283.8	190.0	–33.1
Organic matter accrual rate, horizontal	326.5	225.1	–31.1
Accrual of all material, vertical	817.3	520.3	–36.3
Accrual of all material, horizontal	1252.0	661.6	–47.2
Total inorganic accrual rate, vertical	533.5	330.3	–38.1
Total inorganic accrual rate, horizontal	925.5	436.5	–52.8
Inorganic sedimentation rates, vertical (corrected for inorganic production by diatoms)	249.7	140.3	–43.8
Inorganic sedimentation rates, horizontal (corrected for inorganic production by diatoms)	599.0	211.4	–64.7
Organic sedimentation rates, vertical	30.5	69.3	127.2
Organic sedimentation rates, horizontal	73.2	104.4	42.6
Net organic matter production	253.3	120.7	–52.3

*Data for 1961 are from King and Ball (1966).

†Data for 1979 are based on data from zones I, II, III, and V zince Zone IV was not sampled in 1979.

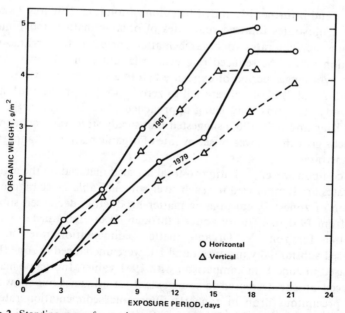

Figure 2. Standing crop of organic matter accrued on the artificial substrates during the summers of 1961 and 1979. (Data for 1961 are from King and Ball, 1966).

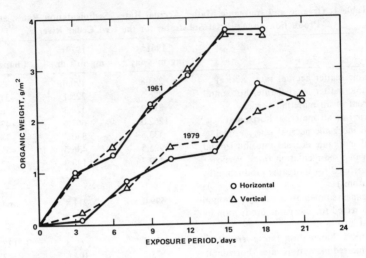

Figure 3. Standing crop of organic matter accrued on the artificial substrates during the summers of 1961 and 1979 corrected for organic sedimentation rates. (Data for 1961 are from King and Ball, 1966).

This trend is further illustrated by colonization rates (Figures 2 and 3). Figure 2 illustrates total accrual rates of organic matter, and Figure 3 illustrates accrual rates of organic matter corrected for organic sedimentation rates. The reduced rate of accrual noted in 1979 was accompanied by a longer period of exposure before a steady state was reached (Figures 2 and 3). Accrual rate fell to zero in 15 days in the 1961 study, but steady state was only being approached after 21 days in the 1979 study. King and Ball (1966) suggested that steady state was reached when aufwuchs growth rate was offset by rate of organic matter sloughing from the substrates.

All components contributing to total accrued material on the artificial substrates are summarized for each river zone in Table 2 (see Figure 1 for location of zones). Total organic matter production decreased substantially from 1961 to 1979 for zones I through III but remained essentially the same for zone V. Organic matter sedimentation rates in 1979 increased substantially in zones I and III, were unchanged in zone II, and decreased in zone V in comparison with 1961 values. Inorganic material produced on site decreased to some extent for all zones but showed the most precipitous drop in zone III. Inorganic sedimentation rates decreased substantially on all horizontal substrates and on vertical sub-

Table 2. Summary of Components Contributing to the Total Accrued Material on the Artificial Substrates*

River zone and placement	Inorganic sedimentation rate, mg m⁻² day⁻¹		Inorganic material produced on site, mg m⁻² day⁻¹		Organic sedimentation rate, mg m⁻² day⁻¹		Total organic matter produced, mg m⁻² day⁻¹	
	1961	1979	1961	1979	1961	1979	1961	1979
Zone I, horizontal	318.8	293.3	298.8	220.0	154.8	169.7	144.0	50.3
Zone I, vertical	60.3	166.6	173.3	146.7	29.3	96.4	144.0	50.3
Zone II, horizontal	461.1	180.0	299.3	213.3	49.8	55.2	249.5	158.1
Zone II, vertical	233.2	93.3	274.7	186.6	25.2	28.6	249.5	158.1
Zone III, horizontal	762.8	153.3	267.6	166.7	-128.3	89.1	395.9	77.6
Zone III, vertical	519.1	73.4	308.6	133.3	-87.3	55.7	395.9	77.6
Zone IV, horizontal*	942.6		385.8		103.3		282.5	
Zone IV, vertical*	352.1		321.1		38.6		282.5	
Zone V, horizontal	524.0	253.3	379.8	273.3	42.6	-42.4	337.2	315.7
Zone V, vertical	52.3	173.3	341.4	286.7	4.2	-29.0	337.2	315.7
Zone VI, horizontal†		146.7		160.0		-22.0		182.0
Zone VI, vertical†		13.3		180.0		-2.0		182.0

*Data for 1961 are from King and Ball, 1966.
†Zone IV, an impoundment, was sampled only during 1961; zone VI, the headwaters zone of the Red Cedar River, was sampled only during 1979.

strates in zones II and III but increased on vertical substrates in zones I and V.

This pattern of organic matter production and sedimentation reflects changes in sources of pollution for the various river zones. Although zone VI was not sampled in 1961, it is the only zone with no sewage treatment plant input and no urban runoff. It reflects agricultural runoff and runoff from fairly extensive wetlands. Thus, its inorganic sedimentation rates are the lowest of any zone on the river (Table 2). Organic sedimentation rates are actually negative and reflect the sloughing of organic matter produced on the substrate with little or no organic sedimentation from offsite. There are few wooded areas along the stream in zone VI, and much of the stream has been channelized, with the result that there is little organic matter input into the stream from surrounding areas.

Zone V extends below Fowlerville, a community of about 2000 people, and receives domestic effluent from a continuous discharge lagoon system. This lagoon system, which is for treatment of previously discharged raw sewage, and other related improvements in discharge of industrial pollutants from a plating plant have alleviated some of the septic zone that occurred below Fowlerville in earlier studies. The lagoon probably has not removed appreciable amounts of phosphorus, however. Lagoon systems exhibit poor phosphrous removal efficiencies (King and Burton, 1979). Also, only minor reductions in exports of organic matter into the river are likely to have occurred since lagoons often export considerable organic matter in the form of algal biomass (King, 1976). The lagoon system should trap sediments and cause substantial decreases in inorganic sediment input into the river. Data for zone V supports these suspected trends; inorganic sedimentation rates are down substantially on the horizontal substrate, where we would expect to measure them most accurately, but total organic matter production remains relatively high in comparison with other zones and is essentially unchanged from 1961 (Table 2).

Zone IV, the impoundment at Williamston, MI, had high inorganic and organic sedimentation rates in 1961 in comparison with other zones, as would be expected of an impoundment (Table 2). No comparative data are available from 1979.

Zone III exhibited the most marked decrease in both organic matter production and inorganic sedimentation rates of any zone (Table 2). This perhaps reflects the upgrading of the Williamston sewage treatment plant from a primary to a tertiary facility. Zones I and II exhibited the same trends (Table 2), perhaps because of changes in organic inputs from upstream, as well as improvements in septic tank overflows and stormwater drainage in the Meridian Township–East Lansing region.

Macrophyte Production

One of the most striking changes in the Red Cedar River was in the macrophyte community during the last 20 yr. Before 1958 macrophyte production was very low, with relatively few dense beds of macrophytes (Ball and Bahr, 1975). Construction of an interstate highway parallel to the river resulted in massive inputs of silt into the river in late 1961. The silt input, coupled with very low discharge caused by several years of low rainfall over the basin, resulted in an explosive increase of macrophytes in the river—from essentially zero to values in excess of 170 g dry weight/m^2 in 1962 in Zone II. Substantial increases occurred in all zones except the impounded zone IV (Ball and Bahr, 1975). This increased production was dominated by *Sagittaria* and *Vallisneria*. Production rates peaked in 1964 at about 1.2 g dry weight m^{-2} day^{-1} for zone I, 0.8 g dry weight m^{-2} day^{-1} for zone II, 2.3 g dry weight m^{-2} day^{-1} for zone IV, and 1.6 g dry weight m^{-2} day^{-1} for zone V (Ball and Bahr, 1975). Assuming a 125-day growing season (Vannote, 1963), these rates translate to biomass production rates of 150, 100, 285, and 200 g dry weight m^{-2} yr^{-1} for zones I, II, III, and V, respectively.

Since the studies were conducted in the early 1960s, extremely high flows have occurred in the Red Cedar River, with the highest discharge on record (168 m^3/s) recorded during April 1975. According to U. S. Geological Survey records, 1968, 1969, 1975, and 1976 were years with much greater than average discharge. These high flows flushed the accumulated silt from the river and scoured out the macrophyte beds that were established in the early 1960s as a consequence of high siltation and low flows. In 1979 biomass production rates were very low, and there were relatively few dense beds of macrophytes. This very patchy distribution included beds of *Potamogeton*, *Elodea*, *Sparghanium*, *Fontinalis*, *Vallisneria*, *Sagittaria*, and other species. The sparse distribution and patchiness made sampling difficult, but estimates derived from the seventeen transects sampled per zone indicated an average biomass production of 0.003 g dry weight/m^2 for zone I, 2.5 g dry weight/m^2 for zone II, 2.2 g dry weight/m^2 for zone III, 1.9 g dry weight/m^2 for zone V, and 2.9 g dry weight/m^2 for zone VI. Production in limited areas within zones is considerably higher, as evidenced by the 1976 estimate of 67 g dry weight/m^2 for a portion of zone II (Mattingly, 1978). Nevertheless, biomass production of the entire river appears to have decreased to values typical of pre-1958 conditions. This decrease could be related to changes in point-source inputs, but it seems much more likely to be related to flushing of silt out of the river as a consequence of the high flows described.

Water-Quality Changes

The most comprehensive study of water quality in the Red Cedar before this study was that conducted in 1958–1959 water year by Vannote (1961). Thus, this study conducted during the 1978–1979 water year is compared to Vannote's work.

Vannote calculated that 2286 kg of phosphorus per year were transported past station 4 at Webberville Road (see Figure 1 for station locations) in 1958–1959 and 9763 kg of phosphorus per year were transported past station 14 at Farm Lane bridge. During 1979, 5466 kg P were estimated to have been transported past Dietz Road (station 5) on the basis of water chemistry samples taken at Dietz Road and discharge values obtained from the U. S. Geological Survey for a gauging station about 0.8 km upstream. Since there are no tributaries entering the Red Cedar River between stations 4 and 5 in this agricultural area of the basin, loads at the two stations would be expected to differ very little. Thus, phosphorus export from the upstream area of the Red Cedar River appears to have increased 2.4-fold over 1961 values. At the downstream site (station 14), phosphorus exports increased from 9763 to 11,612 kg/yr, a 1.2-fold increase over 1961.

The reasons for this disparity are clear if we examine trends in phosphorus concentration along a downstream gradient (Table 3). In

Table 3. Mean Total Phosphorus Concentrations for Main River Stations*

Station description and number	Concentration in 1958–1959†, mg P/l	Concentration in 1978–1979†, mg P/l
Van Buren Road (1)	0.036 ± 0.018	0.075 ± 0.053
Gregory Road (2)		0.148 ± 0.123
Gramer Road (3)	0.079 ± 0.035	0.102 ± 0.066
Webberville Road (4)	0.069 ± 0.022	0.108 ± 0.069
Dietz Road (5)		0.111 ± 0.132
Williamston, below dam (6)	0.065 ± 0.026	0.112 ± 0.106
Williamston, above sewage outfall (7)	0.055 ± 0.024	
Zimmer Road (8)		0.111 ± 0.076
M-43 (9)		0.101 ± 0.069
Dobie Road (10)	0.068 ± 0.018	
Okemos Road (11)		0.090 ± 0.054
Indian Hills (12)		0.090 ± 0.052
Hagadorn Road (13)	0.087 ± 0.037	
Farm Lane Bridge (14)	0.122 ± 0.053	0.110 ± 0.109
Harrison Avenue (15)		0.111 ± 0.102

*Data for 1958–1959 are from Vannote, 1961.
†Values are ±1 standard deviation.

1958-1959, concentrations were much lower in upstream areas than in downstream areas. In 1978-1979, however, concentrations increased from station 1 to station 2 (above and below Fowlerville) and then remained constant downstream of this area. Vannote (1961) demonstrated that phosphorus export by most of the tributaries entering the Red Cedar River was directly correlated to discharge and could be predicted by

$$Y = -4.93 + 0.93X \tag{1}$$

where Y is phosphorus transport (kg/yr) and X is stream discharge (m^3/yr). Since stream discharge of the Red Cedar was 1.6 times higher in 1978-1979 than in 1958-1959, discharge and phosphorus export by tributaries would be expected to rise accordingly.

Phosphorus input from domestic sewage was expected to decline because of upgraded sewage treatment facilities at all municipalities and because of a phosphate detergent ban. This decline should have resulted in less input from each sewage treatment facility and a reduced downstream gradient of increased concentrations. Fowlerville upgraded its sewage treatment facility to continuous discharge lagoons; it had previously dumped its raw sewage into a marsh that drained into the river. This marsh may have been effective at phosphorus removal in 1958-1959, so that phosphorus loading from Fowlerville may have actually increased with the present direct addition of the effluent to the river. Indeed, there were substantial increases from Fowlerville (stations 1 to 2) in both 1958-1959 and 1978-1979, but the 1978-1979 increases were to much higher levels.

Increased populations in the municipalities in the Red Cedar River area would result in greater phosphorus loading and might offset some of the gains derived from better sewage treatment. For example, the city of Williamston has upgraded its sewage treatment from primary to tertiary, but increased population has led to plans to enlarge or to build a new sewage treatment facility.

In addition, the marked increase in urbanization throughout the basin has likely led to increased phosphorus loading (Burton et al., 1977). Thus, gains produced by better sewage treatment may have been offset by continued population growth, urbanization, and increased erosion in the upper part of the Red Cedar watershed. Thus, phosphorus loading may have increased substantially in these upstream areas, whereas the lower portion of the river has exhibited only limited increases in phosphorus concentration and loading.

The limited increase in phosphorus in the lower portions of the river appears to reflect the upgraded sewage treatment at Williamston and the alleviation of septic tank overflows and storm drain problems in the East Lansing-Meridian Township (Okemos) area.

Phosphorus export calculated on a unit area basis is the same for both the upstream and downstream stations (Table 4). This similarity is also exhibited by soluble reactive phosphorus, all forms of nitrogen, and chloride and alkalinity (Table 4). Therefore, the Red Cedar River maintains its water quality in terms of chemical loading from the upstream to downstream location even with the discharge of the Williamston sewage treatment plant (serving a population of 2700 people with tertiary treatment) and the stormwater runoff from the Williamston, Meridian Township, East Lansing, and Michigan State University areas downstream. Exports from the continuous discharge lagoon system of Fowlerville (population of 2000) and from the seepage lagoon system of Webberville (population 1000), plus tributary loads upstream of Williamston, contribute as much load per unit area as do the much more heavily urbanized downstream areas.

The 1958–1959 loads reported by Vannote (1961) were only slightly lower at the downstream site (0.106 kg P ha^{-1} yr^{-1}), and this difference could well have resulted from the much lower discharge in 1958–1959 and, therefore, lower tributary and storm-water inputs. Upstream loading did exhibit a 2.4-fold increase over the past 20 yr as discussed previously.

CONCLUSIONS

Several changes have occurred in the Red Cedar River during the past 20 years which can be related to human activities within the watershed. Net production of organic matter has decreased substantially to levels less

Table 4. Load of Nutrients at Two Sites in the Red Cedar River in 1979

	Farm Lane Bridge, East Lansing,* kg ha^{-1} yr^{-1}	Upstream of Williamston,† kg ha^{-1} yr^{-1}
Total phosphorus	0.126 ± 0.027	0.129 ± 0.024
Soluble reactive phosphorus	0.036 ± 0.005	0.035 ± 0.002
Organic nitrogen	1.250 ± 0.131	1.437 ± 0.164
Inorganic nitrogen	7.493 ± 1.277	6.356 ± 1.247
Nitrate nitrogen	6.954 ± 1.227	5.847 ± 1.194
Nitrite nitrogen	0.140 ± 0.017	0.147 ± 0.020
Ammonium nitrogen	0.399 ± 0.073	0.362 ± 0.062
Chloride	69.083 ± 4.754	69.267 ± 4.485
Alkalinity	296.072 ± 14.571	312.232 ± 15.688

*Drainage area is 92,000 ha.
†Drainage area is 42,240 ha.

than half of 1961 values for aufwuchs production and to only a small fraction of early 1960s values for macrophyte production. The decrease in aufwuchs production appears to be related to decreased dissolved organic inputs associated with upgraded sewage treatment facilities. It did not occur in zone V below Fowlerville, which is served by a lagoon system. Nutrient concentrations were in excess of plant needs for both study periods at all sites. No substantial change in the composition of the periphytic algae occurred for the river; however, colonization rates declined, as did inorganic sedimentation rates.

The dramatic decline in macrophyte production appears to be related to higher flows and flushing of the sediments that entered the river in the early 1960s as a result of highway construction.

Water quality appears to have been degraded substantially in the upper portion of the river but to be essentially unchanged in the lower areas even in the face of increased upstream loading. The downstream gradient of increasing concentrations of phosphorus, characteristic of the river in 1958–1959, no longer existed in 1978–1979.

Since the periphytic algal community is largely unchanged, the observed reduction in aufwuchs accrual rate was probably caused by a decreased abundance of periphytic bacterial and fungal forms associated with upgraded sewage treatment and the resultant decrease in discharge of dissolved organics. Any gain in plant nutrient loading to the river associated with upgraded wastewater treatment has been offset by increased activities within the basin and perhaps the present higher discharge rates. Despite the increased phosphorus loading to the upper portion of the river, the similarity in chemical load per hectare throughout the basin suggests the importance of nonpoint runoff to the Red Cedar River.

Thus it appears that upgraded wastewater treatment over the past 20 yr has resulted in decreased addition of biodegradable organics to the river, but because of increased urbanization in the basin, no net reduction in the nutrient load carried by the Red Cedar River has been realized.

ACKNOWLEDGMENTS

This work was funded by Grant A-099-MICH from the U. S. Department of Interior, Office of Water Research and Technology. Special thanks for technical assistance are given to C. Annett, W. Larsen, and C. Faulkner and the many other students who contributed to this project. Special thanks are also due to Lois Wolfson, who identified the periphyton.

REFERENCES

Ball, R. C. and T. G. Bahr. 1975, Intensive Survey: Red Cedar River,Michigan, in B. A. Whitton (Ed.), *River Ecology*, pp. 431–460, Blackwell Scientific Publications, Oxford, England.

———, K. J. Linton, and N. R. Kevern, 1968, *The Red Cedar River. Report I. Chemistry and Hydrology*, Publications of the Museum, Michigan State University, Biological Series, 4(2): 29–64, East Lansing.

———, N. R. Kevern, and K. J. Linton, 1969a, *The Red Cedar River. Report II. Bioecology*, Publications of the Museum, Michigan State University, Biological Series, 4(4): 105–160, East Lansing.

———, M. E. Stephenson, and T. W. Hardgrove, 1969b, *Continuous Automated Monitoring of Chemical and Physical Characteristics of the Red Cedar River*, Institute of Water Research, Technical Report 8, Michigan State University, East Lansing.

Brehmer, M. L., 1958, A Study of Nutrient Accrual, Uptake, and Regeneration as Related to Primary Production in a Warm-Water Stream, Ph.D. Thesis, Michigan State University, East Lansing.

———, R. C. Ball, and N. R. Kevern, 1968, *The Biology and Chemistry of a Warmwater Stream*. Institute of Water Research, Technical Report 3, Michigan State University, East Lansing.

———, R. C. Ball, and N. R. Kevern, 1969, *Nutrients and Primary Production in a Warm-water Stream*, Institute of Water Research, Tech. Report 4, Michigan State University, East Lansing.

Burton, T. M., R. R. Turner, and R. C. Harriss, 1977, Nutrient Export from Three North Florida Watersheds in Contrasting Land Use, in D. L. Correll (Ed.), *Watershed Research in Eastern North America*, Vol. 1, pp. 323–342, Smithsonian Institution Press, Chesapeake Bay Center for Environmental Studies, Edgewater, MD.

Environmental Protection Agency, 1979, Methods for Chemical Analysis of Water and Wastes, Report EPA-600/4-79-020, Environmental Monitoring and Support Laboratory, Cincinnati, OH.

Garton, R. R., 1968, Effect of Metal Plating Wastes on the Ecology of a Warmwater Stream, Ph.D. Thesis, Michigan State University, East Lansing.

Grzenda, A. R., 1960, Primary Production, Energetics, and Nutrient Utilization in a Warm-Water Stream, Ph.D. Thesis, Michigan State University, East Lansing.

———, R. C. Ball, 1968, Periphyton Production in a Warm-Water Stream, Agricultural Experiment Station, Quarterly Bulletin, 50(3): 296–303, Michigan State University, East Lansing.

———, R. C. Ball, and N. R. Kevern, 1968, *Primary Production, Energetics, and Nutrient Utilization in a Warm-Water Stream*, Institute of Water Research, Technical Report 2, Michigan State University, East Lansing.

———, and M. L. Brehmer, 1960, A Quantitative Method for the Collection and Measurement of Stream Periphyton, *Limnol. Oceanogr.*, 5: 190–194.

Kevern, N. R., 1961, The Nutrient Composition, Dynamics, and Ecological Significance of Drift Material in the Red Cedar River, M.S. Thesis, Michigan State University, East Lansing.

King, D. L., 1964, An Ecological and Pollution-Related Study of a Warm-Water Stream, Ph.D. Thesis Michigan State University, East Lansing.

_____, 1976, Changes in Water Chemistry Induced by Algae, in E. F. Gloyna, J. F. Malina, Jr., and E. M. Davis, (eds.), *Ponds as a Wastewater Treatment Alternative*, Water Resources Symposium 9, Center for Research in Water Resources, University of Texas, Austin.

_____, and R. C. Ball, 1964, *The Influence of Highway Construction on a Stream*, Agricultural Experiment Station Research Report 19, Michigan State University, East Lansing.

_____, and R. C. Ball, 1966, A Qualitative and Quantitative Measure of *Aufwuchs* Production, *Trans. Am. Microsp. Soc.*, 85(2): 232–240.

_____, and R. C. Ball, 1967, Comparative Energetics of a Polluted Stream, *Limnol. Oceanogr.*, 12(1): 27–33.

_____, and T. M. Burton, 1979, A Combination of Aquatic and Terrestrial Ecosystems for Maximal Reuse of Wastewater, in *Water Reuse—From Research to Application, Proceedings*, Vol. 1, Water Reuse Symposium, Washington, DC, Mar. 25–30, 1979, American Water Works Association Research Foundation, Denver, CO.

Linton, K. J., and R. C. Ball, 1965, A Study of the Fish Populations in a Warm-Water Stream, Michigan State University Agricultural Experiment Station Quarterly Bulletin, 48(2): 255–285.

Mattingly, R. L., 1978, An Ecological Comparison of the Red Cedar River (Ingham County, Michigan) in the Early 1960's with 1976, M.S. Thesis, Michigan State University, East Lansing.

Peters, J. C., 1959, An Evaluation of the Use of Artificial Substrates for Determining Primary Production in Flowing Water, M.S. Thesis, Michigan State University, East Lansing.

Vannote, R. L., 1961, Chemical and Hydrological Investigations of the Red Cedar River Watershed, M.S. Thesis, Michigan State University, East Lansing.

_____, 1963, Community Productivity and Energy Flow in an Enriched Warm-Water Stream, Ph.D. Thesis, Michigan State University, East Lansing.

10. SIGNIFICANCE OF FLOODPLAIN SEDIMENTS IN NUTRIENT EXCHANGE BETWEEN A STREAM AND ITS FLOODPLAIN

Mark M. Brinson, H. David Bradshaw, and Russell N. Holmes
Department of Biology
East Carolina University
Greenville, North Carolina

ABSTRACT

This study describes some of the mechanisms by which nitrogen and phosphorus levels are altered in the surface water of a floodplain swamp and examines how this may ultimately affect concentration of these nutrients in the stream channel during hydrologic events that drain floodplain water into the stream. A year-long comparison of stream channel and floodplain water of the lower Tar River, North Carolina, established that hydrologic phases resulting in water flow from floodplain to stream channel would dilute nitrate (and usually ammonium) concentrations in stream water. In contrast, floodplain drainage water would normally augment filterable reactive phosporus in the stream. Seasonal experiments were conducted on sediment-water exchanges of nitrate, ammonium, and phosphate in the floodplain to examine some of the mechanisms responsible for alterations in nutrient concentrations of surface water. Nitrate loss exceeded ammonium loss from surface water to which these ions were added above background levels. Loss rates of both ions were greater in July and November than in February, the coldest month. Nitrogen-15 tracer experiments confirmed a one-way pathway of nitrate from the surface water; this was made possible by denitrification in the sediments. Ammonium diffusion was bidirectional between surface water and sediments. Background concentrations of inorganic nitrogen in interstitial water of surface sediments underwent greatest changes during warm season dry down. Loss of $^{32}PO_4$ from surface water was mostly by abiotic mechanisms during all seasons and was entirely so during the coldest part of the year.

INTRODUCTION

Ecologists tend to establish boundaries of lotic ecosystems at the banks of stream channels. Yet, when low-gradient streams exceed bank-full stage, as they do approximately every year (Leopold et al., 1964), the lotic ecosystem boundary rapidly penetrates floodplains. In the southeastern United States, bottomland hardwood forests accommodate these flood events by providing the lotic ecosystem with a conduit for downstream water flow and a surface area for water storage. There are striking differences, however, between the contact surfaces of water in the stream channel and the floodplain component. Coastal plain stream channels commonly have sandy, unstable substrates where stable loci for sedentary aquatic organisms are limited to fallen branches, snags, and root surfaces from riparian vegetation (Benke et al., 1979). In contrast, the floodplain forest floor is characterized by relatively higher structural complexity, which is contributed by stable sediments, a litter layer of decomposing leaves and woody material, anastomosing tree roots, and complex floodplain topographic features (levees, oxbows, meander scroll ridges, swales, etc.).

Some of the ecological functions attributed to floodplain ecosystems include sinks for nutrients and sediments (Kitchens et al., 1975; Mitsch et al., 1979; Yarbro 1979), water storage for downstream flood ameliora-tion, aquifer recharge/discharge areas (Bedinger, 1979), and productive fish and wildlife habitat (Fredrickson, 1979; Wharton, 1978). Each of these functions is keyed to the occurrence of flood events and to the lateral extension of lotic ecosystem boundaries into floodplains. In this chapter we focus on nutrient coupling between the stream channel and the floodplain and examine the extent to which floodplains may serve as a source or a sink for nitrogen and phosphorus.

Nutrient exchanges between floodplains and stream or river channels may affect the quality of water exported downstream to lakes and estuaries. The coupling between a stream channel and its floodplain can be separated into three hydrologic phases (Figure 1): (a) Normal channel-contained flow when the stream receives seepage or surface flow from the floodplain which originates from precipitation or lateral runoff from

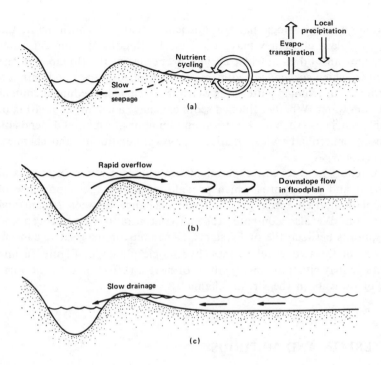

Figure 1. Three hydrologic phases of a floodplain. (a) Normal channel-contained flow, (b) Flood event; overbank flow, (c) Post-flood drainage and impoundment.

adjacent uplands; (b) over-bank flow from the stream channel; this contributes water to the floodplain which flows downstream across the floodplain surface; and (c) postflood drainage from the floodplain to the stream channel.

Because of local peculiarities in hydrologic patterns and nutrient cycling dynamics, it is often difficult to generalize on the degree of coupling between the stream channel and the floodplain. Hydrologic patterns and floodplain morphology influence the proportion of upstream discharge that actually enters the floodplain. For streams that have narrow floodplains, and thus a low capacity to accommodate throughflow, there is a relatively small potential for stream-floodplain nutrient exchange. Likewise, regulated streams that flood infrequently or for only brief periods will have a small potential for nutrient exchange. In contrast, many unaltered, low-gradient streams in the coastal plain of the southeastern United States and the lower Mississippi Valley have a high exchange potential because they have broad floodplains, protracted periods of inundation, and a large capacity for water storage.

For a given hydrologic pattern, nutrient exchange between the stream and floodplain ultimately may be limited by floodplain sediment-water exchanges and other nutrient cycling properties of the floodplain. The extent to which suspended sediments are deposited in the floodplain is a function of hydrologic patterns as well as the suspended sediment content of stream water. Whether flooded sediments have a net influx or efflux of soluble nutrients (such as nitrate, ammonium, or phosphate) depends on a host of microbial transformations, physical conditions, and chemical interactions.

We report here several experiments on sediment-water exchanges of nitrogen and phosphorus which provide insight on the importance of floodplain nutrient cycling in nutrient exchange between stream and floodplain. We also compare seasonal concentrations of nitrogen and phosphorus between the two systems. Depending on the relative concentrations of the two water masses, hydrologic phases in Figure 1a and Figure 1c may either augment, dilute, or have no effect on the concentration of nutients in the stream channel.

MATERIALS AND METHODS

Study Site

The study site is an alluvial swamp forest located in the Atlantic coastal plain in Pitt County, North Carolina ($35°35'$N, $77°10'$W), which is described in some detail by Brinson (1977) and Brinson et al. (1980). Experiments were conducted on the Tar River floodplain approximately 200 m from the river's edge. Density and basal area for trees in the study site were measured in January 1977. For stems >2.5 cm in diameter at breast height (dbh), density was 2730/ha, with a corresponding basal area of 69.0 m^2/ha. There were 2681 stems/ha <2.5 cm dbh and >1.0 m in height. The rather uniform canopy height of 25 m is attributable to clear-cutting about 30 years ago. *Nyssa aquatica* L. (water tupelo) dominates the canopy, with a few scattered *Taxodium distichum* (L.) Richard (bald cypress). The dominant understory species is *Fraxinus caroliniana* Mill., many of which are <2.5 cm dbh. The herbaceous layer is discontinuous, consisting mostly of *Saururus cernuus* L. On the basis of dry weight, surficial sediments contain 14 to 19% organic carbon, 1 to 2% total nitrogen, 20 to 175 mg/kg exchangeable NH_4-N, and 63 mg/kg extractable phosporus. Totals for other elements (mg/kg) are calcium, 1440; magnesium, 2660; sodium 3300; potassium 14,000; and iron, 18,700.

Sampling and Analysis of Tar River and Floodplain Water

Water samples were collected from both the main channel in the Tar River and from a location on the floodplain approximately 150 m from the stream. Sampling frequency was once every 2 weeks from September 1978 through August 1979. During a period of over-bank flooding and flow through the floodplain, Feb. 27 to Mar. 11, the sampling was every 2 days. Filterable reactive phosphorus was analyzed by the molybdate method (Environmental Protection Agency, 1976); ammonium was determined by the indophenol method (Scheiner, 1976); and nitrate was measured by ultraviolet absorption (Brown and Bellinger, 1978) after color removal with $Al(OH)_3$ and filtration (American Public Health Association, 1975). Water levels in the floodplain were continuously recorded, and those in the river were measured on days of collection.

Nitrogen Exchange Experiments and Seasonal Dynamics

The three sets of methods described include: (1) nitrogen enrichment experiments conducted seasonally in which unlabeled nitrate and ammonium were added to the surface water in chambers that contained the sediment–water column, (2) a nitrogen-15 tracer experiment conducted in the spring season only, and (3) observations of interstitial nitrate and ammonium concentrations of the surficial sediments over 17 months.

Nitrogen Enrichment

To isolate areas used for fertilization experiments, we forced six 20-cm-diameter epoxy-coated stovepipes into the soil to prevent lateral movement and mixing with adjacent surface water. Pipes were placed 1 week before the beginning of the experiments to allow sediment conditions to restabilize. Experiments were run July 4, 1976, November 3, 1976, and February 19, 1977. Ammonium and nitrate, as aqueous solutions of NH_4Cl and KNO_3, were added separately in equal amounts to sets of three pipes. Samples were taken at the time of addition and after 1, 3, 6, and 10 days. Water levels were monitored to correct for small changes in volume which occurred over the term of the experiment. Initial concentrations in the chambers ($\bar{x} \pm SE$) for the July 4, November 3, and February 19 experiments were 12.7 ± 2.7, 19.0 ± 2.6, and 10.8 ± 1.1 mg NH_4–N/l and 2.4 ± 0.2, 8.5 ± 1.3 and 5.0 ± 0.3 mg NO_3–N/l, respectively.

Concurrently, samples of surface water (~ 2 l) were isolated in polyethylene containers. Then KNO_3 was added to one group and NH_4Cl

to the other. Initial concentrations in the containers $(x^- \pm SE)$ for the July 4 (no replicates), November 3, and February 19 experiments were 18.0, 32.3 ± 2.7, and 20.0 ± 2.6 mg NH_4-N/l and 3.9, 14.7 ± 1.3, and 7.5 ± 0.5 mg NO_3-N/l respectively. Samples were taken from the containers whenever the pipes were sampled. Both pipe and surface water samples were analyzed for either nitrate or ammonium, depending on which component had been added, using the methods previously described.

Nitrogen-15 Tracer Experiments

Polyvinyl chloride chambers 31.3 cm i.d. were driven approximately 30 cm into the soil so that surface water could circulate though two opposing holes during the equilibration period. An elevated glass plate covered the chamber; this allowed air circulation and light penetration but excluded litterfall and precipitation. One week later on April 20, 1978, inorganic nitrogen was added to the chambers. For each liter of water present in a chamber, 2.5 mg each of ammonium nitrogen and nitrate nitrogen was added after lateral holes were plugged. Three treatments were made in duplicate: (1) $^{15}N-NH_4$ and $^{14}N-NO_3$, (2) $^{14}N-NH_4$ and $^{15}N-NO_3$, and (3) $^{14}N-NH_4$ and $^{14}N-NO_3$ (control). The third treatment was used to determine background levels of ^{15}N.

After nitrogen additions, 500 ml of surface water was taken at 2, 4, 8, 16, 24, and 48 h, taking care that the water column was homogenously sampled. Samples were fixed in the field with 1 ml of 40 mg/ml $HgCl_2$ and were stored at 4°C in the laboratory. Also, samples for determining dissolved oxygen concentration were taken from the control chambers and from an unenclosed area near the chambers and were analyzed by the sodium azide modification of the Winkler method (Golterman and Clymo 1969). After 5 days decomposing leaves on the soil surface and the top 10 cm of soil were collected from each chamber. Sediment was washed from leaves into soil samples with a small amount of deionized water. Large woody parts, mostly twigs and roots, were hand separated from the soil. Each wet soil sampled was weighed and mixed with a gloved hand, and a 500-ml subsample was then homogenized in a blender and stored at 4°C. Leaves and woody parts were dried in a forced-air oven at 85°C and stored in a desiccator.

For each soil sample, approximately 500 mg of wet soil, weighed to the nearest 0.1 mg, was transferred to a Kjeldahl digestion flask and digested according to Bremner (1965). Ammonia was then steam distilled from the digestate into a 50-ml volumetric flask containing 4 ml of $0.1N$ H_2SO_4. A silver condensor was used to prevent retention of ^{15}N in the system between distillations. A 5-ml aliquot of the distillate was used to determine ammonium concentration (Scheiner, 1976). The remaining 45 ml of distillate of each sample was evaporated to about 20 ml on a hot plate and redistilled into a 12 by 75-mm glass tube containing 1 ml of $0.1N$ H_2SO_4. The tubes were then dried in an $80°C$ oven and stored. The $^{15}N/^{14}N$ ratio of each of these samples was determined by using a model 21-620 Consolidated Electronics Corporation mass spectrometer. Nitrogen was converted from diammonium sulfate to nitrogen gas by adding alkaline hypobromide maintained in an oxygen-free condition by purging with argon. Moisture contents of soil samples were determined to relate data to dry soil. Leaves and woody matter were analyzed in the same fashion, except that no moisture calculations were required.

Exchangeable ammonium and nitrate in soil were determined by using nonacidified $2N$ KCl as the exchange solution (Bremner and Keeney, 1966). Ammonia was removed from the KCl filtrate by steam distillation by using MgO to raise the pH. A second distillation followed, and nitrate was recovered as ammonium in the distillate after reduction by Devarda alloy in the distillation flask. Although the second distillation included nitrite as well as nitrate, we feel safe in assuming that nitrite does not exist in measurable concentrations in the soil because of its extremely low levels in surface and interstitial water. The distillate was analyzed for ammonium concentration and ^{15}N was by the methods already described. All surface-water samples were treated in the same fashion as the KCl filtrate.

Interstitial Water

Two soil samples were collected each week over a 17-month period along separate transects during the growing season and every 2 weeks during dormancy. For each sample, soil from the top 5 cm at five stationary sites was mixed. Care was taken not to remove soil from a previously disturbed area. Samples were mixed in the laboratory with a gloved hand, and interstitial water was separated by centrifugation. Ammonium and nitrate were analyzed as previously described. Percent coverage of the soil by surface water was determined at each collection on the basis of the proportion of total sites covered.

Soil-Water Exchange of Phosphorus

Radiotracer experiments were conducted in chambers like those described for the nitrogen enrichment experiment. Of the six chambers used, three were treated with 40 ml of formalin (40% formaldehyde, vol/vol) to inhibit biological activity. Experiments began on April 17, 1976 (spring); October 31, 1976 (fall); and January 2, 1977 (winter). Experiments were initiated by spraying 15 μCi of $H_3{}^{32}PO_4$ onto the surface water in each of three chambers. Thirty-milliliter surface samples of water were taken at 0, 1, 3, 7, 24, 72, and 120 h. From each sample, 10-ml aliquots were filtered through Gelman type-A/E glass-fiber filters to separate dissolved and particulate fractions. One-milliliter quantitites of filtered and unfiltered samples were added to 10 ml of Aquasol (New England Nuclear). At 120 h the surface water was removed from the chambers. All leaves were put in paper bags, and a soil sample was taken to 1-cm depth. Soil and leaf samples were dried (85°C), weighed, and ground with mortar and pestle, and 10-mg aliquots were added to 12 ml of Aquasol and water to form a gel suspension. A Packard 3320 liquid scintillation counter was used for counting. Counting efficiency was determined by channel ratios as recommended by Kobayashi and Maudsley (1974), and quench corrections were made. Activity (cpm) in surface water, decomposing leaves, interstitial water, and sediment was determined, and its distribution among compartments was calculated as percent of total activity. Concentrations of stable phosphorus were not measured.

RESULTS AND DISCUSSION

Effect of Tar River Floodplain On Nutrients In Overbank Flow

Concentrations of NH_4–N, NO_3–N, and filterable reactive phosphorus (FRP) in water samples collected simultaneously from the Tar River channel and from the floodplain surface 150 m from the stream are shown in Figure 2. Our objective in comparing these samples was to determine whether the swamp served as a nutrient source or sink for the river. Evidence for an effect on the river water can be assessed by two observations. First, local precipitation can accumulate and displace surface water in the floodplain toward the river at any time of the year (Figure 1a), either diluting or augmenting the nutrient concentrations in the river. This is most likely to occur during the dormant season when the demand by evapotranspiration for water within the swamp is low and precipitation would lead to a net outflow from the swamp. The other line of evidence would be a change in the nutrient concentration of the water impounded in the swamp following overbank flooding from the river (Figure 1c). A decrease in nutrient concentration of the quiescent water derived from river overflow would suggest nutrient uptake, whereas an

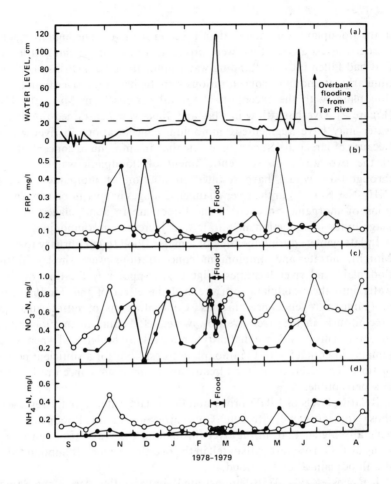

Figure 2. Seasonal concentrations of nitrate, ammonium, and filterable reactive phosphorus in the Tar River channel and in the surface water of the floodplain. Stream water overflows into swamp when stage height in the channel exceeds 20 cm above forest floor, indicated by the broken line in part a. O, Tar River channel. ●, Tar Swamp surface water.

increase in concentration would suggest nutrient supply to the river if flooding were repeated and floodplain surface water was entrained in downstream flow.

Nitrate concentrations in the floodplain were lower than those of the river for most of the year (Figure 2). Thus, when local precipitation displaced floodplain water to the river, there was a high probability that nitrate in the river water would be diluted. Several periods were exceptions to this, most notably in November when the nitrate concentrations in the floodplain were higher than those in the river. The high inorganic nitrogen levels at that time were probably a result of leaching of leaves that had fallen into the standing water in the floodplain (Brinson, 1977). Similarly, ammonium concentrations were higher in the river than in the floodplain, with the exception of several periods from March to July (Figure 2). Overbank flow in May and June which entrained floodplain water may have augmented ammonium concentrations in river water. Because of larger differences in nitrate than in ammonium concentrations of the two water masses, entrainment of floodplain water by river through-flow would have resulted in dilution of inorganic nitrogen ($NH_4-N + NO_3-N$) in the river channel. Changes in post-flood concentrations of inorganic nitrogen in the floodplain in June suggest similar magnitudes of reduction in nitrate and increases in ammonium.

In late winter, during the time of a major flood (late February to early March), nitrate and ammonium concentrations were similar in the floodplain and river localities (Figure 2). Apparently the flow of river water into the floodplain overwhelmed the effect of the floodplain on inorganic nitrogen concentrations. Concentrations of nitrate dropped precipitously in the river overflow water left behind in the floodplain, however. This is evidence that nitrate removal can be detected when water remains in contact with the floodplain forest floor for a sufficient period of time. The decrease in the already low ammonium concentration was less perceptible.

Relative levels of FRP in river and floodplain water were the reverse of those for inorganic nitrogen. Concentrations in the floodplain were generally higher than in the river throughout the year. Also, concentrations of FRP rose rather than fell in the overflow water left impounded in the floodplain after the flood.

Kitchens et al. (1975) demonstrated similar influences of the Santee River Swamp in South Carolina on overflow waters during the winter and early spring. They reported decreases in nitrate concentrations from the stream channel to the floodplain interior, but trends in FRP and ammonium were less well-defined. Gradients of total phosphorus were clearer than those for FRP; this suggests that particulate phosphorus may have been sedimenting on the floodplain. Likewise, Yarbro (1979) reported a significant flux of particulate phosphorus to the Creeping Swamp floodplain.

These observations suggest that the floodplain supplies FRP to the Tar River, whereas it serves as a sink for inorganic nitrogen. This does not take into account particulate forms of nitrogen and phosphorus which are likely controlled by physical considerations. Particulate forms of nitrogen and phosphorus were so variable in the floodplain waters that we could see great differences in suspended material from one place to another and from one sampling period to another.

Nitrogen Exchange Experiments and Seasonal Nitrogen Dynamics

Nitrogen Enrichment

Greater proportions of nitrate than ammonium were removed from the surface water during the three 10-day experiments (Figure 3). At the end

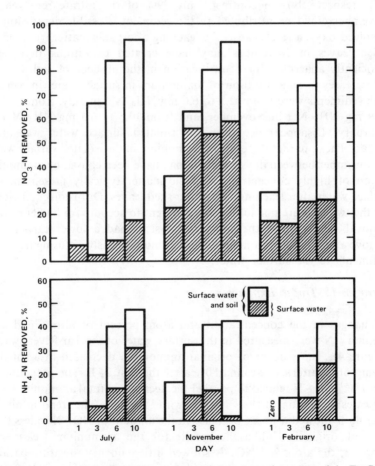

Figure 3. Removal of nitrate and ammonium added to the surface water of the Tar River floodplain by surface water and soil. Shaded portion is removal attributed to surface water in the absence of soil.

of 10 days 86 to 94% of the added nitrate disappeared, in comparison with 40 to 46% of the ammonium. That removal rates for both ions were slightly greater in July and November than in February, the coldest month, suggests temperature dependency. Initial concentrations of inorganic nitrogen were difficult to control because of the uncertainty of the water volumes of the chambers. Average concentrations at the beginning and end of the 10 days were 2.4 and 0.20 mg NO_3-N/l in July, 8.5 and 0.73 mg NO_3-N/l in November, and 5.0 and 0.78 NO_3-N/l in February. Most of the nitrate removed in 10 days from the surface water was a result of the presence of the sediment in July and February (82 and 71%, respectively). In contrast, only 39% of the nitrate removed in November could be attributed to the sediment. Possible depression of dissolved oxygen levels caused by leaching of organic matter from newly fallen leaves in November may have created conditions that were sufficiently anaerobic for denitrification in the absence of soil.

Decreases in concentration of ammonium in surface water in contact with sediments were from 12.7 to 6.6 mg NH_4-N/l in July, from 18.7 to 10.9 mg NH_4-N/l in November, and from 10.6 to 6.3 mg NH_4-N/l in February. Disappearance from the isolated surface water was more erratic. The apparent increases in ammonium in the isolated surface water in November between days 6 and 10 may have been caused by ammonification of highly concentrated organic matter typically present in the surface water after leaching of autumn-shed leaves. Diffusion coefficients for the two ions are not sufficiently different to account for higher nitrate removal rates. Either the soil–water system had a lower capacity to remove ammonium, or ammonium was simultaneously diffusing from the sediments to the surface water.

Nitrogen-15 Tracer Experiment

Changes in the concentration and atom percent of nitrate and ammoniun ^{15}N were measured in the surface water of the Tar River swamp (Figure 4). The percent of original ammonium decreased in a roughly asymptotic pattern to around 80% after the first 24 h. The reduction of atom % $^{15}NH_4$-N tended to parallel the decrease in total ammonium; this indicates a dilution of the $^{15}NH_4$-N with $^{14}NH_4$-N. The trend in nitrate over 48 h showed a linear decrease in the amount of NO_3-N to less than 50% of original. Unlike the results for the ammonium treatments, however, the atom % $^{15}NO_3$-N showed little tendency toward a parallel decrease. This is evidence for a unidirectional pathway for nitrate to the sediments, where it was denitrified. By comparison, the pathway for ammonium was bidirectional, and thus nonlabeled ammonium was generated from the sediments and diffused to the surface water.

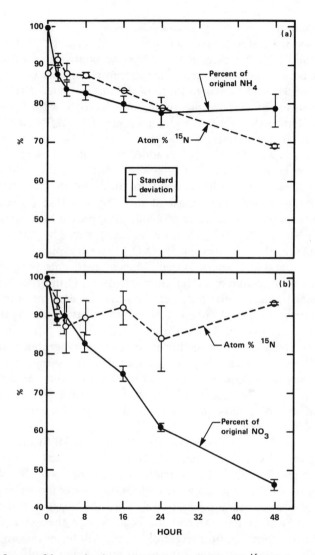

Figure 4. Losses of inorganic nitrogen and changes in atom % ^{15}N of the surface water to which (a) ^{15}N–NH$_4$ and (b) ^{15}N–NO$_3$ were added.

Between the 48-h and the 5-day sample collection, a flood overtopped the chambers, flushing out an unknown amount of ^{15}N. As a result, the amount of recovered ^{15}N was quite low, 7.3% of the original quantity in the ^{15}NH$_4$–N treatments and 2.6% of the original quantity in the ^{15}NO$_3$–N treatments when all compartments (surface water forms, leaf detritus, and exchangeable forms in sediment) were summed.

The amount of exchangeable $^{15}NH_4$ and $^{15}NO_3$ measured in the sediments at the end of 5 days (data not shown) suggested that mass flow alone in the $^{15}NH_4$–N treatment could have accounted for the quantity of $^{15}NH_4$–N present in the sediment since concentrations per unit volume of sediment water were similar to those of the surface water. Exchangeable nitrate in this treatment was present in amounts too low for ^{15}N analysis, however, despite the fact that equal amounts of $^{15}NH_4$–N and $^{14}NO_3$–N were added to the surface water.

In chambers where $^{15}NO_3$–N was added to the surface water, detectable amounts of exchangeable $^{15}NH_4$–N were present in the sediment, but concentrations were less than one-third of those in chambers where $^{15}NH_4$–N was added. Thus most of the $^{15}NO_3$–N that arrived in the sediment from the surface water probably disappeared by denitrification. The remainder of the $^{15}NO_3$–N may have been reduced by pathways other than denitrification. Reduction of $^{15}NO_3$–N to organic nitrogen is one possibility that has been suggested for lake sediments (Keeney et al., 1971). We were unable to detect enrichment of ^{15}N in organic nitrogen above background levels, however, because the concentration of ^{14}N organic nitrogen was two orders of magnitude greater than exchangeable inorganic fractions. Another pathway is reduction of nitrate to ammonium; this is supported by the appearance of exchangeable $^{15}NH_4$–N in the treatment to which $^{15}NO_3$–N was added to the surface water. This pathway has been demonstrated in anaerobic soils incubated for short periods of time (Stanford et al., 1975; Caskey and Tiedje, 1979) and would be an ecologically advantageous mechanism for nitrogen conservation by the system.

Analysis of decomposing leaves for ^{15}N enrichment at the end of the experiments showed that their ^{15}N concentration was approximately ninefold higher for the $^{15}NH_4$–N treatments than for $^{15}NO_3$–N treatments; this suggests that ammonium is the preferred source of inorganic nitrogen for immobilization. In comparison (on a weight basis) with the labeled nitrogen present in the water at the beginning of the experiment, the leaves show a concentration factor of less than 10 for nitrate and greater than 50 for ammonium. Thus leaf detritus represents a highly reactive site for inorganic nitrogen accumulation, but the pathway accounted for only a small proportion of the nitrogen cycled, in comparison with transfers to the sediments. Woody material from the sediments, consisting mostly of dead twigs and branches, showed no enrichment of ^{15}N above background.

We conclude that loss of nitrate from surface water by diffusion to the sediments is a one-way pathway made possible by denitrification of nitrate to dinitrogen gas or nitrous oxide. In view of the ambient conditions and the long incubation times of our experiments, molecular nitrogen is probably the dominant end product (Firestone et al., 1980). By comparison, ammonium diffusion is bidirectional between surface water and sediments since there is no permanent sink in the sediments as there is for nitrate. These pathways have been demonstrated for soil–water columns in the laboratory (Patrick and Tusneem, 1972). The high rates of ammonium loss from the water (Figure 4) are probably caused by the cation exchange system in the soil absorbing excessive concentrations added to the surface water in these experiments. The 2 to 10 mg N/l that were added to conduct the experiments were much higher than background concentrations of interstitial ammonium (Figure 5), thus creating an unnaturally high diffusion gradient. Nitrification limits the rate of ammonium loss from these systems (Reddy et al., 1976), but there appears to be a capacity for some excess storage by exchangeable pools of the sediments and by immobilization. Exchangeable ammonium saturation of the soil is approximately 900 mg NH_4-N/kg dry sediment (Bradshaw,

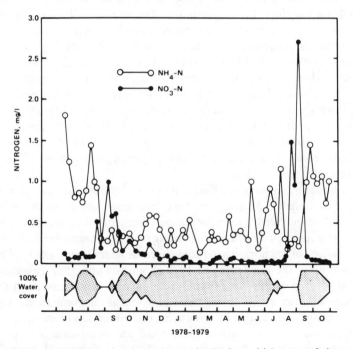

Figure 5. Changes in inorganic nitrogen pools in the interstitial water of the surficial sediments of the Tar River floodplain. Percent water cover is the estimated proportion of the sediments covered by water.

1977). Given the high concentrations of organic carbon (14 to 19%) and the low redox potential ($Eh_7 = 0$ to -300 at 5-cm depth) of the sediments in the swamp, the moderate temperatures during the experiment in the spring, and the geometry of the experimental chambers, nitrate loss rates (Figure 4) were probably limited by diffusion (Phillips et al., 1978). Since roots were severed when the chambers were forced into the soil, uptake by vegetation was unlikely.

Interstitial Water

The results from the experiments described and the reported nitrogen transformations that occur in sediments are supported by observations of seasonal changes of nitrate and ammonium pools in the interstitial water of surficial sediments (Figure 5). When surface water disappeared because of high evapotranspiration rates in the floodplain during the warm season, nitrate concentrations increased. This occurred in August and September of both years and is probably an annual phenomenon in the Tar River swamp and other seasonally flooded swamps. During this period nitrate loss likely continues because of diffusion to deeper anaerobic zones and anaerobic microsites near the surface.

Ammonium concentrations tended to be higher during the warmer months, except when nitrate concentrations reached temporary seasonal highs. Higher rates of ammonium production from decomposition and ammonification might be expected under warmer temperatures during the growing season. The inverse relationship between nitrate and ammonium during dry down indicates a period of intense nitrification under unflooded, oxidizing conditions. When surface water returned to cover the sediments, nitrate again became depleted to levels below those of ammonium.

The apparent pulse of nitrification during dry down in late summer and early fall (Figure 5) suggests that this transformation in nature is ultimately controlled by evapotranspiration from the forest. Reflooding of the soil surface or diffusion of nitrate to deeper anaerobic layers would result in loss of the nitrate produced. Because of the unusually long period of flooding in the late spring of 1979 (Figures 2 and 5), nitrification was probably lower at that time than during most growing seasons. Alternate flooding and dry down during the warm season, in comparison with the continuously flooded conditions during the winter, are more conducive to nitrogen losses to the atmosphere than are continuously aerobic or anaerobic conditions (Reddy and Patrick, 1975). In general, temperature and water level appear to affect microbial activity strongly; this, in turn, increases or depletes inorganic nitrogen pools. Since the experiments

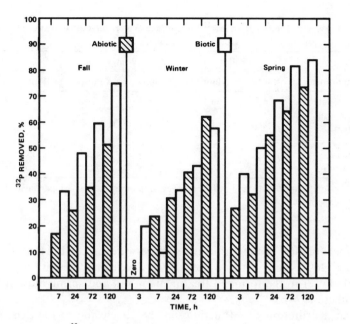

Figure 6. Percent of ^{32}P removed from the surface water of the Tar River floodplain under biotic (control) and abiotic (formalin-treated) conditions.

previously described (Figures 3 and 4) were conducted in water deep enough to allow sequential removal of surface water samples for analysis, there was no opportunity to observe nitrogen transformations induced by dry down.

Phosphorus Exchange

Loss of ^{32}P–PO$_4$ added to surface water was compared between chambers treated with formalin (abiotic) and untreated chambers (biotic). Removal rates were consistently lower in the abiotic than in the biotic chambers in the fall and spring (Figure 6). The winter experiment showed no consistent differences between treatments, and amounts of ^{32}P–PO$_4$ removal in the biotic chambers were lower for all time intervals than those during the spring and fall. The apparent temperature dependence of the biotic portion of removal (untreated minus formalin treated) suggests that microbial activity was responsible.

Colder water temperatures in the winter (1°C) may have been responsible for the lower rates; water temperatures in the fall were 14°C and in the spring 23°C. In the biotic systems, average particulate ^{32}P represented

Table 1. Percent Total Activity of ^{32}P in Surface Water, Interstitial Water, Decomposing Leaves and Sediment at 120 h

Compartment	Fall		Winter		Spring	
	Biotic	Abiotic	Biotic	Abiotic	Biotic	Abiotic
Surface water	24.5	46.1	41.9	47.9	15.4	25.7
Decomposing leaves	5.3	1.6	4.3	0.6	—	—
Interstitial water	4.5	6.1	3.3	4.5	0.9	1.1
Sediment	65.7	46.2	50.5	47.0	83.7*	73.2*

*Decomposing leaves and sediments were combined as one compartment during the spring experiment.

a larger fraction of the ^{32}P in the surface water in the fall (50%) and the spring (36%) than in the winter (23%). Particulate activity was significantly lower in the formalin treatments in fall and spring but not winter (Holmes, 1977).

The distribution of ^{32}P activity at the conclusion of the experiments showed that the sediments were the major sink of ^{32}P removed from the water column (Table 1). Most of the loss from surface water was probably by diffusion to the sediments and absorption there, largely as a result of abiotic processes. However, biotic processes demonstrated in the spring and fall were responsible for measurable portions of phosphorus exchange, although the experiments revealed little about underlying mechanisms. Higher particulate activity for surface water in biotic than in abiotic systems, and in spring and fall than in winter suggest that suspended bacteria and algae were partly responsible. Decomposing leaves showed the highest activity (cpm/g) for ^{32}P of all compartments measured separately (in fall and winter only) and was considerably depressed in the abiotic treatments in comparison with biotic systems. Because the mass of leaves was so much smaller than other compartments, this accounted for no more than about 5% of the total distribution in the chambers (Table 1).

Phosphorus efflux from the sediments may have occurred, as it did for ammonium. Loss of ^{32}P from surface water does not imply net uptake of phosphorus by the sediments. Yarbro (1979) demonstrated that one of the principal transformations of the phosphorus cycle in a swamp floodplain was the conversion of filterable reactive phosphorus to filterable unreactive phosphorus, a process that would have gone undetected by our procedures. On the basis of annual input–output budgets, Yarbro concluded that sedimentation of particulate phosphorus and uptake of filterable reactive phosphorus by algae accounted for most of the phosphorus removal.

Significance of Tupelo–Cypress Swamps in Nutrient Cycling

Associations of water tupelo and bald cypress similar to those in the study area are limited in distribution to riverine floodplains of the southeastern United States and the Mississippi Valley. Aside from unforested areas of open water, these are among the most poorly drained features of the floodplain ecosystem and are commonly found in flats, oxbows, sloughs, and other floodplain depressions. Depending on local geomorphic features, tupelo–cypress communities may dominate the floodplain, as they do in the lower Tar River, or may represent a small proportion ($\sim$ 10%) of floodplain, as they do in the Beidler Tract of the Congaree Swamp in South Carolina (Gaddy et al., 1975). Wilson (1962) estimated that North Carolina has 400,000 ha of "wooded swamps," a wetland type composed largely of cypress–tupelo communities. Other species associations are located either on better-drained soils or at higher elevations on the floodplain, e.g., the overcup oak–bald cypress–water hickory and the cherry-bark oak–sweet gum–swamp chestnut oak communities (Wharton, 1978). Thus, of the floodplain communities collectively designated as southeastern "bottomland hardwoods," tupelo–cypress swamps receive most frequent and longest periods of flooding and are likely to have the highest degree of stream–floodplain nutrient coupling. Within the floodplain, this association represents topographic lows to which local runoff and groundwater move and in which overbank flow is trapped after floods. Despite the high potential that was demonstrated for denitrification, high recycling rates of nitrogen in litterfall (Brinson et al., 1980) and high levels of nitrogen in the soil (1 to 2%) suggest that they are nitrogen rich. This can be explained by mechanisms of nitrogen conservation which apparently override mechanisms of nitrogen loss. These mechanisms are conceptualized in Figure 7 as (1) a recycling loop in which ammonium undergoes a number of transformations, including uptake by plants, leaching as throughfall, mineralization, immobilization, and cation exchange; and (2) an export pathway where ammonium is nitrified aerobically and the resulting nitrate is denitrified anaerobically. Immobilization of inorganic nitrogen, particularly ammonium, by heterotrophic microorganisms after autumn leaf fall is sustained for several months throughout the winter season until temporary storage by filamentous algal mats and uptake by trees in the spring initiate other pathways. Reduction of nitrate to ammonium, although not shown in Figure 7, may be quantitatively important as a mechanism for

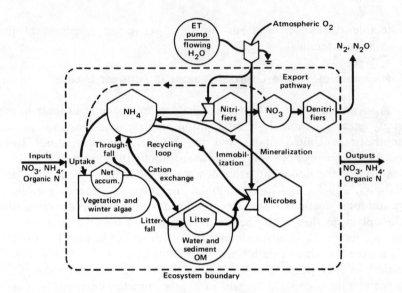

Figure 7. Nitrogen cycle showing pathways and storages in a tupelo–cypress ecosystem. Symbols after Odum (1971).

nitrogen conservation by the ecosystem. The net effect of these and other pathways associated with ammonium (Figure 7) appear to dominate over nitrification and, thus, result in nitrogen conservation by the system. Although high potentials for denitrification are present year round (Figure 3), aerobic conditions necessary for the production of nitrate pools do not occur until the "evapotranspiration pump" (Figure 7) results in seasonal dry down. Sources of nitrate, such as those imported from the stream channel to the floodplain during flooding (Figure 2), may, however, be incorporated into the export pathway of the nitrogen cycle. Thus the tupelo–cypress swamp that we examined is a nitrogen sink for the Tar River, and the atmosphere is simultaneously a nitrogen sink for the floodplain.

The balance for phosphorus in the floodplain cannot be resolved from our results. Abiotic factors probably play a larger role for phosphorus (Figure 6) than for nitrogen cycling. Levels of FRP in interstitial water [$\bar{x}$ = 0.30 mg P/l with no significant difference with depth or season (Holmes, 1977)] are similar to those of floodplain surface water and greater than those in the stream channel (Figure 2). High FRP levels are expected in both compartments of the floodplain because reducing conditions exist in the sediment and at the sediment–water interface. Thus the return of water to the stream channel by hydrologic events (Figures 1a and 1c) tends to enrich the stream with FRP. However, Yarbro (1979) and Mitsch et al. (1979) have demonstrated the importance of sedimentation as a phosphorus source for river swamps. If tupelo–cypress swamps are to be considered either a source or sink of phosphorus for streams, then both sediment deposition and fluxes of dissolved forms of phosphorus must be evaluated.

ACKNOWLEDGMENTS

We would like to thank Joseph Elkins, Jr., Deborah Noltemeier, Martha Jones, and Randy Creech for assistance in the field or the laboratory. Richard Volk of North Carolina State University kindly made available the mass spectrometer for our use. Emilie Kane and Nancy Edwards drafted some of the figures. This research was supported by grants from the North Carolina Science and Technology Committee, the Institute for Coastal and Marine Resources of East Carolina University, and funds provided by the Office of Water Research and Technology (B-114-NC), U. S. Department of the Interior, through the Water Resources Research Institute of the University of North Carolina.

REFERENCES

American Public Health Association, 1975, *Standard Methods for the Examination of Water and Wastewater.* American Public Health Association, New York.

Bedinger, M. S., 1979, *Forests and Flooding with Special References to the White River and Ouachita River Basins, Arkansas,* Water Resources Investigations, Open-File Report 79-68, U. S. Geological Survey, Washington, D.C.

Benke, A. C., D. M. Gillespie, and F. K. Parrish, 1979, *Biological Basis for Assessing Impacts of Channel Modification: Invertebrate Production, Drift, and Fish Feeding in a Southeastern Blackwater River,* Environmental Resources Center, Report 06-79, Georgia Institute of Technology, Atlanta.

Bradshaw, H. D., 1977, *Nitrogen Cycling in an Alluvial Swamp Forest,* Master's Thesis, East Carolina University, Greenville, NC.

Bremner, J. M., 1965, Inorganic Forms of Nitrogen, In C. A. Black (Ed.), *Methods of Soil Analysis, Part 2,* pp. 1179-1237, American Society of Agronomy, Madison, WI.

———, and D. R. Keeney, 1966, Determination and Isotope-Ratio Analysis of Different Forms of Nitrogen in Soils. 3. Exchangeable Ammonium, Nitrate, and Nitrite by Extraction-Distillation Methods, *Soil Sci. Soc. Am. Proc.,* 30: 577-582.

Brinson, M. M., 1977, Decomposition and Nutrient Exchange of Litter in an Alluvial Swamp Forest, *Ecology*, 58: 601-609.

———, H. D. Bradshaw, R. N. Holmes, and J. B. Elkins, Jr. 1980, Litterfall, Stemflow, and Throughfall Nutrient Fluxes in an Alluvial Swamp Forest, *Ecology*, 61: 827-835.

Brown, L., and E. G. Bellinger, 1978, Nitrate Determination in Fresh and Some Estuarine Waters by Ultraviolet Light Absorption: A New Proposed Method, *Water Res.*, 12: 223-229.

Caskey, W. H., and J. M. Tiedje, 1979, Evidence for *Clostridia* as Agents of Dissimilatory Reduction of Nitrate to Ammonium in Soils, *Soil Sci. Soc. Am., J.*, 43: 931-936.

Environmental Protection Agency, 1976, *Methods for Chemical Analyses of Water and Wastes,* Environmental Monitoring and Support Laboratory, Office of Research and Development, Cincinnati.

Firestone, M. K., R. B. Firestone, and J. M. Tiedje, 1980 Nitrous Oxide from Soil Denitrification: Factors Controlling Its Biological Production, *Science*, 208: 749-751.

Fredrickson, L. H., 1979, Lowland Hardwood Wetlands: Current Status and Habitat Values for Wildlife, in P. E. Greeson, J. R. Clark, and J. E. Clark (Eds.) *Wetland Functions and Values: The State of Our Understanding*, pp. 296-306, American Water Resources Association, Minneapolis, MN.

Gaddy, L. L., T. S. Kohlsaat, E. A. Laurent, and K. B. Stansell, 1975, *A Vegetation Analysis of Preserve Alternatives Involving the Beidler Tract of the Congaree Swamp,* South Carolina Wildlife and Marine Resources Department, Columbia, SC.

Golterman, H. L., and R. S. Clymo (Eds.), 1969, *Methods for Chemical Analysis of Fresh Water,* IBP Handbook No. 8, Blackwell Scientific Publications, Oxford, England.

Holmes, R. N., 1977, Phosphorus Cycling in an Alluvial Swamp Forest in the North Carolina Coastal Plain, Master's Thesis, East Carolina University, Greenville, NC.

Keeney, D. R., R. L. Chen, and D. A. Graetz, 1971, Importance of Denitrification and Nitrate Reduction in Sediments to the Nitrogen Budgets in Lakes, *Nature*, 233: 66.

Kitchens, W. M., J. M. Dean, L. H. Stevenson, and J. H. Cooper, 1975, The Santee River Swamp as a Nutrient Sink, in F. G. Howell, J. B. Gentry, and M. H. Smith (Eds.), *Mineral Cycling in Southeastern Ecosystems,* ERDA Symposium Series, CONF-740513, pp. 349-366, NTIS, Springfield, VA.

Kobayashi, Y., and D. V. Maudsley, 1974, Recent Advances in Sample Preparation, In P. E. Stanley and B. A. Scoggins (Eds.), *Liquid Scintillation Counting—Recent Developments,* pp. 189-205, Academic Press, Inc., New York.

Leopold, L. B., M. G. Wolman, and J. P. Miller, 1964, *Fluvial Processes in Geomorphology.* W. H. Freeman and Company, San Francisco.

Mitsch, W. J., C. L. Dorge, and J. R. Weimhoff, 1979, Ecosystem Dynamics and a Phosphorus Budget of an Alluvial Cypress Swamp in Southern Illinois, *Ecology*, 60: 1116-1124.

Odum, H. T., 1971, *Environment, Power, and Society*, John Wiley & Sons, Inc., New York.

Patrick, W. H., Jr., and M. E. Tusneem, 1972, Nitrogen Loss from Flooded Soil, *Ecology*, 53: 735–737.

Phillips, R. E., K. R. Reddy, and W. H. Patrick, Jr., 1978, The Role of Nitrate Diffusion in Determining the Order and Rate of Denitrification in Flooded Soils. II. Theoretical Analysis and Interpretation, *Soil Sci. Soc. Am., J.*, 42: 272–278.

Reddy, K. R., and W. H. Patrick, Jr., 1975, Effect of Alternate Aerobic and Anaerobic Conditions on Redox Potential, Organic Matter Decomposition, and Nitrogen Loss in a Flooded Soil, *Soil Biol. Biochem.*, 7: 87–94.

————, W. H. Patrick, Jr., and R. E. Phillips, 1976, Ammonium Diffusion as a Factor in Nitrogen Loss from Flooded Soils, *Soil Sci. Soc. Am., J.*, 40: 528–533.

Scheiner, D., 1976, Determination of Ammonia and Kjeldahl Nitrogen by Indophenol Method, *Water Res.*, 10: 31–36.

Stanford, G., J. O. Legg, S. Dzienia, and E. C. Simpson, Jr., 1975, Denitrification and Associated Nitrogen Transformations in Soils, *Soil Sci.*, 120: 147–152.

Wharton, C. H., 1978, *The Natural Environments of Georgia*, Georgia Department of Natural Resources, Atlanta.

Wilson, K. A., 1962, *North Carolina Wetlands: Their Distribution and Management*, North Carolina Wildlife Resources Commission, Raleigh.

Yarbro, L. A., 1979, Phosphorus Cycling in the Creeping Swamp Floodplain Ecosystem and Exports from the Creeping Swamp Watershed, Ph.D. Thesis, University of North Carolina, Chapel Hill.

11. THE INFLUENCE OF HYDROLOGIC VARIATIONS ON PHOSPHORUS CYCLING AND RETENTION IN A SWAMP STREAM ECOSYSTEM

Laura A. Yarbro*

Curriculum in Ecology
University of North Carolina
Chapel Hill, North Carolina

ABSTRACT

The fluxes and retention of filterable reactive phosphorus (FRP), filterable unreactive phosphorus (FUP), and particulate phosphorus (PP) in a floodplain swamp in eastern North Carolina varied in response to the seasonality of inundation and yearly variations in runoff from the upstream watershed. High runoff and floodplain inundation typically occurred during the cool season when decomposition rates were low. During the flooded period, the floodplain forest floor had a high absorption capacity for runoff-derived FRP that was both temperature and water FRP concentration dependent. The FUP, on the other hand, was released from the forest floor in approximate proportion to the volume of floodwaters. During the warm season, when decomposition rates were faster and standing water concentrations of phosphorus were relatively high, little or no runoff occurred; thus there was no significant loss of phosphorus from the ecosystem. Loss of FUP from the ecosystem and imports of PP were correlated with yearly variations in runoff. The

---

*Present Address: Harbor Branch Foundation, RR1, Box 196, Ft. Pierce, FL 33450.

PP exports remained constant during the 2 years of the study; this resulted in large net retention during the year of high runoff. Overall, 30 to 56% of total phosphorus inputs were retained by the swamp ecosystem during the 2 years of study.

INTRODUCTION

Wetlands, because they are semi-aquatic and are usually located between aquatic and terrestrial ecosystems, have been recommended for and studied as potential filters for excessive phosphorus inputs from waste treatment plants, agriculture, and mining operations. However, a survey of research demonstrated that wetlands exhibit a wide range in types of phosphorus budgets, extending from net ecosystem retention to quasi-equilibrium or net ecosystem loss (Yarbro, 1979). For such elements as phosphorus, which has no gaseous form significant to ecosystem biogeochemistry, the hydrology of the wetland ecosystem is the driving force behind material transport into and out of the ecosystem (Bormann and Likens, 1979; Gosselink and Turner, 1978). The purpose of this paper is to relate the variations in fluxes and retention of several forms of phosphorus in a floodplain swamp ecosystem to the seasonality of inundation and to annual variations in stream runoff. I will then show that this type of wetland might be a good filter of anthropogenic phosphorus inputs.

MATERIALS AND METHODS

Study area

Creeping Swamp, a hardwood floodplain forest, is located along a small third-order stream of the same name in the Coastal Plain of North Carolina. The watershed is in Pitt, Beaufort, and Craven counties, and drainage is part of the Neuse River system. A portion of the swamp-stream ecosystem about 8.2 km long and 3.2 km^2 in area was selected for intensive study (Figure 1). Within this area, the swamp floodplain was 300 to 1000 m wide.

The floodplain canopy is dominated by *Nyssa sylvatica* var. *biflora*, *Acer rubrum*, *Fraxinus caroliniana*, *Nyssa aquatica*, and *Liquidambar styraciflua*, with *F. caroliniana* and *N. aquatica* most abundant along the stream channel. Herbaceous growth is sparse, but shrubs (*Leucothoe axillaris* and *Vaccinium corymbosum*) are somewhat more abundant. The

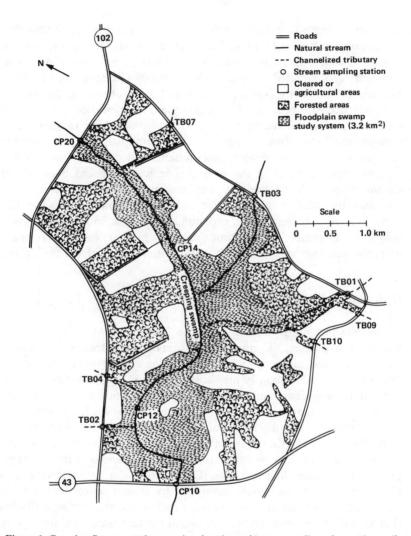

Figure 1. Creeping Swamp study area showing the main stream, tributaries, and sampling stations.

inorganic clay and silt soils of the floodplain are overlain by a leaf litter layer.

The climate is temperate and moist, with cool winters, warm summers, and moderate precipitation. Long-term average precipitation is 122 cm/yr, with a slight seasonal maximum during summer (Sumsion, 1970). Greatest stream runoff and maximum inundation occur during winter and early spring; water depths in the stream may reach 1 m, and velocities are usually 0.05 to 0.15 m/s (Mulholland, 1979). During the flooded season, discharge from the swamp varies widely depending on precipitation and is dominated by storm flow periods (Kuenzler et al., 1977; Yarbro, 1979). In the spring, runoff and water levels in the swamp gradually decrease because of increased plant transpiration in the watershed and within the swamp itself. Usually by mid-May to June water remains only in the stream channel, and the floodplain is dry. As summer progresses the stream becomes intermittent, and often by mid-October the only surface water in the swamp is in widely spaced pools in the stream channel. With leaf fall and cessation of transpiration, rains replenish the shallow groundwater until streamflow resumes (in late November or early December), and inundation of the floodplain occurs in mid to late December. At any time during the growing season, very heavy or prolonged rains may cause short-term inundation of the floodplain.

Hydrologic Measurements

Precipitation was measured daily at a North Carolina Forest Service fire tower in Wilmar, NC, about 5 km from the study area. Streamflow in the main channel of Creeping Swamp entering and leaving the study area (Figure 1) was continuously monitored by the U. S. Geological Survey. For tributary streams and ditches entering the study area and for the upstream main channel station (CP-20) after August 1977, discharge data were obtained from discharge water-level relationships developed by Mulholland (1979). Annual runoff values from ungauged tributaries and ditches were calculated by using slopes from regressions of the discharge of the downstream gauged station (CP-10) with measured discharge of a given tributary. These slopes, multiplied by annual runoff from CP-10, were used to estimate annual runoff for each of the tributaries, assuming that annual runoff was relatively constant per unit watershed for a given year. This was borne out by comparing annual runoff values for gauged watersheds of widely differing sizes in the Coastal Plain (U. S. Geological Survey, 1978; 1979). The hydrologic budget was based on a water year, i.e., October 1 through September 30.

The duration, extent, and frequency of floodplain inundation in the study area were estimated by recording depths of water at 5-m intervals across several transects of the floodplain when it was completely inundated. By using the varying water depths across each transect and the water level at CP-10, I generated a regression model that predicted the fraction of floodplain inundated as a function of water level at CP-10.

Phosphorus Water Chemistry

Streams entering and leaving the study area were sampled biweekly in water year (WY) 1977 (October 1976 to October 1977) and monthly in WY 1978 (October 1977 to October 1978) when significant flow occurred (> 0.05 m^3/s). Four sets of daily stream samples were taken to monitor changes in water quality during storm runoff. Daily sampling usually began on the first day of a large frontal rainstorm and continued until 3 to 4 days after peak flow.

Water samples taken from the center of stream channels were separated into filtered and unfiltered fractions in the field. Unfiltered samples were poured into clean, acid-washed polyethylene bottles and stored on ice. Water was filtered through acid-washed membrane filters (Gelman Metricel GA-6, 0.45 μm pore size) with glass-fiber prefilters (Gelman type A-E), placed in polyethylene bottles, and put on ice. Separate filtration devices were used for stations having high phosphorus concentrations. Samples were frozen, and approximately 2 weeks intervened between sample collection and analysis.

Phosphorus was differentiated by filtration and chemical analysis into filterable (molybdate) reactive phosphorus (FRP), total filterable phosphorus (TFP), and total phosphorus (TP). The FRP has been called orthophosphate by other researchers, but my studies (Yarbro, 1979) and those of others (e.g., Kuenzler and Ketchum, 1962; Rigler, 1968; Lean, 1973; Downes and Paerl, 1978; Stainton, 1980) show that the measurement of molybdate-reactive phosphorus greatly overestimates the amount of orthophosphate in natural waters. The FRP was measured either by the automated stannous chloride procedure on a Technicon Autoanalyzer (Environmental Protection Agency, 1974) or by the method of Strickland and Parsons (1972). Autoanalyzer analyses of FRP at 660 nm wavelength were corrected for sample color by using sample blanks to which mixed reagent minus ammonium molybdate had been added. Total fractions were measured by persulfate digestion and the automated stannous chloride method (Environmental Protection Agency, 1974) or by the persulfate digestion method of Menzel and Corwin (1965) followed by the method of Strickland and Parsons (1972). Reproducibility between the two methods of analysis was tested with standards and swamp water samples; no detectable differences were found. Standards and blanks were included with each analysis, and recoveries were checked intermittently by using internal standards. Particulate phosphorus (PP) and filterable unreactive phosphorus (FUP) concentrations were obtained by difference. The FUP has been called dissolved organic phosphorus (DOP) by other researchers.

Annual stream fluxes of phosphorus were obtained by multiplying annual runoff by annual flow-weighted mean phosphorus concentrations. Fluxes into and from the 3.2 km² floodplain area were normalized for area. Area-based fluxes allowed easy comparisons but did not represent uniformity of processes in the floodplain.

Bulk precipitation was collected for phosphorus measurement by using polyethylene funnels mounted on wooden frames 1 to 1.5 m above the ground and connected to plastic bottles by rubber tubing. Nylon mesh (0.25 mm) and glass wool excluded debris. Three collectors were placed in clearings near the study area. Volumes were measured every 2 weeks from August 1976 through August 1978. Samples for phosphorus analysis, collected every 4 weeks over the same period, consisted of intercepted rainfall of the previous 2-week period. Collection bottles were acid-washed, and samples were preserved in the field with $HgCl_2$ (American Public Health Association, 1975). Phosphorus data from the three rain collectors were averaged and then volume weighted. Annual weighted mean concentrations of precipitation phosphorus were multiplied by annual precipitation volumes to calculate annual fluxes.

Sedimentation of Particulate Phosphorus

Sedimentation was assessed by measuring the increment of dry weight and phosphorus on tared squares of constant area (4 by 4 cm) of *Quercus michauxii* leaves placed in hardware-cloth holders and located in triplicate on the floodplain at four different elevations. These leaves decomposed slowly and provided a natural substrate for sedimenting materials. Leaves were collected immediately after leaf fall. Before they were placed in the swamp, the leaf squares were soaked in deionized water for 48 h to remove easily leachable materials and then were air-dried and weighed. Leaf squares were placed in the swamp for 2 week periods from January to May 1978, the period of floodplain inundation in WY 1978. After 2 weeks the leaf squares were removed from the floodplain and placed in plastic bags. In the laboratory, squares were air-dried for a week, weighed, oven-dried at 80°C for 48 h and reweighed to determine air and dry weights. Replicates for each elevation were then pooled and ashed to determine ash weights; phosphorus analysis of the ash followed the same protocol as litter analyses (see below) except that $Mg(NO_3)_2$ was omitted. Studies using standards demonstrated about a 5% reduction in phosphorus recovery in samples not treated with $Mg(NO_3)_2$. Replicate control leaf squares were submitted to the same laboratory treatments to account for air- to dry-weight changes, leaf ash, and phosphorus content. Only on rare occasions did some leaf squares lose weight while on the swamp floor; in those few cases weight losses were a very small percentage of the total weight of the squares.

Biomass and Soil Phosphorus

Phosphorus in herbaceous vegetation, shrubs, and vines was estimated by dry ashing (Likens and Bormann, 1970; Lee et al., 1965) subsamples of harvests made by Mulholland (1979) in 1976 and 1977. Bryophyte phosphorus was measured in samples taken from 15 1-m^2 plots across a transect of the floodplain. The standing stock of forest floor litter was estimated bimonthly in WY 1978 by harvests of 15 0.25-m^2 plots across a floodplain transect. Phosphorus analysis followed the same dry-ashing procedure. Standing stocks of phosphorus in canopy vegetation were estimated using literature phosphorus concentrations (Duvigneaud and Denaeyer de Smet, 1970; Likens and Bormann, 1970; Cromack and Monk, 1975) and biomass data of Mulholland (1979).

Soil phosphorus was estimated twice during the dry season. At 15 sites along a floodplain transect, five cores of soil were taken using a 2.5-cm corer and separated into surface (0 to 5 cm) and subsurface (6 to 25 cm) subsamples. At each site, depth subsamples were pooled. Each subsample was weighed wet, dried at 80°C for 48 to 72 h and reweighed. The phosphorus content was measured by dry ashing at 500°C for 4 h according to the methods of Legg and Black (1955), Saunders and Williams (1955), and Lee et al. (1965). This method resulted in quantitative measurement of organic and sorbed soil phosphorus but not phosphorus in compounds of clay minerals.

Canopy Losses of Phosphorus

Throughfall was measured using fifteen collectors similar to precipitation collectors placed along a transect across the swamp floodplain. Field sampling, duration, and frequency of collections and laboratory analyses were the same as for precipitation. On each collection date average phosphorus concentrations of the fifteen samples were volume weighted and then summed to obtain annual weighted mean concentrations of throughfall phosphorus. The average annual throughfall volume was multiplied by the annual weighted mean phosphorus concentrations to calculate annual fluxes.

Qualitative estimates of phosphorus concentrations in stemflow were made by placing polyurethane collars around ten swamp trees [diameter at base height (DBH) = 24.6 cm ± 6.04 SD]. These collars directed stemflow into funnels that emptied into large covered plastic garbage cans. Stemflow was collected bimonthly from March 1977 through June 1978. Glass wool in the funnels excluded debris, and $HgCl_2$ was used as a preservative. Total and reactive phosphorus concentrations were measured on the autoanalyzer as described. Quantitative fluxes were estimated using measured concentrations and literature estimates of the quantity of stemflow relative to precipitation (Helvey and Patric, 1965).

Litter smaller than 0.3 m in length (leaves, small twigs, fruits, and flowers) of vegetation greater than 1.5 m in height was sampled by 0.2-m^2 circular litter collectors over a 3-year period from March 1975 through March 1978. During the first 2 years of collection, baskets were located across three transects of the swamp floodplain with 9 or 10 baskets on each transect. A single transect of fifteen baskets was sampled during the third year. Individual collectors were emptied into plastic bags every 3 to 4 weeks except during peak leaf-fall in October and early November when litter was collected at 1 to 2 week intervals. In the laboratory samples were oven-dried at 80°C for at least 48 h and then weighed. Phosphorus determinations on harvested materials, litter, and other detrital materials followed ashing procedures suggested by Likens and Bormann (1970) and Lee et al. (1965).

Forest Floor Water Exchanges of Phosphorus

Phosphorus fluxes between the swamp forest floor and overlying floodwaters were measured by following short-term changes (1 to 5 hr) in phosphorus concentrations and $^{32}PO_4$ activity in water in in situ chambers, which isolated portions of the forest floor and associated floodwaters. Changes in $^{32}PO_4$ activity and FRP concentrations were measured simultaneously to estimate both net and gross fluxes of phosphorus. Measurements were made throughout the winter and early spring of 1977, 1978, and 1979 over a wide variety of temperatures, phosphorus concentrations, and floodplain elevations and with the addition of biological inhibitors. A detailed description of these methods is given in Yarbro (1979) and is not presented here.

RESULTS

Hydrology

Hydrologic fluxes in the Creeping Swamp watershed varied widely between WY 1977 and WY 1978 (Figures 2 and 3). Annual precipitation

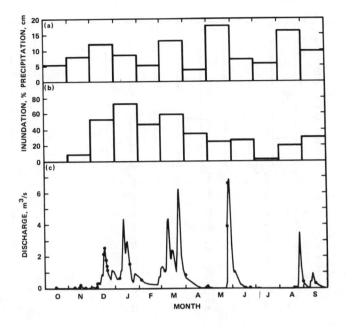

Figure 2. Precipitation (a) annual total, 112 cm; inundation (b); and surface water discharge (c), annual total runoff, 23.9 cm, for Creeping Swamp during WY 1977.

varied about 10 cm around the long-term mean of 122 cm (Sumsion, 1970). Runoff from the downstream station, CP-10, varied by nearly a factor of 3 between years and ranged from 21 to 46% of annual precipitation. This difference is explained by a few high-intensity storms in WY 1978.

The monthly distribution of precipitation showed no consistent variation in either year (Figs. 2a and 3a). Floodplain inundation, however, had a distinct seasonal pattern in both years, with greatest inundation during winter and spring months and minima in July and August of both years (Figures 2b and 3b). More floodplain was inundated for a longer period of time in WY 1978. Daily stream discharge was greater in the cool months and was characterized by sharp peaks following storms. There were four peaks in WY 1977 when discharge exceeded 4 m³/s. In WY 1978, there were eleven such peaks, three of which exhibited extremely high discharge (> 19 m³/s).

A mass-balance calculation of water entering and leaving the floodplain study area showed that more water left as surface runoff at CP-10 and as evapotranspiration than entered as stream runoff and precipitation

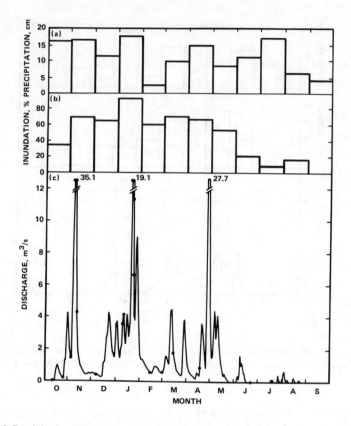

Figure 3. Precipitation (a) annual total, 135 cm; inundation (b); and surface water discharge (c), annual total runoff, 61.8 cm, for Creeping Swamp during WY 1978.

(Table 1). Tributary runoff was estimated in two ways: (1) regression with CP-10 discharge, as described in the methods section, and (2) by multiplying annual CP-10 runoff by the tributary-drained area (33 km²), assuming runoff was constant over the area. Unmeasured runoff was that coming from the uplands (11.8 km²) surrounding the floodplain within the study area (Figure 1), which was not measured at the stream stations. This input was estimated by multiplying the annual runoff measured at CP-10 by the area of these uplands. Evapotranspiration in the swamp was estimated by using the fraction of total annual precipitation for the whole watershed that did not leave CP-10 as runoff and multiplying it by the volume of precipitation entering the floodplain. A watershed-based evapotranspiration estimate may not represent what actually occurred in the floodplain, but no data were available to evaluate this. Annual deep

Table 1. Water Budget for the Creeping Swamp Floodplain*

Source	Water flux, 10^6 m^3/yr	
	1977	1978
Inputs		
Precipitation	3.58	4.32
CP-20	6.80	17.6
Tributaries		
Estimate 1	5.2 to 7.6	14 to 17
Estimate 2	7.9	20
Unmeasured upland runoff	2.8	7.3
Total	18–21	43–49
Outputs		
CP-10	19.1	49.4
Floodplain evapotranspiration	2.8	2.3
Total	22	52

*Runoff estimates for CP-10 and CP-20 were obtained from the U. S. Geological Survey (1978; 1979) and were corrected for watershed area.

ground water losses, which Winner and Simmons (1977) estimated to be quite small (2% of annual precipitation), were not included in the budget. The difference in inputs and outputs might be accounted for by unmeasured groundwater inputs to the floodplain. Winner and Simmons (1977) concluded that the portion of the swamp floodplain between CP-20 and CP-10 (Figure 1) was an area where water from the Castle Hayne aquifer discharged into the swamp.

Phosphorus in Precipitation and Stream Waters

Phosphorus in bulk precipitation was primarily in the reactive form (0.034 to 0.042 mg/l), and concentrations were consistently higher than FRP concentrations in undisturbed streams (Table 2). Mean stream concentrations showed large variations between streams and from year to year. Wastes from a poorly managed hog farm upstream of the polluted tributary, TB-02, markedly affected phosphorus concentrations in WY 1977 in TB-02 and at CP-10, downstream of the confluence of TB-02 and the Creeping Swamp main channel. In WY 1978, when the hog farm was not operating, phosphorus concentrations in these streams, except for PP, were much lower. In the natural streams, TB-03 and CP-20, phosphorus concentrations were extremely low, and PP tended to constitute the largest proportion of total phosphorus.

Table 2. Annual Weighted Mean Concentrations of Phosphorus (mg/1) in Bulk Precipitation and Selected Stream Waters During WY 1977 and 1978

	Bulk precipitation	Downstream, CP-10	Upstream, CP-20	Polluted tributary, TB-02	Natural tributary, TB-03
1977					
FRP	0.042	0.064	0.003	0.962	0.004
FUP		0.012	0.006	0.048	0.005
PP		0.047	0.014	0.258	0.015
TP	0.054	0.123	0.025	1.268	0.024
1978					
FRP	0.034	0.005	0.006	0.173	0.003
FUP		0.013	0.009	0.055	0.007
PP		0.017	0.013	0.337	0.009
TP	0.045	0.036	0.028	0.566	0.018

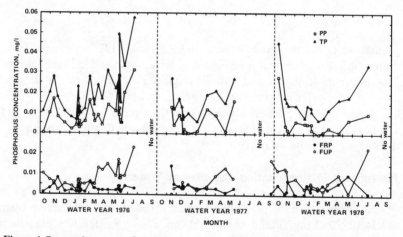

Figure 4. Seasonal patterns of phosphorus concentration at the mid-swamp station, CP-14.

The seasonal variations in concentrations of phosphorus fractions were examined at the mid-swamp station, CP-14 (Figure 4). This station more closely approximated natural conditions than did CP-10 or CP-20, where weirs caused pooling of the water. Seasonally, FRP showed no consistent variations throughout 3 years and was the phosphorus form with lowest concentration. In contrast, FUP tended to be higher in summer and autumn and was usually higher than FRP. The PP concentrations were variable but were low in January and February of each year. Although runoff was highest in WY 1978, PP concentrations were less variable and lower than in the other 2 years.

Phosphorus Cycling in the Swamp Floodplain Ecosystem

I constructed a yearly phosphorus budget for the 3.2-km^2 floodplain swamp located in the Creeping Swamp study area for WY 1977 and WY 1978. The vertical limits of the ecosystem were the tops of the trees and a depth of 25 cm in the soil. I partitioned the ecosystem into five components, pathways between components, and pathways leading to and from the ecosystem. Components were (1) the swamp floodwaters, which were further subdivided into FRP, FUP, PP, and LPP (particulate P > 0.25 mm); (2) trees and saplings; (3) herbaceous vegetation, shrubs, vines, and bryophytes; (4) the swamp forest floor, including the seasonally abundant filamentous algae; and (5) the mineral soil. The overall budget is presented in Figure 5.

Inputs

The source of phosphorus to the ecosystem was hydrologic, and surface water phosphorus inputs were primarily in the forms of FRP and PP. Inputs of LPP, estimated from measurements made by Mulholland (1979) and analyses of harvested drift materials, were small (< 1%) compared with total surface water phosphorus inputs. Inputs in bulk precipitation were 6 to 7% of surface water inputs, 40% of which fell on the dry forest floor (Table 3). Sixty-one to seventy-seven percent of surface water inputs were brought into the ecosystem via the polluted tributary, TB-02. Total inputs to the swamp ecosystem were slightly greater in WY 1978 than in WY 1977.

Export

Losses from the ecosystem were also hydrologic, occurring almost entirely in surface water. The loss of LPP remained quite small but was slightly greater than LPP imports. Total exports decreased from WY 1977 to WY 1978 in contrast with an increase in imports over the same period, resulting in a twofold increase in the amount of TP retained by the ecosystem from WY 1977 to WY 1978.

In WY 1977 FRP and PP comprised the largest fraction of total exports, whereas in WY 1978 FUP and PP were the major components. This was caused by a nearly fivefold decrease in FRP exports and nearly a

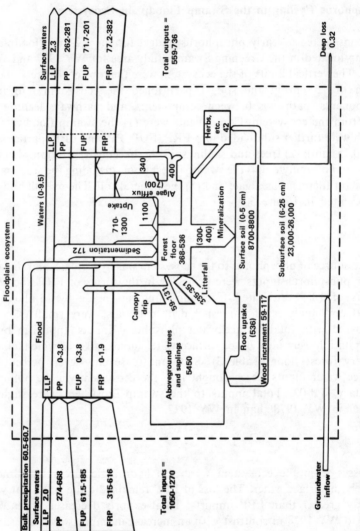

Figure 5. Phosphorus budget for the Creeping Swamp floodplain ecosystem. Units are mg P/m²-yr (fluxes) and mg P/m² (standing stocks).

Table 3. Differences in Hydrologic Characteristics and Phosphorus Fluxes Between Wet and Dry Seasons During WY 1977 and 1978 in the Creeping Swamp Floodplain*

	WY 1977		WY 1978	
	Wet	Dry	Wet	Dry
	Dec. 76–June 77	Oct. 76–Nov. 76 July 77–Sept. 77	Nov. 77–May 78	Oct. 77 June 78–Sept. 78
Hydrologic Characteristics				
Annual precipitation, %	60	40	61	39
Annual runoff, %	91	9	94	6
Average floodplain area inundated	46%	10%	69%	17%
Phosphorus Fluxes				
Inputs, mg P m^{-2} year^{-1}				
Precipitation	38	22	39	21
Stream Runoff	950	95	1190	80
Outputs, mg P m^{-2} year^{-1}				
Stream runoff	670	66	525	33
Internal fluxes, mg P m^{-2} year^{-1}				
Forest floor FRP uptake	1090	210	600	110
Sedimentation			146	26
Litter fall	92	220	83	67
Throughfall	41	19		
Stemflow	1.3	.8	2.4	1.6

*The wet season consisted of those consecutive months when the floodplain was > 30% inundated. Phosphorus fluxes are based on weighted seasonal means of phosphorus concentrations. Fluxes that are area-inundated and season dependent (forest floor uptake and sedimentation) were also adjusted for relative inundation in each season.

threefold increase in FUP exports from the swamp between the 2 years. Lessening of the influence of the hog farm upstream of TB-02 (imports of FRP in WY 1978 were about half of imports in WY 1977) probably lowered FRP exports. Furthermore, runoff in WY 1978 was almost three times greater than in WY 1977; if water-volume-dependent leaching controlled FUP release from the forest floor, then the dramatic increase in FUP release could be explained by the increased volume of water leaving the swamp ecosystem in WY 1978. During both water years there was a net export of FUP from the swamp ecosystem. In WY 1977 there was a small net export of PP from the swamp. In WY 1978 the trend was reversed, with a net retention of PP during this relatively wet year. Even with greatly increased imports and runoff, the swamp ecosystem maintained a relatively constant level of PP exports. Therefore the potential effects of increased erosion and increased PP concentrations in surface waters may have been decreased by the swamp ecosystem.

Internal cycling

Phosphorus cycling in the floodplain was dominated by fluxes through the floodwaters (Figure 5). Standing stocks of phosphorus in floodwaters, calculated from average concentrations at CP-14 (Figure 4) and inundation volumes, varied from 0 mg P/m^2 when there was no water in the swamp to an average of 9.5 mg P/m^2 during greatest floodplain inundation. The small size of the floodwater phosphorus compartment indicated that the residence times of surface water inputs in the floodwaters must have been extremely short. In WY 1978 a net transfer of 172 mg PP m^{-2} yr^{-1} occurred between floodwaters and the forest floor as sedimentation; this was slightly greater than half the estimated amount of surface water PP retained by the swamp. The fact that the greatest contributor of PP, tributary TB-02, was downstream from the sedimentation measurement site may account for the underestimate. The single greatest transfer of phosphorus from the floodwaters to the forest floor was in the form of FRP, and most of the transfer could be attributed to algal uptake. Measurements of the forest floor uptake made by using $^{32}PO_4$ overestimated what actually occurred. Algal uptake estimates were conservative and indicated that algal dynamics were very important in phosphorus fluxes between the forest floor and floodwaters. Based on an annual algal productivity of 16.5 g C/m^2 (Mulholland, 1979) and an assumed Redfield (1958) molar ratio of 106 C : 1 P in algal biomass, about 400 mg P/m^2 were incorporated into algal biomass during a bloom period. The difference between uptake and incorporation was probably returned to the floodwaters either as FRP or FUP.

Estimated FUP releases from the forest floor were 10 to 20 times greater than the net release of FUP from the ecosystem in surface waters. The FUP release measurements were based on changes in FUP concentrations in water overlying the swamp floor and are, therefore, not subject to the same criticism as FRP estimates with ^{32}P. This implies that the FUP fraction was also recycled between the forest floor and floodwaters.

Cycling of phosphorus between the forest floor and vegetation included a canopy return of 390 to 500 mg P m^{-2} yr^{-1}, with litterfall phosphorus comprising the largest portion (70 to 73%). Throughfall inputs were slightly larger than precipitation inputs, and stemflow phosphorus inputs were < 1% of total canopy return. Forty-three to sixty-two percent of litter-fall inputs occurred during October and November, whereas most throughfall and stemflow inputs occurred during the winter and spring months. Comparison of canopy inputs and the standing stock of phosphorus in the forest floor litter suggested a residence time of a year or a little more for forest floor phosphorus.

DISCUSSION

Seasonal Differences in Floodplain Phosphorus Cycling

Separating each water year and associated phosphorus fluxes into wet and dry seasons facilitates examination of the hydrologic influence on floodplain phosphorus cycling (Table 3). Each wet season lasted 7 months (58% of year) and covered a slightly different time period in each year. The distribution of precipitation between wet (60%) and dry (40%) seasons was the same each year. In sharp contrast, runoff was distinctly seasonal; > 90% occurred during the wet season. As expected, more floodplain was inundated during the wet season as well.

Phosphorus fluxes, especially inputs and outputs, occurred predominately during the wet season (Table 3). Precipitation, throughfall, and stemflow inputs were greater during the wet season because the amount of precipitation was greater and because phosphorus concentrations were higher. The wet season spanned the spring months when tree pollen production and site preparation for crops contributed phosphorus to the atmosphere. Litter fall was the only measured flux that was greater during the dry season. Most litter fell during the autumn when the floodplain was dry. Despite a predominance of fluxes during the wet season, the floodplain ecosystem retained 30 to 56% of all inputs annually. The floodplain retained more phosphorus in WY 1978 and retained a greater percentage of inputs (56%) despite greatly increased runoff. Although PP concentrations and inputs were higher in WY 1978, export of PP from the ecosystem remained the same. Long and extensive inundation with slowly flowing waters may have provided the opportunity for sedimentation of PP. The FRP and FUP concentrations, particularly in the polluted tributary, TB-02, were lower in WY 1978, and this may have also increased floodplain retention capabilities.

Studies of phosphorus fluxes in other wetlands have revealed three dominant mechanisms driving phosphorus exports: (1) temperature-dependent increases in community respiration rates, (2) autumnal senescence of macrophytes, and (3) spring runoff. These losses often offset phosphorus retention during other portions of the year, resulting in net ecosystem losses of phosphorus. Tidal marshes, which are usually flooded regularly throughout the year, export large amounts of phosphorus during the warm summer months when respiration rates are high (Bender and Correll, 1974; Settlemeyre and Gardner, 1975; Valiela et al., 1978; Axelrad et al., 1976; Heinle and Flemer, 1976; Stevenson et al., 1976; Woodwell and Whitney, 1977). Marshes and bogs adjacent to lakes and streams export phosphorus in the autumn after the onset of macrophyte senescence (Lee et al., 1975; Turner et al., 1976; Spangler et al., 1976; 1977; Prentki et al., 1978; Richardson et al., 1978; Tilton and Kadlec, 1979). Wetlands in the northern temperate zones are ice-covered for several months during the winter and metabolically fairly inactive, resulting in poor assimilation of phosphorus inputs (Richardson et al., 1978). A large proportion of the annual runoff occurs during spring thaw when large ecosystem losses of phosphorus have been recorded (Spangler et al., 1976; 1977; Richardson et al., 1978; Tilton and Kadlec, 1979). Floodplain swamps in general appear to retain net amounts of phosphorus (e.g., Kuenzler et al., 1977). Retention has been attributed to sedimentation (Mitsch et al., 1979) or to the removal of dissolved phosphorus from floodwaters during floodplain inundation (Butler, 1975; Hartland-Rowe and Wright, 1975; Kitchens et al., 1975; Day et al., 1976; Boyt et al., 1977; Kemp, 1978).

If phosphorus fluxes were dependent on hydrologic fluxes and if most of the flux and retention of phosphorus occurred during the wet season, why then was Creeping Swamp retaining phosphorus over an annual period when other wetland types were not? The answer may lie in the timing of hydrologic fluxes with respect to the metabolism of the swamp community. The floodplain forest floor community of Creeping Swamp was very active during the flooded cool season. The inundated forest floor supported a large and productive community of macroinvertebrates and fish (Sniffen, 1981). The ecosystem was lotic in character, and floodwaters remained aerobic (Kuenzler et al., 1977). Net uptake of floodwater FRP was consistently measured during this period and uptake was positively dependent on temperature and FRP concentration. Forest floor litter increased in phosphorus content early in the flooded season and then slowly began releasing phosphorus through decomposition (Yarbro, unpublished data). Filamentous algae on the forest floor removed FRP from floodwaters; however, the algae were present only during the winter and early spring. Leafing out of the canopy and drawdown of floodwaters resulted in the senescence and death of the algal bloom. When the algae decomposed, mineralized phosphorus either was released to floodwaters or was leached into the forest floor. Since typically the floodplain was drying during these periods, leaching is more likely to have occurred. Thus the filamentous algae may have acted as a trap and served as temporary storage for phosphorus brought in by winter floodwaters until the onset of the plant growing season. The spring wildflower *Erythronium americanum* has been observed to play a similar role in the Hubbard Brook forest (Muller and Bormann, 1976).

In summary, during the time when runoff was greatest and potential for phosphorus losses appeared large, the ecosystem retained phosphorus by incorporation into algae; uptake by the forest floor, including the leaf litter; and sedimentation. During the summer months, when litter decomposition rates were greatest, runoff was very low. In the autumn, during leaf fall, little or no runoff occurred. Relatively high phosphorus concentrations during these months in slowly flowing or stagnant waters pointed to the potential for phosphorus losses had there been significant runoff.

The Creeping Swamp floodplain ecosystem retained phosphorus over a water year because of active uptake and PP sedimentation during floodplain inundation and because little or no runoff occurred during summer and autumn, when higher respiration rates and leaf fall may have resulted in greater potential phosphorus mobility. Measurement of phosphorus fluxes during storm flow resulting from a tropical storm in summer or early autumn would test this hypothesis.

ACKNOWLEDGMENTS

I thank Pat Mulholland, Bob Sniffen, and Paul Carlson for field assistance and helpful suggestions. Shirley Wasson and Thomas Smith helped with laboratory analyses. Willow Baker of the of the North Carolina Forest Service provided precipitation data. Mark Brinson, Judy Meyer, and Kathy Ewel criticized the manuscript. The study was supported by a National Science Foundation Predoctoral Fellowship at the University of North Carolina, Chapel Hill, and by grants to E. J. Kuenzler from the North Carolina Water Resources Research Institute (B-084-NC and B-110-NC). Support during manuscript preparation was provided by Harbor Branch Foundation, Fort Pierce, FL. Harbor Branch Foundation Contribution No. 222.

REFERENCES

American Public Health Association, 1975. *Standard Methods for the Examination of Water and Wastewater*, New York.

Axelrad, D. M., K. A. Moore, and M. E. Bender, 1976, *Nitrogen, Phosphorus and Carbon Flux in Chesapeake Bay Marshes*, Virginia Water Resources Research Center, Bulletin 79, Blacksburg.

Bender, M. E., and D. L. Correll, 1974. *The Use of Wetlands as Nutrient Removal Systems*, Publication No. 29, Chesapeake Research Consortium, Baltimore, MD.

Bormann, F. H., and G. E. Likens, 1979. *Pattern and Process in a Forested Ecosystem*, Springer-Verlag, New York.

Boyt, F. L., S. E. Bayley, and J. Zoltek, Jr., 1977, Removal of Nutrients from Treated Municipal Wastewater by Wetland Vegetation, *J. Water Pollut. Control Fed.*, 49: 789-799.

Butler, T. J., 1975, Aquatic Metabolism and Nutrient Flux in a Southern Louisiana Swamp and Lake System, M.S. Thesis, Louisiana State University, Baton Rouge.

Cromack, K., Jr., and C. D. Monk, 1975, Litter Production, Decomposition, and Nutrient Cycling in a Mixed Hardwood Watershed and a White Pine Watershed, in F. G. Howell, J. B. Gentry, and M. H. Smith (Eds.), *Mineral Cycling in Southeastern Ecosystems*, pp. 609-624 ERDA Symposium Series CONF-740513, NTIS, Springfield, VA.

Day, J. W., Jr., T. J. Butler, and W. H. Conner, 1976, Productivity and Nutrient Export Studies in a Cypress Swamp and Lake System in Louisiana, in M. Wiley, (Ed.), *Estuarine Processes, Vol. II*, pp. 255-269, Academic Press, Inc., New York.

Downes, M. T., and H. W. Paerl, 1978, Separation of Two Dissolved Reactive Phosphorus Fractions in Lakewater, *J. Fish. Res. Board Can.*, 35: 1636-1639.

Duvigneaud, P., and S. Denaeyer de Smet, 1970, Biological Cycling of Minerals in Temperate Deciduous Forests, in D. E. Reichle (Ed.), *Analyses of Temperate Forest Ecosystems*, pp. 199-225, Springer-Verlag, New York.

Environmental Protection Agency, 1974, *Methods for Chemical Analysis of Water and Wastes*, Cincinnati.

Gosselink, J. G., and R. E. Turner, 1978, The Role of Hydrology in Freshwater Wetland Systems, in R. E. Good, D. F. Whigham, R. L. Simpson (Eds.), *Freshwater Wetlands*, pp. 63-78, Academic Press, Inc., New York.

Hartland-Rowe, R., and P. B. Wright, 1975, Effects of Sewage Effluent on a Swampland Stream, *Verh. Intnat. Verein. Limnol.*, 19: 1575–1583.

Heinle, D. R., and D. A. Flemer, 1976, Flows of Materials Between Poorly Flooded Tidal Marshes, *Mar. Biol.*, 35: 359–373.

Helvey, J. D., and J. H. Patric, 1965, Canopy and Litter Interception of Rainfall by Hardwoods of the Eastern U.S., *Water Resour. Res.*, 1: 193–206.

Kemp, G. P., 1978, Agricultural Runoff and Nutrient Dynamics for a Swamp Forest in Louisiana, M.S. Thesis, Louisiana State University, Baton Rouge.

Kitchens, W. M., Jr., J. M. Dean, L. H. Stevenson, and J. H. Cooper, 1975, The Santee Swamp as a Nutrient Sink, in F. G. Howell, J. B. Gentry, and M. H. Smith (Eds.), *Mineral Cycling in Southeastern Ecosystems*, ERDA Symposium Series CONF-740513, pp. 349–366, NTIS, Springfield, VA.

Kuenzler, E. J., and B. H. Ketchum, 1962, Rate of Phosphorus Uptake by *Phaedactylum tricornutum*, *Biol. Bull.*, 123: 134–145.

——, P. J. Mulholland, L. A. Ruley, and R. P. Sniffen, 1977, *Water Quality in North Carolina Coastal Plain Streams and Effects of Channelization*, North Carolina Water Resources Research Institute, Report No. 127, Raleigh.

Lean, D. R. S., 1973, Phosphorus Dynamics in Lakewater, *Science*, 179: 778–780.

Lee, G. F., E. Bentley, and R. Amundson, 1975, Effects of Marshes on Water Quality, in A. D. Hasler (Ed.), *Coupling of Land and Water Systems*, pp. 105–127, Springer-Verlag, New York.

——N. L. Clesceri, and G. P. Fitzgerald, 1965, Studies on the Analysis of Phosphates in Algal Cultures, *Int. J. Air Water Pollut.*, 9:715–722.

Legg, J. O., and C. A. Black, 1955, Determination of Organic Phosphorus in Soils. II. Ignition Method, *Soil Sci. Soc. Am., Proc.* 19:139–143.

Likens, G. E., and F. H. Bormann, 1970, *Chemical Analyses of Plant Tissues from the Hubbard Brook Ecosystem in New Hampshire*, School of Forestry Bulletin No. 79, Yale University, New Haven, CT.

Menzel, D. W., and N. Corwin, 1965, The Measurement of Total Phosphorus in Seawater Based on the Liberation of Organically Bound Fractions by Persulfate Oxidation, *Limnol. Oceanogr.*, 10:280–282.

Mitsch, W. J., C. L. Dorge, and J. R. Weimhoff, 1979, Ecosystem Dynamics and a Phosphorus Budget of an Alluvial Cypress Swamp in Southern Illinois, *Ecology*, 60: 1116–1124.

Mulholland, P. J., 1979, Organic Carbon Cycling in a Swamp-Stream Ecosystem and Export by Streams in Eastern North Carolina, Ph.D. Thesis, University of North Carolina, Chapel Hill.

Muller, R. N., and F. H. Bormann, 1976, Role of *Erythronium americanum* Ker. in Energy Flow and Nutrient Dynamics of a Northern Hardwood Forest Ecosystem, *Science*, 193: 1126–1128.

Prentki, R. T., T. D. Gustafson, and M. S. Adams, 1978, Nutrient Movement in Lakeshore Marshes, in R. E. Good, D. F. Whigham, and R. L. Simpson (Eds.), *Freshwater Wetlands*, pp. 169–194, Academic Press, Inc., New York.

Redfield, A. C., 1958, The Biological Control of Chemical Factors in the Environment, *Am. Sci.*, 46: 205–222.

Richardson, D. J., D. L. Tilton, J. A. Kadlec, J. P. M. Chamie, and W. A. Wentz, 1978, Nutrient Dynamics of Northern Wetland Ecosystems, in R. E. Good,

D. F. Whigham, and R. L. Simpson (Eds.), *Freshwater Wetlands*, pp. 217–241, Academic Press, Inc., New York.

Rigler, F. H., 1968, Further Observations Inconsistent with the Hypothesis that the Molybdenum Blue Method Measures Orthophosphate in Lakewater, *Limnol. Oceanogr.* 13: 7–13.

Saunders, W. M. H., and E. G. Williams, 1955, Observations on the Determination of Total Organic Phosphorus in Soils, *J. Soil Sci.*, 6: 254–267.

Settlemeyre, J. L., and L. R. Gardner, 1975, A Field Study of Chemical Budgets for a Small Tidal Creek—Charleston Harbor, S.C., in T. M. Church (Ed.), *Marine Chemistry in the Coastal Environment*, ACS Symposium Series, No. 18, pp. 152–175, American Chemical Society, Washington, DC.

Sniffen, R. P., 1981, The Abundance of Aquatic Invertebrates in the Floodplain of a Seasonally Inundated Stream Swamp, Ph.D. Thesis, University of North Carolina, Chapel Hill.

Spangler, F. L., C. W. Fetter, Jr., and W. E. Sloey, 1977, Phosphorus Accumulation-Discharge Cycles in Marshes, *Water Resour. Bull.*, 13: 1191–1201.

————, W. E. Sloey, and C. W. Fetter, Jr., 1976, Artificial and Natural Marshes as Wastewater Treatment Systems in Wisconsin, in D. L. Tilton, R. H. Kadlec, and C. J. Richardson (Eds.), *Freshwater Wetlands and Sewage Effluent Disposal*, pp. 215–240, Proceedings of a National Symposium, University of Michigan, Ann Arbor.

Stainton, M. P., 1980, Errors in Molybdenum Blue Methods for Determining Orthophosphate in Freshwater, *Can. J. Fish. Aquat. Sci.*, 37: 472–478.

Stevenson, J. C., D. R. Heinle, D. A. Flemer, R. J. Small, R. A. Rowland, and J. F. Ustach, 1976, Nutrient Exchanges Between Brackish Water Marshes and the Estuary, in M. Wiley (Ed.), *Estuarine Processes, Vol. II*, pp. 219–240, Academic Press, Inc., New York.

Strickland, J. D. H., and T. R. Parsons. 1972. *A Practical Handbook of Seawater Analysis*, Fisheries Research Board of Canada, Ottawa.

Sumsion, C. T., 1970, *Geology and Ground-Water Resources of Pitt County, North Carolina*, Ground Water Bulletin No. 18, U. S. Geological Survey, Washington, DC.

Tilton, D. L., and R. H. Kadlec, 1979, The Utilization of a Fresh-Water Wetland for Nutrient Removal from Secondarily Treated Waste Water Effluent, *J. Environ. Qual.*, 8: 328–334.

Turner, R. E., J. W. Day, Jr., N. Meo, P. M. Payonk, T. B. Ford, and W. G. Smith, 1976, Aspects of Land-Treated Waste Application in Louisiana Wetlands, in D. L. Tilton, R. H. Kadlec, and C. J. Richardson (Eds.), *Freshwater Wetlands and Sewage Effluent Disposal*, Proceedings of a National Symposium, pp. 145–170, University of Michigan, Ann Arbor.

U. S. Geological Survey, 1978, *Water Resources Data for North Carolina*, Water-data Report NC-77-1, Washington, DC.

————, 1979, *Water Resources Data for North Carolina*, Water-data Report NC-78-1, Washington, DC.

Valiela, I., J. M. Teal, S. Volkmann, D. Shafer, and E. J. Carpenter, 1978, Nutrient and Particulate Fluxes in a Salt Marsh Ecosystem: Tidal Exchanges and Inputs by Precipitation and Groundwater, *Limnol. Oceanogr.*, 23: 798–812.

Winner, M. D., Jr., and C. E. Simmons, 1977, *Hydrology of the Creeping Swamp Watershed, North Carolina, with References to Potential Effects of Stream Channelization*, Water Resources Investigation 77-26, U. S. Geological Survey, Washington, DC.

Woodwell, G. M., and D. E. Whitney, 1977, Flax Pond Ecosystem Study: Exchanges of Phosphorus Between a Salt Marsh and the Coastal Waters of Long Island Sound, *Mar. Biol.*, 41: 1–6.

Yarbro, L. A., 1979, Phosphorus Cycling in the Creeping Swamp Floodplain Ecosystem and Exports from the Creeping Swamp Watershed, Ph.D. Thesis, University of North Carolina, Chapel Hill.

12. ORGANIC CARBON SUPPLY AND DEMAND IN THE SHETUCKET RIVER OF EASTERN CONNECTICUT

E. A. Matson

Department of Biology
East Carolina University
Greenville, North Carolina

R. L. Klotz

Department of Biology
State University of New York
Cortland, New York

ABSTRACT

Autochthonous and allochthonous supplies of organic carbon were compared in the Shetucket River (annual discharge, 10^9 m^3) of eastern Connecticut during 1974. The biomass of the sixth-order river segment was dominated by epilithic, unicellular algae, and heterotrophic activity was at least 75% microbial. Imported and exported dissolved organic carbon were measured, and gross primary production and total community respiration were estimated using the upstream-downstream oxygen technique. Light-limited in situ gross primary production (2000 kcal m^{-2} yr^{-1}) provided the equivalent of 40% of the carbon respired in the river segment, and calculated net primary production (60% of gross) supplied about 14% of the calculated carbon assimilated by heterotrophs. The ratio of net primary production to heterotrophic assimilation is prefered here when expressing the dependence of heterotrophs on autochthonous carbon supplies. Even though primary production was two orders of magnitude greater than in small, shaded New England brooks, total community respiration was only twice as great. The supply of alloch-thonous carbon per square meter of river bottom was at least 4×10^5 times greater than autochthonous production per square meter, and total carbon import and export were equal.

INTRODUCTION

It is apparent that system studies that separate a river from its watershed are unproductive (Davis, 1899); the couple between the two communities is a tight one. As rivers enlarge downstream the degree of autotrophy and heterotrophy is controlled by hydrologic, geologic, and biologic factors. Decreasing watershed slope along the river continuum may result in longer soil retention time for runoff (Ward, 1975). Prior processing of labile organic matter by the soil flora (Alexander, 1961; McDowell and Fisher, 1976) may result in the accumulation of refractory organic material which is discharged into rivers via runoff and ground-water or interflow (Beck et al., 1974).

Impoundments, agricultural runoff, and municipal sewage alter lotic trophic structure (Fisher, 1977; Marzolf, 1978) and may cause "serial discontinuities" in river characteristics (King and Ball, 1967; Ward and Stanford, this volume). Rivers in low-terrain areas receive substantial inputs of carbon from swamps and marshes (Mulholland and Kuenzler, 1979), and impoundments may accumulate a diverse carbon pool subject to storm washout and/or drawdown for downstream export. Labile organic matter that does enter the river is rapidly used, and perhaps the refractory compounds persist downstream (Hynes, 1975).

Algae release important amounts of their photosynthate as dissolved organic compounds (Fogg, 1977), and many of these are rapidly assimi-lated by heterotrophs (Brylinsky, 1977; Iturriaga and Hoppe, 1977; Ward et al., 1980). Therefore, as autochthonous production increases in wider, sunlit, downstream reaches, organic matter from algae and other plants may become increasingly important to river heterotrophs as the refrac-tory percentage increases.

In this regard, it is difficult to distinguish between allochthonous and autochthonous carbon sources in the downstream reaches of larger rivers. Organic material imported to the study section may have been autoch-thonously produced immediately upstream from the products of decom-position of terrestrial debris. Thus arbitrarily drawn boundaries may restrict the development of the system theory. Nonetheless, the concept of lotic heterotrophy (Hynes, 1963) has been supported by studies in low-order shaded streams (Kaushik and Hynes, 1971; Hall, 1972; Peterson and Cummins, 1974; Sedell et al., 1974; Small, 1975; McDowell and Fisher, 1976) which import most of their recognizable organic matter in the form of vegetation debris. Generally, primary production in these streams is low (e.g., Fisher and Likens, 1973), but several examples of small autotrophic systems have been given (Odum, 1957; Minshall, 1978). However, the amount of respiration per square meter of stream is relatively constant, regardless of the presence of an active autotrophic flora. For example, respiration in several New England streams increases only several-fold while autochthonous production increases by several orders of magnitude along a continuum from first- and second-order brooks (Fisher and Likens, 1973) to fourth- and sixth-order rivers (Fisher, 1977; this study). Cummins (1980) reviewed several hypotheses that may explain these observations.

We have studied the relative importance of allochthonous and autochthonous organic carbon sources in a sixth-order, serially impounded river of a heavily forested watershed. Although autochthonous production was expected to be significant in this river and to result in the export of net production to downstream reaches, allochthonous import to our study area was operationally defined as any organic carbon imported to the area regardless of its immediate source. Estimates of stream heterotrophic community assimilation were calculated to determine the magnitude of any autochthonous deficit (the excess of assimilation over net primary production) which might exist. The magnitude of this deficit in a larger river along the stream continuum might support hypotheses on lotic heterotrophy and the interdependence of terrestrial and aquatic communities.

MATERIALS AND METHODS

Study Area

The effluent from a sewage treatment plant (STP) was diluted to 1.4% ($\pm$ 0.91%, N = 25, 1974) of station 5 discharge and entered the confluence of the Willimantic and Natchaug Rivers where they combine to form the Shetucket (Figure 1). Klotz (1977) reviewed the results of a 3-year study of the effects of this effluent, which had been upgraded from primary to secondary (activated sludge), on the biology and chemistry of this river system. Our river-segment study areas were located at least 2 km downstream of the confluence, where nutrient and conductivity levels had returned to values typical of those upstream of the effluent. Pertinent study area data are given in Tables 1 and 2.

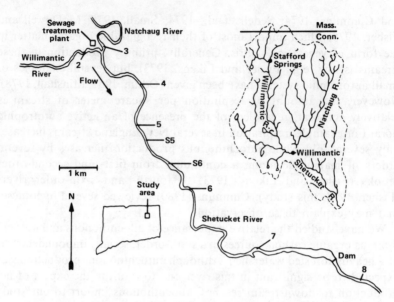

Figure 1. The Shetucket River Watershed (inset) and study area.

Allochthonous Organic Carbon Supplies

The import of organic carbon from upstream was estimated from 42 samples taken at stations 5 and s5 over an annual and diurnal sampling regime. Water for dissolved organic carbon (DOC) analysis was filtered through precombusted ~0.45 μm pore size glass-fiber filters, and aliquots of the filtrate were combusted to CO_2 in a tube furnace at 550°C, in line with an infrared CO_2 analyzer, with 500 μM benzoic acid standards (van Hall et al., 1963). Since concentrations of particulate organic carbon (POC) averaged 10% of the DOC in several adjacent rivers (see discussion), our DOC data were increased by this amount for an estimate of total organic carbon (TOC).

The import of allochthonous TOC to our study area was calculated to obtain units of carbon (kcal) transported over a square meter of river bottom per unit time. This approach allows for comparison with other processes, such as productivity and respiration, which are described in the same terms (e.g., kcal m^{-2} day^{-1}), and avoids problems associated with different size study areas. It is assumed, somewhat erroneously, that (1) the river bottom is flat, and (2) the material transported over it is available for biological processing (i.e., the question is, how much of

Table 1. Characteristics of the Shetucket River*

Watershed area	1330 km^2
Land use (State of Connecticut)	74% mixed deciduous forest
	16% agricultural
	10% residential and urban
River water chemistry	
pH	~6 to 8
Conductivity	50 to 150 μS/cm
Carbonate alkalinity	5 to 25 g/m^3 ($\overline{x}$ = 15)
Turbidity (JTU)	2 to 10 ($\overline{x}$ = 7)
BOD$_5$	0.7 to 5.8 g/m^3($\overline{x}$ = 1.7)
NO$_3$–N	36 to 140 μM
O–PO$_4$	16 to 22 μM
Total suspended solids	0.7 to 250 g/m^3($\overline{x}$ = 26)
Dissolved oxygen	65 to 122% saturation
Dissolved inorganic carbon	160 to 770 μM
Dissolved organic carbon	Station 4: 430 ± 110 μM(N = 39)
	Station 5: 480 ± 120 μM(N = 42)
	Station 6: 480 ± 250 μM(N = 67)
	Station 7: 540 ± 210 μM(N = 22)
River hydrology/morphometry	
Gradient	1.33 m/km
Bottom texture	Fine silt to large boulders
Channel form	Rectangular, with vertical banks
Precipitation	118 cm/yr (45 year $\overline{x}$, 115 cm)
Runoff	~60% of precipitation
Discharge	1.2 to 200 m^3/s($\overline{x}$ = 23)
Velocity	0.25 to 1.0 m/s($\overline{x}$ = 0.72)
Depth	0.1 to 2.5 m ($\overline{x}$ = 0.64)
Width	40 to 60 m ($\overline{x}$ = 50)

*Data are from Klotz, 1977.

mass transport actually encroaches the bottom?). Nonetheless, the data for these units are easily obtained and applied to all lotic systems.

First, the concentration of material (kcal/m^3) is multiplied by river discharge (m^3/s) for an estimate of mass transport (kcal/s). Next, the area is defined as a meter of stream length multiplied by the width for units of a square meter. Mass transport estimates (whole river flux) are divided by the area of this 1-m-long stream segment for units of kilocalories per square meter per second. For example, in a stream where:

TOC = 50 kcal/m^3
Q = 20 m^3/s
W = 30 m

Table 2. Diurnal Study Data, 1974

Item	Mar. 6	Apr. 30	May 21	June 11	July 9	Aug. 6	Sept. 19	Oct. 31*	Dec. 12
Stations	4–6	4–6	4–6	5–6	5–6	s5–s6	s5–s6	s5–s6	4–s6
Length, m	3430	3430	3430	2680	2680	1020	1020	1020	2480
Depth, m	0.77	0.63	0.47	0.29	0.28	0.37	0.46	0.87	0.90
Mean Q, m^3/s	27	20	13	6.2	6.4	4.5	6.1	12–36	49
Flow time, min	82	92	106	105	98	70	62	35	38
Water temperature, C	7	17	18	24	26	24	19	11	2
Insolation†	450	630	1500	2100	1500	3200	980	870	200
O_2 concentration, μM	370–400	280–330	240–310	180–290	210–280	210–310	220–280	330–390	410–420
O_2 % saturation	96–106	91–108	79–106	65–111	79–111	76–122	73–100	94–108	95–100
CO_2 concentration, μM	165–210	230–280	220–260	270–350	240–320	280–360	220–300	230–280	160–175
N each per Station	6	6	10	16	9	12	13	15	8
O_2 reaeration (daytime), kcal m^{-2} light cycle^{-1}	0.79	1.71	2.59	3.64	2.57	2.58	2.25	—	—‡
O_2 diffusion (nighttime), kcal m^{-2} dark cycle^{-1}	0.66	0.84	1.52	1.65	1.45	1.43	0.49	—	—‡
Gross primary productivity (GPP), kcal m^{-2} day^{-1}	5.72	5.75	7.40	15.8	8.31	4.0	3.26	4.77	—‡
Total community respiration (TCR), kcal m^{-2} day^{-1}	10.8	10.8	19.0	22.2	13.5	3.4	4.5	43.9	—‡
Biomass§									
Algae (N = 9)	14.5	11.3	10.6	8.0	16.4	16.7	13.8	21.9	10.0
Insects (N = 22)	2.4	6.5	2.8	5.0	1.8	1.7	2.4	0.4	2.4
Bacteria (N = 27)	0.59 ± 0.94								
Fish (N = 15,000)	2.1								

*During the October diurnal, severe weather caused large changes in many parameters, and the values given are means or ranges or, where dash (-) is shown, were not calculated.

†Values are insolation to the river bottom (kcal m^{-2} day^{-1}).

‡In December there were no differences between up and downstream concentrations.

§Data for bacteria and fish are annual means.

$D = 1$ m
$V = 0.67$ m/s

mass transport per square meter is:

$$[(50 \text{ kcal}/m^3)(20 \text{ } m^3/s)] / [(30 \text{ m})(1 \text{ m})] = 33 \text{ kcal } m^{-2} s^{-1}$$

The same result is obtained with Odum's (1957) velocity times concentration formula. Odum's formula requires a depth adjustment (a dimensionless m/m proportion), but ours does not because the depth term is included in the discharge estimate.

With our calculation method, the size of the study area actually investigated is irrevelant. We define the system as width multiplied by 1 m of length. All material transported downstream in a 1-m-wide river passes over the bottom; thus, e.g., 1/50th of the material passes over the bottom of a river 50 m wide. This assumes, of course, that the depth is constant for the entire width of the river. This is not common, but errors introduced by this fallacy are insignificant in comparison with those resulting from infrequent sampling and qualitative uncertainties about the TOC.

Autochthonous Organic Carbon Supplies

In situ gross primary productivity (GPP) was estimated in nine diurnal studies between March and December 1974 according to the two-station, upstream-downstream oxygen methods described by Odum (1956), with reaeration and diffusion corrections taken from Churchill et al. (1962). A large data pool from several stations was reduced to describe the O_2 rate of change among the stations listed in Table 2. We made a total of 540 measurements of O_2 (field probe and Winkler titrations) and CO_2 (infrared analyzer). For the purposes of this synthesis, we will discuss only the O_2 data obtained from the water column at the stations listed in Table 2. Other CO_2 and O_2 data from benthic and water column light/dark chambers and water column CO_2 data are discussed elsewhere. Several 7-day continuous recordings of O_2 and temperature were made in August and October to examine the photosynthetic response to weekly weather variations.

Autotrophic biomass was composed almost entirely of epilithic, unicelluluar algae (Klotz et al., 1976; Heisey, 1975). Algae were quantified by the rock scraping method of Douglas (1958), and live cells were distinguished with Nomarski optics. The mean dry weight per cell of *Acnanthes deflexa* and *Chlorella* spp. (the two dominant taxa) were obtained from laboratory culture (Sorokin, 1973) and converted to caloric equivalents from data in Cummins and Wuycheck (1971).

Biological Carbon Demand

Estimates of total community respiration (TCR) were calculated according to Odum (1956), and respiration and assimilation were partitioned among several community groups as described in the following discussion.

Algal respiration was assumed to be 40% of GPP (Whittaker, 1975; Likens, 1973). Therefore net primary production (NPP) was 60% of GPP, and algal respiration was subtracted from TCR for an estimate of total heterotrophic respiration. Respiratory estimates for aquatic insects (Olson and Reuger, 1968) and fish (Beamish, 1964) were corrected for seasonal temperature changes and multiplied by biomass data obtained simultaneously by others (below). The sum of algal, insect, and fish respiration was subtracted from TCR, and the remainder was attributed to the bacteria, yeasts, fungi, protozoa, and micrometazoa, hereafter collectively referred to as microbes or bacteria.

The production of heterotrophic biomass by macroorganisms (insects and fish) was estimated by using a respiration-to-assimilation ratio of 0.67. This is a reasonable compromise (range of 0.5 to 0.8) among several published values for invertebrates (Odum, 1956; Edwards, 1973; Wu and Levins, 1978) and fish (Edwards, 1973; Small, 1975). Significant alterations of this ratio for macroorganisms do not affect conclusions about heterotrophic assimilation partitioning or carbon demand.

Estimates of the respiration-to-assimilation ratio for microbes are difficult to obtain and interpret. Under "optimum" laboratory conditions, bacteria may respire between 1 and 100% of their carbon substrate, depending on growth conditions and the type of substrate (Doelle, 1975). In field studies of organic ^{14}C uptake, microbial respiration has ranged from 1 to 44% of assimilation (Hobbie and Crawford, 1969; Iturriaga and Hoppe, 1977; Wiebe and Smith, 1977; Meyer-Reil, 1978). Odum (1957) used a respiration-to-assimilation ratio of 0.91, and we choose 0.50 as a compromise (Payne, 1970). Thus, for the macroorganisms, assimilation is 1.5 times respiration and, for the microbes, 2 times respiration. Subsequently, heterotrophic production was calculated by subtracting respiration from assimilation, ignoring metabolic (but not digestive) excretion.

Bacteria, some fungi, and yeasts in the sediment and water column of the Shetucket River were enumerated by Matson et al. (1978). Dry weights of representative heterotrophic, mesophilic bacteria (*Bacillus*, *Micrococcus*, and *Pseudomonas* spp.) were obtained from log-phase cultures on glass fiber filters and compared with direct counts in a hemacytometer.

Aquatic insects in this system were described by Klattenberg (1974) and

Costello (1976), both of whom collected colonizers of artificial substrata at biweekly intervals. We used their data as first-order estimates of the standing crop biomass for the purposes of this study. The vast majority of colonizing insects belonged to the Tendipedidae and Hydropsychidae. Insect and microbial biomass were converted to calories by using Cummins and Wuycheck's (1971) estimates.

The fishes of the Shetucket watershed have been studied by Goldstein (1975), who identified and weighed over 15,000 specimens. Total numbers and community structure were determined by seining, gill netting, angling, and direct observation (snorkeling). Wet weights of the dominant fish, *Catostomas commersonii* (the white sucker), which made up ~95% of total fish biomass, were converted to dry weight (20%) and to calories (5 kcal/g dry weight) by using data from Small (1975).

Storage of Organic Carbon

The standing crop of biomass and organic carbon in the top centimeter of sediments (U. S. Geological Survey, 1971–1975, CHN analyzer) was combined to estimate total carbon storage between stations. This is probably a significant underestimate since deeper sediments also contain organic carbon that can be reworked by the hyporheos. Also, some sediment is stored and probably decomposed transiently between flood events, with mean residence times less than 1 year.

Export of Organic Carbon

A total of 67 water samples from stations s6 and 6 was analyzed for DOC, and rates of TOC export were calculated in the same way as import rates. A t-test for paired data and an f-test were used to compare variances among upstream and downstream DOC data and to determine whether import and export were equal.

Physical Data

Daily insolation rate data obtained from the University of Connecticut Agricultural Station located about 10 km from our study area were used to calculate surface and benthic light supplies and photosynthetic efficiencies. A submersible photometer was used to measure light transmission to the river bottom at different times of the day and the year.

The energy released as water fell through the study-area gradient was calculated as the mass of water times acceleration due to gravity times the difference in elevation (Hall, 1972). This work done on the river bottom serves to modify and maintain the habitat, scour biomass from rock surfaces, and remove metabolic products.

The energy input resulting in changes in daily water temperatures was calculated by multiplying the predawn water temperature by study-area water volume (Table 2) and subtracting this from the daily maximum water temperature times volume.

Unless otherwise indicated, we converted biological and detrital data into caloric equivalents by using the figures recommended by the International Biological Programme (Winberg, 1971); i.e., 1 g O_2 = 3.6 kcal, 1 g dry weight of organic matter = 4.5 kcal, and 1 g organic carbon = 10 kcal.

RESULTS

Organic Supplies

Allochthonous Import

Rates of TOC import ranged from 0.2 to 6.0 × 10^6 kcal m^{-2} day^{-1}, for an annual total of 8.3 × 10^7 kcal/ m^2 (Figure 2a). At upstream stations s5 and 5, DOC concentrations averaged 480 ± 120 μM, and river discharge ranged from 1.0 to 200 m^3/s (Table 1). At least half of the total river discharge and TOC import occurred between November and March at temperatures generally less than 5° C. In 1971, 1972, 1973 and 1974, 53, 50, 63, and 66% respectively, of total annual discharge occurred in this period. Greater than 85% of the variation in DOC export downstream is attributable to discharge alone. Irregular midwinter thaws often result in short-term high export phenomena.

Autochthonous Supplies

Live algae on the rocks remained at densities of ∼ 10^{11} cells/ m^2 all year (Klotz, 1977). Correspondingly, diurnal changes in concentration of O_2 and CO_2 were observed during all diurnal studies (Table 2). At 2° C in December, however, there was no observed difference between upstream and downstream concentrations. Diffusion and reaeration of O_2 often significantly dampened otherwise large diurnal oscillations in O_2 saturation. For example, in August nighttime reaeration (2.58 kcal m^{-2} dark cycle^{-1}) was 76% of 24-h TCR (3.4 kcal m^{-2} day^{-1}) and daytime O_2 diffusion was 35% of GPP. These percentages were lower during all other diurnal studies.

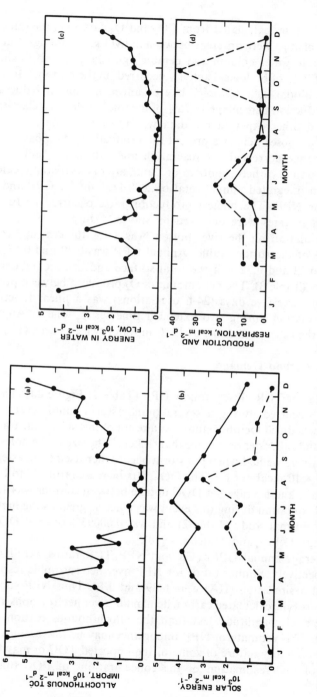

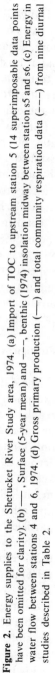

Figure 2. Energy supplies to the Shetucket River Study area, 1974. (a) Import of TOC to upstream station 5 (14 superimposable data points have been omitted for clarity). (b) ——, Surface (5-year mean) and ––––, benthic (1974) insolation midway between station s5 and s6. (c) Energy in water flow between stations 4 and 6, 1974. (d) Gross primary production (——) and total community respiration data (–––) from nine diurnal studies described in Table 2.

Daily GPP rates calculated from CO_2 and O_2 data were the same, but the rates of change were different between individual samplings. Furthermore, separate water column and benthic light/dark chambers showed that at least 95% of biological activity occurred on the bottom. However, a transient bloom of a *Chlorella*-like organism in August reduced the benthic production percentage to 75% of the total. A detailed elaboration of these and other experimental data will appear elsewhere.

For our purposes here, we present the productivity data based on O_2 rates of change corrected for reaeration and diffusion (Table 2). We assumed a photosynthetic quotient of 1 and, after conversion to calories, obtained an integrated annual total of 2000 kcal/m^2 for GPP and 1200 kcal/m^2 for NPP. The highest production rates occurred in June, and relatively high rates were observed in May and July.

Total insolation on the river bottom was 3.8×10^5 kcal m^{-2} yr^{-1}, or about 38% of the surface value. Annual GPP was 0.20 and 0.53% and NPP was 0.12 and 0.32% of the total surface and benthic light supply, respectively (Table 3). The net amount of O_2 produced between predawn minima on successive days (24-h oscillations) was a linear function of surface insolation during two extended time series—one in August at 24° C and the other in October at 12° C in water twice as deep (Figure 3).

Biological Carbon Demand

The data for TCR closely follow GPP (Table 2, Figure 2d). The only exception occurred during a severe thunderstorm runoff event in the regularly scheduled October study. As a result of this storm, the transport of leaves and other debris greater than 1.3 cm^2 (the size of the holes in the only containers we had when the storm began) increased from essentially zero to 29×10^4 kcal m^{-2} h^{-1}. The DOC flux increased from 13 to 17×10^4 kcal m^{-2} h^{-1}, and average net DOC uptake between stations was 4.0 kcal m^{-2} h^{-1}. Respiration during this event was twice as great as the highest of any other diurnal study (Table 2) and was 0.00093% of total DOC and particle flux (Klotz and Matson, 1978).

The average annual GPP/TCR and NPP/TCR ratios were 0.40 and 0.24, respectively. Similar ratios for NPP over heterotrophic respiration (HR) and assimilation (HA) were 0.29 and 0.14. Thus NPP supported about 14% of the calculated carbon demand of river heterotrophs and was five orders of magnitude less than the allochthonous carbon import (Table 3). This estimate of NPP importance may be low because of the possibly refractory nature of some of the imported TOC or may be high because of underestimation of the concentrations of POC (discussed below).

Table 3. Calculated Mean Annual Biological Rates and Efficiencies

Item	Algae	Bacteria	Insects	Fish
Respiration (4955),* kcal m^{-2} yr^{-1}	800	4100	40	15
Assimilation (10283), kcal m^{-2} yr^{-1}	2000	8200	60	23
Production (5328), kcal m^{-2} yr^{-1}	1200	4100	20	8
Photosynthetic efficiency				
GPP, % of insolation on river bottom	0.53			
NPP, % of insolation on river bottom	0.32			
Heterotrophic				
Respiration, % of NPP	—	340	3.3	1.3
Production, % of NPP	—	540	1.7	0.63
GPP, % of TOC import	0.0000024			
NPP, % of TOC import	0.0000014			
Heterotrophic				
Respiration, % of TOC import		0.00050	0.0000048	0.0000018
Production, % of TOC import		0.00075	0.0000024	0.00000091

*Data in parentheses are community totals for the year.

Organic Carbon Storage

The U. S. Geological Survey (1975) data averaged 2400 kcal/m^2 in the top centimeter, and ignition losses at 500° C in another study (Matson et al., 1978) provided an estimate of 2000 kcal. Both are underestimates but represent biomass-to-sediment organic carbon ratios of 0.0044 to 0.021, which agree well with terrestrial data (Alexander, 1961)

Export of Organic Carbon

The DOC concentrations (Table 1) and river discharge (U. S. Geological Survey 1975) at stations 4, 5, and 6 were not significantly different. Therefore import and export of DOC (and calculated TOC) were equal. As in other studies (e.g., Manny and Wetzel, 1973), large diurnal variations in DOC concentrations were observed (up to 50% of the annual mean). In the Shetucket River these variations did not coincide with any known or speculated events, such as diurnal variations in photosynthesis or STP effluent volume.

Physical Energy in Water Flow

Water moving between stations 4 and 6 dissipated about 5×10^5 kcal m^{-2} yr^{-1} (Figure 2c), or 250 times GPP and about 1700 times less than TOC import. Sunlight and water flow function together to cause diurnal temperature fluctuations of up to $5°C$/day in summer, or about 3200 kcal m^{-2} day^{-1} (~ 1.6 times GPP).

DISCUSSION

Organic Carbon Supplies

Autochthonous Productivity

In situ autotrophy of the Shetucket River was greater than that in most other unshaded lotic systems studied. Gross photosynthetic efficiency (0.53%) was similar to that observed in Silver Springs, Florida (0.56%; Odum, 1957), the Piedmont streams (0.28%; Nelson and Scott, 1962), and the southern great plains (0.1 to 2.7%; Duffer and Dorris, 1966). The linear relationship between photosynthetic O_2 production and insolation (Fig. 3) agrees with other data from Florida (Odum, 1957) and Virginia (Kelly et al., 1974). In the Shetucket, benthic algae were always exposed to inorganic forms of nitrogen and phosphorus (Table 1). Most of the algae observed in the plankton were displaced benthic forms, which were often stacked six deep on the rocks (Klotz, 1977). We believe that primary productivity in the Shetucket was mostly limited by available space, rather than light, nutrients, or temperature.

Allochthonous Carbon Import

Organic carbon imported from upstream supplied at least 4×10^5 more material per square meter than autochthonous sources for river hetero-trophs. As the allochthonous material flows through the study area, however, it is spatially available for benthic heterotrophs in each square meter for a time proportional to velocity, concentration, and vertical mixing. Furthermore, much of this lotic DOC may indeed be refractory and, therefore, metabolically unsuitable for heterotrophs (Hynes, 1975);

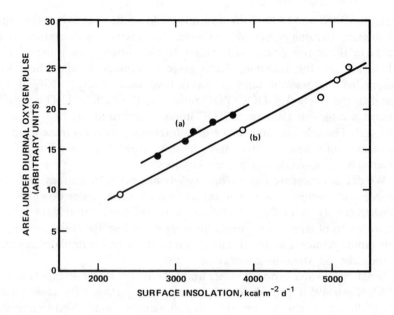

Figure 3. Regression of area under the diurnal O_2 concentration curve on surface insolation for two 5-day continuous O_2 and temperature recordings in (a) August at 24°C at stations midway between s5 and s6, and (b) October at 12°C at station s6. Least-squares linear regressions (a) and (b) have coefficients of +0.99, and the pooled data have a coefficient of +0.96, all significant at $P < 0.01$. Each point is one day.

this results in limits to qualitative availability also. Thus our alloch-thonous supply rates probably greatly overestimate the actual amount subject to assimilation. Even if 99.9% of the imported DOC is "coal," allochthonous sources supplied at least 400 times more organic carbon than the benthic algae. We are not convinced, however, that this much imported carbon is refractory since large diurnal fluctuations may be a result of in situ processes possibly occurring here (Klotz and Matson, 1978) and elsewhere (Manny and Wetzel, 1973).

Our estimates of Shetucket POC concentrations may be low. In the summary of DOC/POC ratios from several watersheds prepared by Moeller et al. (1979), the range was between 0.09 and 70. In two other adjacent eastern Connecticut watersheds, however, the DOC/POC ratios were 12 and 8 (N = 178, Matson and Buck, in review). Both of these watersheds (58 and 18 km^2) are less influenced by man's activities than the Shetucket, and one is relatively pristine. Conversely, the total suspended loads in these two rivers were about tenfold lower than those of the Shetucket. If, for example, the organic percentage of the Shetucket suspended load was the same as that in these other rivers ($\sim 65\%$ of dry weight), the Shetucket DOC/POC ratio would be about 1. This would almost double our estimate of TOC import rates to about 15×10^8 kcal m^{-2} yr^{-1}. The relative importance of autotrophy would decrease proportionally. Unfortunately, thse mathematical projections may not reflect heterotrophic assimilation preferences in situ.

We did not measure input from river banks between stations, but this source may become less important as rivers become wider downstream (Fisher, 1977). Actually, normalization of the bank input data to the entire width of larger rivers probably does not reflect the real behavior of this input. Adjacent material falling into rivers often accumulates in quiet eddies and on streamside snags.

Small headwater brooks (Fisher and Likens, 1973) and springs (Odum, 1957; Teal, 1957) provide examples of lotic systems with identifiable allochthonous input. Indeed, there is no upstream from which to import, although this narrow view separates the stream from its watershed (Davis, 1899; Hynes, 1975). Accrual of groundwater, interflow, and direct runoff progressing downstream combine with autochthonous sources for a diverse dissolved and particulate carbon pool. Organic matter imported to Deep Creek, Idaho, comes from autochthonous material produced upstream (Minshall, 1978) and possibly spiraled (Elwood et al., this volume) into the study area. Despite different origins of the Shetucket allochthonous debris, we consider input from upstream to be a blackbox containing material from many different sources. This includes authochthonous material produced immediately upstream and imported to our study area, even though the production-to-respiration ratios were mostly less than 1.

Organic Carbon Demand

Our TCR data are similar to those from Fort River, which is in a 105-km^2 watershed in adjacent New England terrain (Fisher, 1977). Both of these estimates are only two times greater than that obtained in Bear Brook, despite a 200- to 300-fold increase in autochthonous production (Fisher and Likens, 1973). Table 4 is a comparative summary. Our TCR estimates are probably somewhat low since we have seen evidence of suboxic and anoxic catabolism in some quiet eddies in the Shetucket. Measuring CO$_2$ evolution would give a more direct estimate of carbon

Table 4. Comparison of Three New England Lotic Energy Budgets

System	Watershed area, km²	Mean Q,* m³/s	Allochthonous,† kcal m⁻² yr⁻¹	GPP,‡ kcal m⁻² yr⁻¹	TCR,‡ kcal m⁻² yr⁻¹	GPP/TCR
Bear Brook§ (second order), New Hampshire	1.3	~0.02	50×10^5 (42×10^2)	10 (0.0002)	2040 (0.04)	0.0049
Fort River¶ (fourth order), Massachusetts	105	2.7	23×10^7 (18×10^2)	2700 (0.0012)	5540 (0.0024)	0.50
Shetucket** (sixth order), Connecticut	1330	23	83×10^7 (no data)	2000 (0.00024)	5000 (0.0006)	0.40

*Mean river discharge.
†Calculated by method used in this study. Data in parentheses in this column are meteorologic and subsurface input.
‡Gross primary production and total community respiration as reported. Data in parentheses in these two columns are percentages of allochthonous input.
§Data from Fisher and Likens, 1973.
¶Data from Fisher, 1977.
**Data from this study.

oxidation in habitats such as these, but measurements would be difficult to interpret. The difficulties in the Shetucket River arise from a diurnal oscillation in pH, which shifts the $CO_{2(g)} \rightleftharpoons HCO_3^-$ equilibrium and results in intractable diffusion and reaeration calculations.

Nonetheless, assuming that all Shetucket respiration is oxic, our data provide a community respiratory partitioning of 83% microbial, 16% algae, 0.8% aquatic insects, and 0.3% fish. If we had used algal respiration rates of 20 or 60% or GPP instead of 40%, the bacteria would still be responsible for at least 75% of TCR. Furthermore, if our insect biomass data were low by a factor of 10, insect respiration would amount to a maximum of 10% of TCR. The respiratory partitioning reported here agrees well with data from other New England streams (Fisher, 1977). Aquatic insects, fish, and other macroorganisms turn over insignificant amounts of organic material, but their physical activities, such as particle chewing, shredding, grazing, and habitat modification are important in maintaining and stimulating an active microbial flora (Cummins, 1974). Direct predation and grazing may also stimulate metabolic activity in streams (Porter, 1976). Minshall (1978) contrasts these conclusions with data indicating that invertebrate respiration was greater than microbial respiration at two sites in Deep Creek, Idaho.

At our compromise microbial respiratory efficiency of 0.50, the GPP/HA and NPP/HA ratios are significantly more heterotrophic than the simpler P/R ratio would indicate. If we had used Odum's (1957) microbial efficiency of 0.91, the GPP/HA and NPP/HA ratios would be 0.44 and 0.25, respectively. Actually, many organic ^{14}C uptake data indicate an efficiency closer to 0.1 or 0.2, and these figures would provide a Shetucket heterotrophic assimilation estimate of 20,000 kcal m^{-2} yr^{-1}, or about twice what we report. Thus uncertainties about microbial assimilation in situ significantly affect quantitative and qualitative conclusions concerning the degree of lotic heterotrophy and the importance of autochthonous production.

There are serious problems with all microbial rate measurements (Bull, 1980), and firm statements about lotic trophic dynamics cannot yet be made (McIntire, this volume). If heterotrophic assimilation estimates can be obtained, we suggest using the ratio of NPP/HA rather than the traditional P/R ratio for describing autochthonous deficits in natural communities. The NPP/HA ratio often provides estimates of this deficit that are several-fold greater than those provided by the P/R ratio. This may be extremely important in calculations and projections of lotic capacity for pollutant dissipation, fish production, and expectations for aesthetic improvements. Assimilation includes the entire metabolic carbon demand of the heterotrophic community and is often many times greater than respiration. The importance of assimilation estimates has been made clear in several other expositions (McIntire and Colby, 1978; Howarth and Teal, 1980). Flow and cycling of matter and energy in many communities and ecosystems are complex and poorly understood processes (Pomeroy, 1978; Wiebe, 1978), and lotic systems are no exception. Basic, simple differences in flux calculations, such as those discussed here, may result in very different conclusions about biological rate processes in situ.

ACKNOWLEDGMENTS

We gratefully appreciate discussions with J. D. Buck, P. H. Rich, and F. R. Trainor at the University of Connecticut and A. Kowalczewski of the University of Warsaw, Poland. P. H. Rich generously supplied laboratory space and equipment. Mark Hines, Janice Ibbison, Pat Bubucis, Peter Matson, Loel Meckel, Judy Mickle, Laurie Oracle, and Bob DeGoursey all improved this study.

This work was supported in part by a grant from the Institute of Water Resources, The University of Connecticut, with federal funds provided by the Office of Water Resources, OWRR project No. A-052-CONN. East Carolina University provided travel funds under contract number 90821.

An early version of this paper benefitted from criticisms by J. R. Webster and two anonymous reviewers.

This paper is dedicated to the memory of Fran de Lara.

Contribution No. 130 of the Univeristy of Connecticut Marine Research Laboratory, Noank, CT.

REFERENCES

Alexander, M., 1961, *Introduction to Soil Microbiology*, John Wiley & Sons, Inc., New York.

Beamish, F. W. H., 1964, Respiration of Fishes and Special Emphasis on Standard Oxygen Consumption. II. Influence of Weight and Temperature on Respiration of Several Species, *Can. J. Zool.*, 42: 177–188.

Beck, K. C., J. H. Reuter, and E. M. Perdue, 1974, Organic and Inorganic Geochemistry of Some Coastal Plain Rivers of the Southeastern United States, *Geochim. Cosmochim. Acta*, 38: 361–364.

Brylinsky, M., 1977, Release of Dissolved Organic Matter by Some Marine Macrophytes, *Mar. Biol.*, 39: 213–220.

Bull, A. T., 1980, Biodegradation: Some Attitudes and Strategies of Microorganisms and Microbiologists, in D. C. Ellwood, M. J. Latham, J. H. Slater, J. N.

Hedger, and J. M. Lynch (Eds.), *Contemporary Microbial Ecology*, pp. 107–136, Academic Press, Inc., New York.

Churchill, M. A., R. A. Buckingham, and H. L. Elmore, 1962, *The Prediction of Stream Reaeration Rates*, Tennessee Valley Authority, Division of Health and Safety, Environmental Hygiene Branch, Chattanooga, TN.

Costello, R., 1976, Organic Enrichment of Sediments and Its Effect on Macroinvertebrate Communities with Specific Reference to Aquatic Insects, M.S. Thesis, The University of Connecticut, Storrs.

Cummins, K. W., 1974, Structure and Function of Stream Ecosystems, *Bio-Science*, 24: 631–641.

———, 1980, The Multiple Linkages of Forests to Streams, in R. H. Waring (Ed.), *Forests: Fresh Perspectives from Ecosystem Analysis*, pp. 191–198, Biology Colloquium 40, Oregon State University, Corvallis.

———, and J. C. Wuycheck, 1971. Caloric Equivalents for Investigations in Ecological Energetics, *Mitt. Internat. Ver. Limnol.*, 18: 1–158.

Davis, W. M., 1899, The Geographical Cycle, *Geog. J.*, 14: 481–504.

Doelle, H. W., 1975, *Bacterial Metabolism*, 2nd ed., Academic Press, Inc., New York.

Douglas, B., 1958, The Ecology of the Attached Diatoms and Other Algae in a Small Stony Stream, *J. Ecol.*, 46: 295–322.

Duffer, W. R., and T. C. Dorris, 1966, Primary Productivity in a Southern Great Plains Stream, *Limnol. Oceanogr.*, 11: 143–151.

Edwards, R. R. C., 1973, Production Ecology of Two Caribbean Marine Ecosystems. II. Metabolism and Energy Flow, *Est. Coast. Mar. Sci.*, 1: 319–333.

Fisher, S. G., 1977, Organic Matter Processing by a Stream-Segment Ecosystem: Fort River, Massachusetts, U.S.A., *Int. Rev. Gesamten Hydrobiol.*, 62: 701–727.

———, and G. E. Likens, 1973, Energy Flow in Bear Brook, New Hampshire: An Integrative Approach to Stream Ecosystem Metabolism, *Ecol. Monogr.*, 43: 421–439.

Fogg, G. E., 1977, Excretion of Organic Matter by Phytoplankton, *Limnol. Oceanogr.*, 22: 576–577.

Goldstein, R. M., 1975, Man's Effect on the Fishes of the Upper Shetucket River System, Ph.D. Thesis, University of Connecticut, Storrs.

Hall, C. A. S., 1972. Migration and Metabolism in a Temperate Stream Ecosystem, *Ecology*, 53: 585–604.

Heisey, R. M., 1975, Production and Heavy Metal Concentration of Vascular Aquatic Macrophytes in Three Eastern Connecticut Rivers, M.S. Thesis, The University of Connecticut, Storrs.

Hobbie, J. E., and C. C. Crawford, 1969, Respiration Corrections for Bacterial Uptake of Dissolved Organic Compounds in Natural Waters, *Limnol. Oceanogr.*, 14: 528–532.

Howarth, R. W., and J. M. Teal, 1980, Energy Flow in a Salt Marsh Ecosystem: The Role of Reduced Inorganic Sulfur Compounds, *Am. Nat.*, 116: 862–872.

Hynes, H. B. N., 1963, Imported Organic Matter and Secondary Productivity in Streams, Proceedings of the 16th International Congress on Zoology, Vol. 4, pp. 324–329.

————, 1975, Edgardo Baldi Memorial Lecture: The Stream and Its Valley, *Verh. Internat. Verein. Limnol.*, 19: 1–15.

Iturriaga, R., and H. G. Hoppe, 1977, Observations of Heterotrophic Activity on Photoassimilated Organic Matter, *Mar. Biol.*, 40: 101–108.

Kaushik, N. K., and H. B. N. Hynes, 1971, The Fate of Dead Leaves that Fall into Streams, *Arch. Hydrobiol.*, 68: 465–515.

Kelly, M. G., G. M. Hornberger, and B. J. Cosby, 1974, Continuous Automated Measurement of Rates of Photosynthesis and Respiration in an Undisturbed River Community, *Limnol. Oceanogr.*, 19: 305–312.

King, D. L., and R. C. Ball, 1967. Comparative Energetics of a Polluted Stream, *Limnol. Oceanogr.*, 12: 26–33.

Klattenberg, R. P., 1974, Effects of Primary Sewage Effluent from the City of Willimantic on the Benthic Invertebrate Fauna of the Willimantic/Shetucket Rivers, M.S. Thesis, The University of Connecticut, Storrs.

Klotz, R. L., 1977, *The Effects of Secondarily Treated Sewage Effluent on the Willimantic/Shetucket River*, Report No. 27, Institute of Water Resources, The University of Connecticut, Storrs.

————, and E. A. Matson, 1978, Dissolved Organic Carbon Fluxes in the Shetucket River of Eastern Connecticut, U.S.A., *Freshwater Biol.*, 8: 347–355.

————, J. R. Cain, and F. R. Trainor, 1976, Algal Competition in an Epilithic River Flora, *J. Phycol*, 12: 363–368.

Likens, G. E., 1973, Primary Production: Freshwater Ecosystems, *Human Ecol.*, 1: 347–356.

Manny, B. A., and R. G. Wetzel, 1973, Diurnal Changes in Dissolved Organic and Inorganic Carbon and Nitrogen in a Hardwater Stream, *Freshwater Biol.*, 3: 31–43.

Marzolf, G. R., 1978, *The Potential Effects of Clearing and Snagging on Stream Ecosystems*, Biological Services Program, Fish and Wildlife Service, U.S. Department of the Interior.

Matson, E. A., S. G. Hornor, and J. D. Buck, 1978, Pollution Indicator and Other Microoganisms in River Sediment, *J. Water Pollut. Control Fed.*, 50: 13–19.

McDowell, W. H., and S. G. Fisher, 1976, Autumnal Processing of Dissolved Organic Matter in a Small Wood Land Stream Ecosystem, *Ecology*, 57: 561–569.

McIntire, C. D., and J. A. Colby, 1978, A Hierarchical Model of Lotic Ecosystems, *Ecol. Monogr.*, 48: 167–190.

Meyer-Reil, L.-A., 1978, Uptake of Glucose by Bacteria in the Sediment, *Mar. Biol.*, 44: 293–298.

Minshall, G. W., 1978, Autotrophy in Stream Ecosystems, *BioScience*, 28: 767–771.

Moeller, J. R., G. W. Minshall, K. W. Cummins, R. C. Petersen, C. E. Cushing, J. R. Sedell, R. A. Larson, and R. L. Vannote, 1979, Transport of Dissolved Organic Carbon in Streams of Differing Physiographic Characteristics, *Organ. Geochem.*, 1: 139–150.

Mulholland, P. J., and E. J. Kuenzler, 1979, Organic Carbon Export from Upland and Forested Wetland Watersheds, *Limnol. Oceanogr.*, 24: 960–966.

Nelson, D. J., and D. C. Scott, 1962. Role of Detritus in the Productivity of a Rock Outcrop Community in a Piedmont Stream, *Limnol. Oceanogr.*, 7: 396–413.

Odum, H. T., 1956, Primary Productivity in Flowing Waters, *Limnol. Oceanogr.*, 1: 102–117.

———, 1957, Trophic Structure and Productivity of Silver Springs, Florida, *Ecol. Monogr.*, 27: 55–112.

Olson, T. A., and M. E. Reuger, 1968, Relationship of Oxygen Requirements to Index-Organism Classification of Immature Aquatic Insects, *J. Water Pollut. Control Fed.*, 40: R188–201 (Res., Suppl. 5, Part 2).

Payne, W. J., 1970, Energy Yields and Growth of Heterotrophs, *Ann. Rev. Microbiol.*, 24: 17–52.

Peterson, R. C., and K. W. Cummins, 1974, Leaf Processing in a Woodland Stream Ecosystem, *Freshwater Biol.*, 4: 343–368.

Pomeroy, L. R., 1978, Secondary Production Mechanisms of Continental Shelf Communities, in R. J. Livingston (Ed.), *Ecological Processes in Coastal and Marine Systems*, Marine Science Vol. 10., pp. 163–188, Plenum Press, New York.

Porter, K. G., 1976, Enhancement of Algal Growth and Productivity by Grazing Zooplankton, *Science*, 192: 1332–1334.

Sedell, J. R., F. J. Triska, J. D. Hall, N. H. Anderson, and J. H. Lyford, 1974, Sources and Fates of Organic Inputs in Coniferous Forest Streams, in R. H. Waring and R. L. Edmons (Eds.), *Integrated Research in the Coniferous Forest Biome*, pp. 57–69, Bulletin No. 5, Coniferous Forest Biome, Ecosystem Analysis Studies, International Biological Program.

Small, J. W., Jr., 1975, Energy Dynamics of Benthic Fishes in a Small Kentucky Stream, *Ecology*, 56: 827–840.

Sorokin, C., 1973, Dry Weight, Packed Cell Volume and Optical Density, in J. R. Stein (Ed.), *Handbook of Phycological Methods*, pp. 321–343. Cambridge University Press, New York.

Teal, J. M., 1957, Community Metabolism in a Temperate Cold Spring, *Ecol. Monogr.*, 27: 283–297.

U. S. Geological Survey, 1971–1975, Water *Resources Data for Connecticut* (Water Years 1971 through 1975) Water Data Reports CT-71-1, CT-72-1, CT-73-1, CT-74-1, CT-75-1, Hartford, CT.

van Hall, C. E., J. Safrank, and V. A. Stanger, 1963, Rapid Combustion Technique for the Determination of Organic Substance in Aqueous Solutions, *Anal. Chem.*, 35: 315–319.

Ward, A. K., C. N. Dahm, and K. W. Cummins, 1980, Transformation of Algal Derived Dissolved Organic Material in Northwest Mountain Streams, Abstracts, Annual Meeting, Knoxville, TN., American Society of Limnology and Oceanography, Ann Arbor, MI.

Ward, R. C., 1975, *Principles of Hydrology*, McGraw-Hill Book Company, New York.

Whittaker, R. H., 1975, *Communities and Ecosystems,* 2nd ed., MacMillan Publishing Co., Inc., New York.

Wiebe, W. J., 1978. Anaerobic Benthic Microbial Processes: Changes from the Estuary to the Continental Shelf, in R. J. Livingston (Ed.), *Ecological Processes in Coastal and Marine Systems*, Marine Science Vol. 10, pp. 469-486, Plenum Press, New York.

————, and D. F. Smith, 1977. Direct Measurement of Dissolved Organic Carbon Release by Phytoplankton and Incorporation by Microheterotrophs, *Mar. Biol.*, 42: 213-223.

Winberg, G. G., 1971, *Symbols, Units, and Conversion Factors in Studies of Freshwater Productivity*, International Biological Programme, London.

Wu, R. S. S., and C. D. Levins, 1978, An Energy Budget for Individual Barnacles (*Balanus glandula*), *Mar. Biol.*, 45: 225-235.

SECTION 3

DYNAMICS AND CONTROL OF
SYSTEM COMPONENTS

As indicated by papers in Sections 1 and 2, interesting questions can be asked at the ecosystem level of organization. Yet, in a reductionistic sense, we hope to increase our understanding of components that comprise ecosystems. In this way, we may test whether principles or theories that derive from system level investigations are consistent with mechanistic understandings of individual processes that comprise lotic systems. This section contains papers that focus on one or two components of lotic systems. We have organized them in a sequence that reflects our bias toward aggregation in relation to energy flow or nutrient processing. However, we wish to emphasize relationships among the categories.

PRODUCERS

One question of continuing importance is the relative contribution of allochthonous and autochthonous material to the energy pool available for secondary production in lotic systems. Advective transport complicates the issue; upstream autochthonous inputs are downstream allochthonous inputs. Rogers, et al. compare production by submersed and emergent macrophytes and identify at least two functional groups of macrophytes with regard to turnover of biomass. Similarly, Hill and Webster consider the role of autochthonous production in the organic matter budget of the New River. They point out that the timing, as well as the relative magnitudes of allochthonous and autochthonous production, must be considered when judging the importance of one source versus another. These papers should alert the reader that the prevalent concept of lotic system dependence on external carbon sources may reflect geographical locations of concentrated stream research.

271

CONSUMERS

A major area of research in benthic ecology concerns processes that regulate the structure of benthic invertebrate communities. Peckarsky discusses the concepts of "harsh" and "benign" environments as factors that must be considered along with predation and competition in regulating the structure of benthic insect assemblages. Reice's data suggest that habitat availability overrides vertebrate predation in control of benthic assemblages, although predation might influence patterns of habitat selection and association among resident insects. Ward and Stanford discuss invertebrate diversity in relation to spatial and temporal heterogeneity of lotic environments, especially as influenced by impoundments of major river systems. The observations of Matter et al. suggest that reservoir releases affect invertebrate drift, at least over a 24-hour period including the release.

Problems in measuring secondary production in lotic systems are implied by Allan's assessment of benthic invertebrate consumption by stream fishes. Gatz suggests that optimal foraging behavior is not characteristic of stream fishes and introduces a novel technique for classifying fishes as "searchers" or "pursuers" in their mode of predation.

DETRITUS AND NUTRIENTS

Breakdown, transport and cycling of organic matter downstream continues to receive much attention from stream ecologists. The relative importance of temperature, macroinvertebrates and microorganisms in processing leaf material changes with stream size, according to Paul et al. One extreme is evidenced by Kirby and Webster's research with temporary and permanent streams. Leaf breakdown and POM transport were positively related to abundance of macroinvertebrate shredders. Not surprisingly, shredder abundance was more often greater in the permanent streams that were studied. Fairchild et al. emphasize the role of microorganisms in modifying leaf material and making it palatable to macroinvertebrates. These authors caution that pollutant induced destruction of microbe assemblages could ultimately constrain fish production through food chain relationships that depend on microbial action.

Apart from organic matter processing, nutrient dynamics play a key role in the function of lotic systems. In relation to physical transport, Hill notes that at low-flow conditions, denitrification might significantly deplete stream nitrogen. Elaboration of the relative importance of denitrification in lotic systems requires quantification of process kinetics as discussed by Stammers et al.

13. AQUATIC MACROPHYTE CONTRIBUTION TO THE NEW RIVER ORGANIC MATTER BUDGET

B. H. Hill* and J. R. Webster

Department of Biology
Virginia Polytechnic Institute and State University
Blacksburg, Virginia

ABSTRACT

The contribution of aquatic macrophytes to the energy budget of a 135-km reach of the New River was estimated. Production rates were measured by the harvest method and extrapolated to the entire reach on the basis of measurements of cover made by aerial photography. The estimated macrophyte contribution was compared with measurements of periphyton production and model estimated allochthonous inputs. Macrophytes contributed 13.1% of the total input and 28% of the input generated within the reach. Macrophyte input to the New River trophic dynamics occurs as an autumnal pulse of rapidly decomposed detritus. This pulse forms an important link between spring–summer periphyton production and fall–winter allochthonous-based production.

INTRODUCTION

Recent studies of energy flow in lotic ecosystems indicate that streams are strongly dependent on watershed-derived organic matter (Cummins, 1974; Hynes, 1975; Vannote et al., 1980). However, appreciable in situ production of organic matter can occur under favorable conditions of insolation and nutrient availability (Minshall, 1978). Such conditions are

*Present address: Graduate Program in Environmental Sciences, University of Texas at Dallas, P.O. Box 688, Richardson, TX 75080.

273

likely to be met in higher order streams where shading by riparian vegetation is minimal and nutrient levels are generally high (Vannote et al., 1980). In such streams the ratio of photosynthesis to respiration may be greater than one (Minshall, 1978).

Generally the first producers to appear along the length of a stream system are attached periphyton. As stream size increases, autotrophic production by attached benthic algae often decreases in proportion to contributions by other primary producers. Assuming that planktonic forms are rare in swift-flowing, medium-sized rivers (Hynes, 1970; Wetzel, 1975a), the other important primary producers are aquatic macrophytes. Hynes and Wetzel stated that macrophytes (which include bryophytes, macroalgae, and angiosperms) are, as a whole, poorly adapted to lotic conditions. In spite of this, macrophytes can contribute significantly to energy budgets of some streams. Previous studies have shown that aquatic macrophytes contribute between 1.2 and 30% of stream primary production (Odum, 1957; King and Ball, 1967; Mann et al., 1972; Westlake et al., 1972; Fisher and Carpenter, 1976).

Since aquatic macrophytes are not extensively grazed in most aquatic systems (Westlake, 1965; Fisher and Carpenter, 1976), the only avenues for macrophyte input into stream trophic dynamics are excretion of dissolved organic matter (DOM) by living macrophytes and decay of senescent macrophyte tissue. The excretion of DOM by aquatic macrophytes has been extensively studied in lake ecosystems (e.g., Wetzel, 1975b), but little is known of this phenomenon in lotic ecosysems. Apparently, the major contribution by aquatic macrophytes to stream ecosystems comes via death and decay. Aquatic vegetation has been found to decay considerably faster than terrestrial vegetation (Fisher and Carpenter, 1976; Godshalk and Wetzel, 1978; Hill, 1979). Thus, although autumn-shed tree leaves may be an organic energy supply for many months (e.g., Petersen and Cummins, 1974), macrophytic detritus occurs as an autumn pulse in the energy budget.

The purpose of this study was to estimate the relative contribution of aquatic macrophytes to the organic matter budget of the New River. We hypothesized that aquatic macrophytes, although perhaps only secondary as an annual energy source to streams, may contribute a significant organic matter pulse in late summer and autumn and can provide a readily usable carbon source between high summer production by periphyton and the breakdown of autumn-shed allochthonous litter.

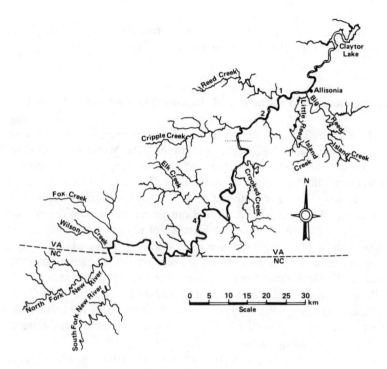

Figure 1. Map of the New River study area. Numbers refer to the sampling locations. The dotted line at the center of the figure separates hardwater (downstream) and softwater (upstream) sections of the river.

METHODS

Site Description

The New River originates in the Appalachian highlands of North Carolina and flows north through Virginia and West Virginia to the Ohio River. It is characterized by a narrow floodplain, steep gradient (2.33 m/km, average), and high velocity (Kanawha River Basin Coordinating Committee, 1971). The river passes through two distinct geologic formations, gneiss and limestone/dolomite, which divide it into soft and hardwater regions. The section of the New River considered in this study extends from the confluence of the North and South Forks of the New River (forming a sixth-order stream) downstream 135 km to Allisonia, VA, at the upper end of Claytor Lake (Figure 1). Average river width in this reach is 167 m, and depths are often less than 1 m. Riparian vegetation covers about 47% of the river bank.

Distribution and Production of Aquatic Macrophytes

The distribution and extent of aquatic macrophyte cover in the study area was determined by aerial photography. The Montana method of 35-mm aerial photography (Meyer and Grumstrup, 1978) was used with Ektachrome daylight color transparency film. The film was exposed on October 16, 1979, at an altitude of 305 m above the river surface. After processing, the slides were projected onto a gridded screen for estimation of percent cover by presence or absence of aquatic macrophytes within the squares of the grid. Total area of macrophyte beds and total river area were determined by measuring these areas on the slides with calibration from U. S. Geological Survey 7.5-minute topographic maps.

Production of *Podostemum ceratophyllum* L., *Justicia americana* (L.) Vahl, and *Potamogeton crispus* L. was determined by harvesting above-ground and belowground biomass at monthly intervals throughout the 1979 growing season. Biomass in 0.25 m^2 plots (0.10 m^2 for *P. ceratophyllum*) was collected (three to five replicates) from four sites (Figure 1), washed, air-dried, weighed, ashed (525°C for 30 min), and reweighed to determine ash-free dry weight (AFDW). Production rates at these sites were determined by differences in biomass on subsequent sampling dates. Losses of biomass caused by physical and biological processes were assumed to be negligible. Data from all four sites were combined to give a single production value for each species to facilitate extrapolation to the whole river.

Periphyton Contributions

Estimates of New River periphyton production were obtained by extrapolating in-stream measurements of ^{14}C uptake by periphyton in the New River at Glen Lyn, VA, 128 km downstream from Allisonia (Figure 1) (Rodgers, 1977). In estimating production from this source, we assumed that periphyton cover was 100% in all areas where aquatic macrophytes were absent and that there were no site differences in periphyton production between Glen Lyn and our study reach. Because of the assumption of 100% coverage, our estimate of the periphyton contribution is undoubtably an overestimate.

Allochthonous Input

Allochthonous particulate organic matter (POM) input as litter fall was estimated by using the New River model developed by Webster et al. (1979). Litter fall was 201.8 g m^{-2} year^{-1} on the stream bank (Hill, 1981), and it decreased linearly to zero at 10 m from the stream bank (Gasith and Hasler, 1976). By solving numerically a partial differential equation relating litter fall to river distance and time, we estimated the upstream and tributary inputs to the study reach and the allochthonous input along the study reach. This estimate of upstream inputs ignores upstream macrophyte and periphyton production. From our observation, ignorance of upstream macrophyte production is probably justified; we have observed few macrophytes in the river upstream from our study reach. We have no information to help us with upstream periphyton production. The model estimate also assumes that allochthonous leaf material is not processed upstream and is, therefore, an overestimate of upstream input. Newbern et al. (1981) estimated that total organic matter transport at a point about halfway through our study reach was 67,400 T/year, of which 24,322 T/year was particulate. This latter value is more than twice the model estimate, 10,962 T/year (see Table 3), which we are using.

Table 1. Mean Monthly Aquatic Macrophyte Biomass in the New River*

Species	June	July	August
Justicia americana			
Aboveground	255.5 ± 111.9	341.5 ± 78.5	447.8 ± 123.4
Belowground	886.9 ± 398.8	1568.6 ± 550.1	2076.7 ± 460.0
Combined	1313.8 ± 328.7	1910.1 ± 615.5	2524.5 ± 515.0
Podostemum ceratophyllum	157.0 ± 50.4	251.8 ± 58.4	318.6 ± 156.5
Potamogeton crispus	350.3 ± 87.9	300.3 ± 94.1	269.2 ± 38.0

*Biomass given in g AFDW/m^2 ± SE.

Table 2. Aquatic Macrophyte Contribution to the New River Study Area

Species	Input, T/AFDW/yr
Podostemum ceratophyllum	1154
Justicia americana	179
Typha latifolia	97
Potamogeton crispus	3
Elodea canadensis	2
Total macrophyte contribution	1435

Table 3. Particulate Organic Matter Inputs to a 135 km Reach of the New River

Source	Input, (T AFDW/yr)	Percent of total input
Allochthonous		
Upstream and tributary	5,893	53.8
Within study area	64	0.5
Autochthonous		
Periphyton	3,570	32.6
Aquatic macrophytes	1,435	13.1
Total POM input	10,962	

Table 4. Breakdown Rates, Sample Size (n), and Coefficient of Determination (r^2) for Five Species of Aquatic Macrophytes in the New River

Species	n	Breakdown rate*	r^2
Podostemum ceratophyllum	26	0.037 ± 0.009	0.74
Elodea canadensis	28	0.026 ± 0.004	0.84
Potamogeton crispus	28	0.021 ± 0.007	0.59
Justicia americana	28	0.016 ± 0.003	0.79
Typha latifolia	28	0.007 ± 0.002	0.64

*Values are rate/d ± SE.

Breakdown of Aquatic Macrophytes

The rate at which aquatic macrophyte organic matter was broken down was measured by the loss of weight from litter bags. Two to five g (air-dried weight) of five species of aquatic macrophytes (*P. ceratophyllum, J. americana, Typha latifolia* L., *P. crispus,* and *Elodea canadensis* Michx.) were placed in nylon mesh bags (15 by 15 cm, with 3-mm octagonal openings). Five bags of each species were placed between two layers of wire mesh to hold the samples to the river bed. Six sets of samples were anchored at each of four sites, and one set was returned immediately to the laboratory to determine handling loss. The others were removed after 2 days and 1, 2, 4, 6, and 8 weeks. Retrieved samples were air-dried, weighed, ashed, and reweighed to determine loss of AFDW. Breakdown rate coefficients were calculated by using linear regression of log-transformed data (Jenny et al., 1949; Olson, 1963). Analysis of covariance (Sokal and Rohlf, 1969) was used to compare breakdown rates.

RESULTS

Aerial photography indicated that aquatic macrophytes covered about 27% (590 ha) of the New River study area. *Podostemum ceratophyllum*, the dominant aquatic macrophyte in the New River, accounted for 25% of the macrophyte cover. Other species measured were *T. latifolia* (1.4%), *J. americana* (0.9%), *P. crispus* (0.03%), and *E. canadensis* (0.03%). Of these species, only *P. ceratophyllum* and *E. canadensis* occurred throughout the study area. *Justicia* and *P. crispus* were restricted to the hardwater section of the river, and *T. latifolia* occurred mostly in two small impounded areas.

Aquatic macrophyte biomass increased rapidly from late spring to midsummer and then appeared to level off (Table 1). Average production rates were: *J. americana*, 23.3 g AFDW m^{-2} day^{-1} (4.7 g AFDW m^{-2} day^{-1} for aboveground biomass only); *P. ceratophyllum*, 3.4 g AFDW m^{-2} day^{-1}; and *P. crispus*, 2.9 g AFDW m^{-2} day^{-1}. Maximum standing crops of these three species were 2500 (450 aboveground), 320, and 300 g AFDW/m^2, respectively. Standing crops for *T. latifolia* and *E. canadensis* were estimated from reported values (McNaughton, 1966; Sculthorpe, 1967; Klopatek and Stearns, 1978) as 2800 (500 aboveground) and 300 g AFDW/m^2, respectively.

The contribution of each macrophyte species to the New River study area was estimated by multiplying the area of coverage by growing season aboveground production or maximum standing crop (*T. latifolia* and *E. canadensis*) (Table 2). Belowground production of *J. americana* and *T. latifolia* was estimated by assuming a belowground biomass turnover of 4.5 years, a rate midway between the values suggested by Westlake (1965) and Sculthorpe (1967). The values in Table 2 can only be considered approximate, especially those for *J. americana* and *T. latifolia*, because of our lack of knowledge concerning belowground dynamics. Because of its wide distribution in the New River, *P. ceratophyllum* was the greatest source of aquatic macrophyte POM, contributing 80% of the macrophyte input. This was followed by *J. americana* (12%), *T. latifolia* (7.7%), *P. crispus* (< 1%), and *E. canadensis* (< 1%) (from Table 2).

Annual periphyton production averaged 0.60 g AFDW m^{-2} day^{-1} (Rodgers, 1977). Extrapolating this value to our study area yielded an estimated organic matter input from this source of 3570 T/year, or roughly twice that of aquatic macrophytes. Upstream and tributary litterfall inputs were estimated to be 5893 T/year, and in situ allochthonous input contributed 64 T/year to our study area (Table 3).

Breakdown of aquatic macrophytes proceeded rapidly at all sites. Weight loss from litter bags was greatest for *P. ceratophyllum*. Since there were no overall site effects ($p < 0.05$), all sites were combined to give an average breakdown rate for each species (Table 4).

DISCUSSION

From our estimates, aquatic macrophytes account for at least 13.1% of the total input of particulate organic matter to our study area on the New River (Table 2). They are responsible for nearly one-third (28%) of the POM generated within the study reach, however (autochthonous production plus direct riparian inputs). We feel that the latter number is more significant for two reasons. First, our estimate of upstream and tributary inputs is an overestimate because it assumes no instream utilization. A large portion of the POM entering the New River upstream of our study area is, in fact, used before it enteres the study area. Second, the material entering from upstream is low quality, partly because of upstream processing but also because terrestrial leaves generally have lower quality than aquatic macrophyte tissue. Because aquatic macrophytes consist mostly of cellulose and other easily degraded compounds, with little lignin (Sculthorpe, 1967), they break down rapidly (Table 4) in comparison with terrestrial leaves (e.g., Petersen and Cummins, 1974).

The timing of the availability of aquatic macrophytes to aquatic food chains is the key to their importance in the energy dynamics of mid-sized streams. Since aquatic macrophytes are not generally used while living, biomass accumulates through the growing season. In autumn, when the plants die, this material is released as a pulse that is rapidly used by aquatic detritivores. Periphyton production occurs throughout spring, summer, and early fall and probably is the most important trophic base during this period. Allochthonous leaf input occurs in fall and is used by detritivores after a period of conditioning (e.g., Barlocher and Kendrick, 1975). Because some leaves condition and breakdown rapidly and others condition and breakdown slowly, there is a continuum of leaf availability lasting through winter and spring (Petersen and Cummins, 1974).

Vannote et al. (1980) speculated that natural stream ecosystems should tend toward a temporal uniformity of energy flow. In this regard Fisher and Carpenter (1976) and Hill (1979) suggested that the autumn pulse of aquatic macrophyte detritus may be the major energy source during the period when periphyton production is decreasing with decreasing insolation and before allochthonous litter input has become important.

Therefore the role of aquatic macrophytes in rivers should be viewed not only with respect to their organic matter pool or annual production but also with respect to the temporal aspects of stream energy budgets.

REFERENCES

Barlocher, F., and B. Kendrick, 1975, Leaf-Conditioning by Microorganisms, *Oecologia,* 20: 359-362.

Cummins, K. W., 1974, Structure and Function of Stream Ecosystems, *Bio-Science,* 24: 631-641.

Fisher, S. G., and S. R. Carpenter, 1976, Ecosystem and Macrophyte Primary Productivity of the Fort River, Massachusetts, *Hydrobiologia,* 47: 175-187.

Gasith, A., and A. D. Hasler, 1976, Airborn Litterfall as a Source of Organic Matter in Lakes, *Limnol. Oceanogr.,* 21: 253-258.

Godshalk, G. L., and R. G. Wetzel, 1978, Decomposition of Aquatic Angio-sperms. II. Particulate Components, *Aquat. Bot.,* 5: 301-327.

Hill, B. H., 1979, Uptake and Release of Nutrients by Aquatic Macrophytes, *Aquat. Bot.,* 7: 87-93.

_____, 1981, Organic Matter Inputs to Stream Ecosystems: Contributions of Aquatic Macrophytes to the New River, Ph. D. Thesis, Virginia Polytechnic Institute and State University, Blacksburg.

Hynes, H. B. N., 1970, *The Ecology of Running Waters,* University of Toronto Press, Toronto.

_____, 1975, The Stream and Its Valley, *Verh. Internat. Verein. Limnol.,* 19: 1-15.

Jenny, H., S. P. Gessel, and F. T. Bingham, 1949, Comparative Study of Decomposition Rates of Organic Matter in Temperate and Tropical Regions, *Soil Sci.,* 68: 419-432.

Kanawha River Basin Coordinating Committee, 1971, *Kanawha River Comprehensive Basin Study.* Vol. I, Main Report, U. S. Department of Agriculture, Washington, DC.

King, D. L., and R. C. Ball, 1967, Comparative Energetics of Polluted Streams, *Limnol. Oceanogr.,* 12: 27-33.

Klopatek, J. M, and F. W. Stearns, 1978, Primary Productivity of Emergent Macrophytes in a Wisconsin Freshwater Marsh Ecosystem, *Am. Midl. Nat.,* 100: 320-332.

Mann, K. H., R. H. Britton, A. Kowalczewski, J. J. Lack, C. P. Matthews, and I. McDonald, 1972, Productivity and Energy Flow at all Trophic Levels in the River Thames, England, in Z. Kajak and A. Hillbrict-Ilkowska (Eds.), *Productivity Problems of Freshwaters,* IBP/UNESCO Symposium, pp. 579-596, PWN Polish Scientific Publishers, Warszawa-Krakow.

McNaughton, S. J., 1966, Ecotype Function in the Typha Community Type, *Ecol. Monogr.,* 36: 297-325.

Meyer, M. P., and P. G. Grumstrup, 1978, *Operating Manual for the Montana 35 mm Aerial Photography System,* Sec. Rev., Remote Sensing Laboratory, College of Forestry and Agricultural Experiment Station, University of Minnesota, St. Paul.

Minshall, G. W., 1978, Autotrophy in Stream Ecosystems, *BioScience,* 28: 767-771.

Newbern, L. A., J. R. Webster, E. F. Benfield, and J. H. Kennedy 1981, Organic Matter Transport in an Appalachian Mountain River, Virginia, U.S.A., *Hydrobiologia,* 83: 73-83.

Odum, H. T., 1957, Trophic Structure and Productivity of Silver Springs, Florida, *Ecol. Monogr.,* 27: 55-112.

Olson, J. S., 1963, Energy Storage and the Balance Between Producers and Decomposers in Ecological Systems, *Ecology,* 44: 322-332.

Petersen, R. C., and K. W. Cummins, 1974, Leaf Processing in a Woodland Stream, *Freshwater Biol.,* 4: 343-368.

Rodgers, J. H., 1977, Aufwuchs Communities of Lotic Systems—Nontaxonomic Structure and Function, Ph.D. Thesis, Virginia Polytechnic Institute and State University, Blacksburg.

Sculthorpe, C. D., 1967, *The Biology of Aquatic Vascular Plants,* Edward Arnold, Ltd., London.

Sokal, R. R., and F. J. Rohlf, 1969, *Biometry,* W. H. Freeman and Co., San Francisco.

Vannote, R. L., G. W. Minshall, K. W. Cummins, J. R. Sedell, and C. E. Cushing, 1980, The River Continuum Concept, *Can. J. Fish. Aquat. Sci.,* 37: 130-137.

Webster, J. R., E. F. Benfield, and J. Cairns, Jr., 1979, Model Predictions of Effects of Impoundment on Particulate Organic Matter Transport in a River System, in J. V. Ward and J. A. Stanford (Eds.), *The Ecology of Regulated Streams,* pp. 339-364, Plenum Press, New York.

Westlake, D. F., 1965, Some Basic Data for the Investigation of the Productivity of Aquatic Vascular Plants, *Mem. Ist. Ital. Idrobiol.,* 18: 229-248.

———, H. Casey, H. Dawson, M. Ladle, R. K. H. Mann, and A. F. H. Marker, 1972, The Chalk Stream Ecosystem, in Z. Kajak and A. Hillbrict-Ilkowska (Eds.), *Productivity Problems of Freshwater,* IBP/UNESCO Symposium, pp. 615-635, PWN Polish Scientific Publishers, Warszawa-Krakow.

Wetzel, R. G., 1975a, Primary production, in B. A. Whitton (Ed.), *River Ecology,* pp. 230-247, University of California Press, Berkeley.

———, 1975b, *Limnology,* W. B. Saunders Co., Philadelphia.

14. PRIMARY PRODUCTION AND DECOMPOSITION OF SUBMERGENT AND EMERGENT AQUATIC PLANTS OF TWO APPALACHIAN RIVERS

John H. Rodgers, Jr., Mark E. McKevitt, Don O. Hammerlund, and Kenneth L. Dickson

Institute of Applied Sciences and Biology Department
North Texas State University
Denton, Texas

John Cairns, Jr.

Center for Environmental Studies and
Biology Department
Virginia Polytechnic Institute and State University
Blacksburg, Virginia

ABSTRACT

Primary production and decomposition rates of macrophytes were estimated by removing sequential standing crop and measuring weight lost over time from in situ litter-bag incubations of plant materials, in the New River, Giles County, Virginia, and the Watauga River, Carter County, Tennessee, respectively. *Heteranthera dubia, Justicia americana, Elodea nuttallii,* and *Typha latifolia* were sampled from the New River (sixth order). *Elodea canadensis* and *Nitella flexilis* were sampled from the Watauga River (fourth order), and *Podostemum ceratophyllum* and *Potamogeton crispus* were collected from both rivers. Distributions were recorded and areal coverage was estimated and mapped for each species.

Standing crops (including roots and rhizomes) ranged from 790.1 g/m^2 ash-free dry weight for *T. latifolia* to 9.8 g/m^2 for *P. crispus* in the New

River. In the Watauga River the maximal standing crops of *P. crispus* and *P. ceratophyllum* were 86.8 and 97.6 g/m², respectively, whereas maximal standing crop of *P. ceratophyllum* in the New River was 12.8 g/m².

Examination of functional loss or decomposition rates revealed two apparent categories: (1) fast decomposition, (containing *E. nuttallii, E. canadensis,* and *P. crispus* (k = 0.12 – 0.21 g g⁻¹ d⁻¹) and (2) a slower category, containing the other species, with *T. latifolia* as the slowest (k = 0.03 – 0.04 g g⁻¹ d⁻¹). Comparing these rates with those reported for allochthonous inputs can further clarify the role of macrophytes in lotic ecosystems.

INTRODUCTION

Production and distribution of macrophyte populations have been extensively studied in lentic aquatic systems, as well as in estuarine systems and tidal and nontidal marshes (Good et al., 1978; de la Cruz, 1973; Keefe, 1972). Relatively little attention has been given to investigating the dynamics and functions of macrophytes in lotic systems, however. Most often emphasis has been placed on changes in lotic systems brought about by resident aquatic plant populations viewed in a negative way. For example, overabundant growths of macrophytes have decreased accessibility of waters for recreational uses (Davis and Brinson, 1976), increased numbers of disease vectors (Van Zon, 1977), altered pH and dissolved ammonia concentrations, influenced temperature gradients (Dale and Gillespie, 1977), destroyed fish populations, interfered with navigation, increased evaporation and slowed flows (Peltier and Welch 1964; Sculthorpe, 1967), increased color content of waters (Novak et al., 1975), and altered the oxygen dynamics of systems (Edwards, 1968; Jewell, 1971). Macrophytes can contribute greatly to an aquatic ecosystem, however, by providing shelter or habitat and eventually food for a variety of fish and other aquatic organisms (Sculthorpe, 1967).

With recent interests in allochthonous inputs into headwater streams and the "detrital dynamic" concept of aquatic ecosystems (Wetzel and Rich, 1973), voluminous information has been accumulated on degradation or processing of plant materials of terrestrial origin. Spurred on by the "river continuum" framework (Cushing, 1976; Cummins, 1974; 1977), numerous investigators have studied allochthonous material input, transport, and processing rates in small woodland streams. Studies of processing or decomposition of autochthonous materials in freshwater have been for the most part limited to lentic systems or estuaries (see, e.g.,

Good et al., 1978). There is considerable evidence that, although lotic ecosystems may function with a detritus-based economy, autochthonous primary production predominates as an energy source on an annual basis for systems that do not have a closed canopy or are not situated in a heavily forested region (Minshall, 1978). Across geographic regions in North America and within a given stream size or order, the relative proportion of autotrophy to heterotrophy varies considerably.

Since Appalachian rivers in the United States are inherently difficult to study because of their size, flow fluctuations, longitudual and cross-channel changes in morphometry, and temporal and spatial dynamics of population distributions, little knowledge exists concerning nutrient fluxes and energy flow. One purpose of this study was to compare the production and relative importance of submerged and emergent macro-phytes of two Applachian rivers—the New River, Giles County, Virginia, and the Watauga River, Carter County, Tennessee. A second objective was to determine the processing or decomposition rates of these aquatic macrophytes to better assess the role of this portion of autochthonous production in these rivers. Aquatic macrophyte processing rates were compared with rates reported for allochthonous inputs to lead to further hypotheses concerning the functional role of the autotrophically derived materials.

STUDY AREAS

Macrophyte samples (Table 1) were collected from a reach of the New River at Glen Lyn, Giles County, Virginia (Figure 1). The New River at Glen Lyn (river mile 95) has an average width of 500 m, is relatively shallow (mean depth, about 1.2m), and contains a series of rapids and pools. River flow ranges from about 28.3 m^3/s (1000 cfs) upward to an annual maximum of about 1416 m^3/s (50,000 cfs), with an average of 142.3 m^3/s (5023 cfs). Flows are controlled by a dam approximately 50 km upstream from this reach. The water is slightly alkaline and carries a considerable supply of nitrogen and phosphorus (Table 2). The substrate in this reach of the New River ranges from cobble and pebbles in riffles to gravel with sand and silt as minor constituents (Hynes, 1970). The average fall in this reach is about 0.4m/km, and the river is sixth order (Strahler, 1957). The river basin is both granitic and sedimentary.

The Watauga River is similar to the New River in alternation of riffles and pools (Figure 2). The study area of this fourth-order stream encompasses a 15-km reach centered around river mile 46. The water has considerable nitrogen and phosphorus (Table 2). The average fall in this

Table 1. Production of Dominant Submerged and Emergent Macrophytes in the
New River, Virginia, and the Watauga River, Tennessee

Species	Maximum standing crop (~NPP), g AFDW/m^2	Standard deviation	Colonized area, % covered	Relative importance rank
New River				
Typha latifolia	790.2	86.9	5	2
Justicia americana	319.3	16.7	18	1
Elodea nuttallii	107.3	11.2	21	4
Heteranthera dubia	106.8	8.1	29	3
Podostemum ceratophyllum	21.6	5.3	24	5
Potamogeton crispus	9.8	2.8	3	6
Watauga River				
Nitella flexilis	201.2	22.5	17	2
Podostemum ceratophyllum	97.6	21.2	41	1
Potamogeton crispus	86.8	4.8	22	3
Elodea canadensis	40.4	4.0	20	4

Table 2. Water Quality Characteristics of the New and Watauga Rivers*

Parameter	New River, 1976 Mean	New River, 1976 Range	Watauga River, 1979 Mean	Watauga River, 1979 Range
Temperature, °C	14.0	0.0–33.0	12.0	2.0–25.0
Dissolved oxygen, mg/l	9.5	5.0–14.0	9.8	6.3–16.0
pH	7.7	7.1–8.0	7.3	6.8–8.5
Alkalinity, mg/l as CaCO$_3$	44	39–54	34	29–71
Hardness, mg/l as CaCO$_3$	64	51–76	35	32–44
Conductivity, μmho/cm	130	90–115	95	70–130
Turbidity, FTU	15	5–90	7	4–68
Total suspended solids, mg/l	8	1–25	6	1–16
Total Kjeldahl nitrogen, mg/l	0.02	0.005–0.04	0.01	0.002–0.02
Nitrate, mgNO$_3^-$/l	0.015	0.01–0.04	0.01	0.005–0.03
Ammonia, mgNH$_4^+$/l	0.04	0.005–0.09	0.01	0.000–0.02
Total phosphate, mg/l	0.1	0.05–0.20	0.1	0.02–0.15
Dissolved ortho-phosphate, mg/l	0.02	0.005–0.04	0.01	0.005–0.02

*Values are means and ranges of monthly samples for the year.

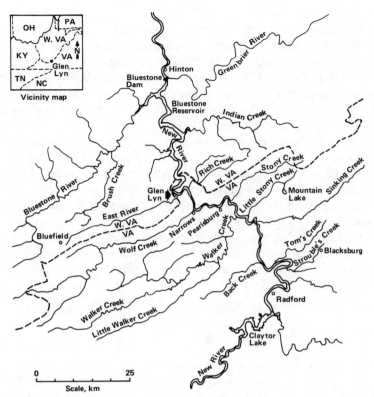

Figure 1. Map of the New River, Virginia, showing sampling site at Glen Lyn, Giles County.

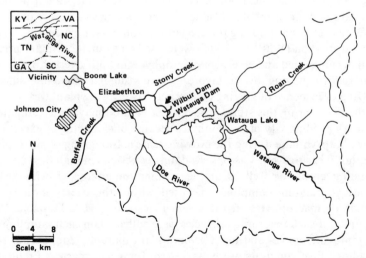

Figure 2. Map of the Watauga River, Tennessee, showing sampling site below Wilbur Dam, Carter County.

reach is about 0.6 m/km. Water is supersaturated with oxygen most of the time. Regulated by upstream hydroelectric dams, flows vary somewhat seasonally and extensively diurnally, ranging from 0.57 m³/s (20 cfs) to 7.25 m³/s (256 cfs). For the most part the Watauga River has a cobbled bottom interspersed with fine sediments. In the river basin, gneisses and schists predominate in the higher elevations, with limestone and dolomite in the valley regions (Rodgers, 1953).

Both watersheds receive runoff from agricultural lands, along with industrial and domestic sewage discharges. There are additional organic and inorganic inputs from urban areas and mining operations. Extensive agriculture in both watersheds constitutes the predominant source of nutrient inputs to these aquatic systems. Because of the dams that act as sediment traps on these regulated lotic systems, the nutrients passed to the downstream reaches are largely in dissolved forms. The sites on the New and Watauga rivers were chosen for study because they appeared to be representative of the rivers and because the sites were readily accessible for sampling.

MATERIALS AND METHODS

After initial reconnaissance and collection of macrophytes from the two rivers, dominant species were identified (Radford et al., 1968; Beal, 1977) and mapped. Dominant species were defined as those occurring more frequently than a single plant per 5-m-wide transect across the river at a sampling location. Both in situ decomposition rates and production estimates for macrophytes (Table 1) were obtained at each site.

Once each month during the growing season, quadrat sampling and the harvest method (Westlake, 1965) were used to estimate primary production of submerged and emergent macrophytes in both rivers. Loss of plant material to grazing by herbivores was considered negligible since no grazing or evidence of grazing on any actively growing plant was observed in the field during this study. The harvest method may underestimate net primary productivity of macrophytes because of sloughing or excretion of new photosynthate during the growing season. Sloughing of macrophytes was not measured but was observed during periods of high flow and peak standing crops, as well as at the termination of the growing season. Stratified random sampling was used within the nearly monospecific stands of macrophytes in the rivers (Westlake, 1969; Dawson, 1976). Samples of each macrophyte species were collected on each sampling date by removing above- and belowground plant material from each 0.25-m² quadrat. Five quadrats were harvested for each species. Preliminary studies of biomass-to-area relationships demonstrated the adequacy of this sample size for estimating standing crops of the macrophytes in the two rivers. Both shoots and roots were collected with a small shovel; collectors used diving masks and snorkels where necessary. Plant material was washed in a 0.3-cm-mesh screen and transferred to a labeled plastic bag. In the laboratory, plants were sorted into roots (including rhizomes) and shoots where appropriate. Wet weight was determined, and sub-samples of plant material were dried to a constant weight at 105°C. Dry weight was determined, and ash-free dry weight (AFDW) was calculated after incinerating at 550°C. For *Podostemum ceratophyllum*, weights were corrected after ashing by removing nonplant material (i.e., pebbles) by screening.

In situ decomposition of the dominant macrophytes of these two rivers was estimated by using nylon net litter bags and measuring weight loss over time of air-dried plant material. Boyd (1970) emphasized that in nature aquatic plants die back gradually; thus decomposition as measured by the litter-bag technique is not a true reflection of how a shoot breaks down. An approximation of the overall relative rate can be obtained, but it may be a conservative estimate (Cummins et al., 1980). Plant material was collected in the fall of the preceeding year at the conclusion of the growing season. The use of litter bags satisfied the basic requirements of containing known weights of plant material, providing a means of placing and securing the plant material in the rivers for a specified period of time, and exposing the plant material to physical, chemical, and biotic factors present in the rivers. Ten grams of air-dried plant material were placed in each bag (bags were 12 by 12 cm, with 3 mm square mesh). Plant material consisted of shoots and leaves, or whole plants of *Podostemum* and *Nitella*. The mesh size chosen was a compromise—it was sufficiently small to prevent large fragments of plants from being washed away and sufficiently large to allow at least a portion of the grazing invertebrates or benthic fauna access to the plant material. Bags were secured in the rivers with steel tie wire. Five bags were removed on each sampling day at 5-day intervals after 20 to 30 days of incubation. One set was removed initially to serve as a transportation control. Litter bags were transported to the laboratory in buckets of water. Weights of plant material remaining were determined, and any associated organisms were preserved for identification and counting. Additional litter bags were incubated with strips of plastic to simulate the habitat presented by the litter bags and plant material in an attempt to assess the role of macroinvertebrates in the decomposition process. The artificial substrates provided information on macroinvertebrate colonization which we could compare with colonization of the various macrophyte tissues.

Assuming that plant-litter weight loss followed an exponential pattern, we calculated instantaneous rate constants (K) as a means of estimating the fractional rate of weight loss in time (see, e.g., Jenny et al., 1949). Computing K involves the negative natural logarithm of the percentage weight remaining after designated periods of time. Calculations and statistical analyses were performed using the Statistical Analysis System (Barr et al., 1976). All tests were performed at the $p = 0.05$ significance level.

RESULTS AND DISCUSSION

Primary Production

Patterns of productivity were similar in the two rivers in that maximum biomass or standing crop was found in September for all macrophytes studied (Figures 3 and 4). The relatively large amount of biomass that remained over winter in the emergents, *Typha* and *Justicia*, followed a pattern previously found by Kvet et al. (1969). The overwintering biomass of the submergents was significantly less. In the New River there was great similarity between the seasonal patterns of production of *Heteranthera* and *Elodea* and between *Podostemum* and *Potamogeton*. In the Watauga River the macroalga *Nitella* (Wood and Imahori, 1965) produced the greatest standing crop. The great fluctuations in river flow and the relatively steep banks probably eliminated emergent vegetation from this river. As noted by Haslam (1978), vegetation and flow are probably in equilibrium, with the species present being those which are tolerant of the normal and extremes of flow. Turbulent flows can abrade and batter leaves and stems, as well as uproot and remove species. Perhaps the greater standing crops of *Podostemum* in the Watauga River than in the New River can be attributed to the presence of larger amounts of cobble substrate. The affinity of *Podostemum* for stable substrates has been noted by other investigators (e.g., Nelson and Scott, 1962). *Potamogeton* appeared to colonize areas where sand predominated in the substrate; however, efforts to correlate standing crops with sediment types were not successful. Macrophytes colonized 42% of the river bottom area in the New River and 34% in the Watauga River.

There were also more predominant species of macrophytes in the New River than in the Watauga River (Table 1). The New River (sixth order) is considerably larger than the Watauga River (fourth order). Haslam (1978) reported that she observed a positive correlation between the number of species and stream or drainage order in studies of North

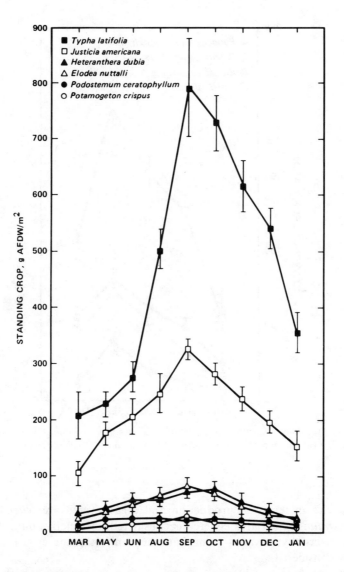

Figure 3. Macrophyte standing crops in the New River. Samples were taken in 1976. Vertical lines represent one standard deviation on either side of the mean (N = 5).

American and English rivers. *Justicia* and *Heteranthera* (emergent and submerged macrophytes, respectively) make a significant contribution to the New River when we consider both production and the proportion of the total colonized area which they cover. In the Watauga River the top

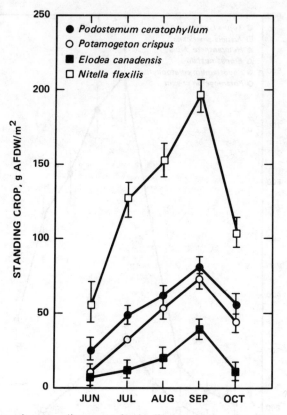

Figure 4. Macrophyte standing crops in the Watauga River. Samples were taken in 1979. Vertical lines represent one standard deviation on either side of the mean (N = 5).

two species in terms of production and area covered are *Nitella* and *Podostemum*. When area covered is considered, the relative importance of the macrophytes varies from that when only standing crop per unit area is considered.

If macrophyte production in these rivers is compared with that reported elsewhere (Table 3), it is apparent that some rivers support greater standing crops. Standing crops in some English rivers are two to three times greater than in these Applachian rivers (Mathews and Kowalczewski, 1969). High winter flows in the New River and great diurnal flow fluctuations in the Watauga River have kept floating-leaved plants from achieving significant standing crops in either lotic system. On the average, emergent vegetation was twice as productive as submergent

Table 3. Estimates of Seasonal Maximal Biomass for Submerged and Emergent Aquatic Macrophytes

Type and genus	Seasonal maximum biomass, g AFDW/m^2	Location	Reference
Submerged			
Berula erecta	395*	River Ivel, England	Edwards and Owens, 1960
Ranunculus pseudofluitans	79–316*	River Ivel, England	Owens and Edwards, 1961
Ranunculus penicillatus	200–380	Bere Stream, England	Dawson, 1976
Potamogeton lucens var. *calcareus*	301*	River Yare, England	Owens and Edwards, 1962
Heteranthera dubia	107	New River	This study
Elodea nuttallii	107	New River	This study
Elodea canadensis	40	Watauga River	This study
Nitella flexilis	201	Watauga River	This study
Podostemum ceratophyllum	22	New River	This study
Podostemum ceratophyllum	98	Watauga River	This study
Podostemum ceratophyllum	93	Middle Oconee River, Georgia	Nelson and Scott, 1962
Potamogeton crispus	87	Watauga River	This study
Potamogeton crispus	8–9	Fort River, Massachusetts	Fisher and Carpenter, 1976
Potamogeton crispus	9.8	New River	This study
Emergent			
Typha latifolia	3712*	Cedar Creek, Minnesota	Bray et al., 1959
Typha latifolia	790	New River	This study
Justicia americana	320	New River	This study

*Calculated from dry weight measurements.

vegetation if sloughing can be discounted. Although measurements were not attempted, sloughing appeared to be significant during periodic high flows. Some plant parts would serve as propagules and perhaps increase the distribution of a species, whereas others would decompose.

Several processes may be important in loss of detritus weight in a litter bag. Processing or decomposition may be the result of a combination of leaching, microbial respiration, fragmentation, and shredding caused by biological and physical activities. Weight of detritus may be increased by colonization and growth of microorganisms. In the New and Watauga rivers, the estimated time interval for 95% loss of the macrophytes studied was from 16 to 100 days (Table 4). The instantaneous rate constant (K) ranged from 0.21 for *Potamogeton* in the New River at 24 to 26°C to 0.03 for *Typha latifolia* in the same river at 11 to 12°C. The temperature dependency of the macrophyte processing rates and agreement with the well known Q_{10} phenomenom indicate that processing is primarily a result of biological activity. Previous values reported for emergent and submerged aquatic plants ranged from 0.0358/d for duckweed (Lambe and Wohler, 1973) to 0.0007/d for *Juncus squarrosus* (Latter and Cragg, 1967). Processing rates measured in this study were somewhat greater than even the rate reported for duckweed. Boyd (1970) reported a processing rate for *Typha* of 0.003/d. In Claytor Lake, an impoundment of the New River, Webster and Simmons (1978) found a rate of 0.01/d for *Typha*, which was somewhat slower than the rates of 0.03 to 0.04/d measured in this study. Physical flow or current may increase fragmentation in the litter bags or increase microbial activity, but Cummins et al. (1980) discounted the role of flow in processing rates. There was no significant difference in processing rates for the same species in the two rivers when similar temperatures are compared.

There is a significant difference in processing rates between the categories of submerged and emergent macrophytes (Table 5). Although available information is scanty, rates in these rivers are greater than those reported for the same species in other aquatic systems. Submerged macrophytes generally lack secondary thickening of cell walls and have lacunae (with the notable exception of *Podostemum*) which would increase their vulnerability to microbial attack (Sculthorpe, 1967). Emergent macrophytes must have sufficient secondary thickening and lignification to support aerial shoots. The mesh size (3 mm) may have contributed somewhat to the rapid processing ratio observed, but Mason and Bryant (1975) found differences in marsh plant decay rates of less than 15% between litter-bag mesh sizes of 4.6 and 0.25 mm.

In comparison with the processing rates reported for fast, medium, and slow categories of allochthonous plant materials by Peterson and

Table 4. Macrophyte Processing Rate (K), Standard Deviation (SD), and Time Interval to 95% Loss in the New and Watauga Rivers

	August to September (24–26° C)			November (11–12° C)			Interval to 95% loss, days(=3/k)
Species	K, $g\ g^{-1}\ d^{-1}$	SD	R*	K, $g\ g^{-1}\ d^{-1}$	SD	R*	
New River							
Potamogeton crispus	0.21	0.011	0.96	0.12	0.014	0.97	25
Elodea nuttallii	0.15	0.009	0.85	0.12	0.010	0.98	25
Podostemum ceratophyllum	0.08	0.010	0.84	0.05	0.008	0.98	60
Justician americana	0.07	0.009	0.99	0.05	0.012	0.95	60
Heteranthera dubia	0.05	0.010	0.93	0.05	0.009	0.95	60
Typha latifolia	0.04	0.008	0.94	0.03	0.007	0.98	100
	August 9–20° C						
Watauga River	K, $g\ g^{-1}\ d^{-1}$	SD	R				
Nitella flexilis	0.19	0.012	0.90				16
Elodea canadensis	0.12	0.008	0.96				25
Potamogeton crispus	0.12	0.014	0.87				25
Podostemum ceratophyllum	0.05	0.009	0.91				60

*The correlation coefficient (R) is highly significant in all cases.

Table 5. Predicted Half-Lives* of Submerged and Emergent Macrophytes

Species	T½ * (0.693/k), days	Location	Reference
Submerged			
Potamogeton crispus	4-6	New River	This study
P. crispus	6	Watauga River	This study
P. pectinatus	35	South African Lake	Howard-Williams and Davies. 1979
Elodea nuttallii	5-6	New River	This study
E. canadensis	6	Watauga River	This study
Podostemum ceratophyllum	9-14	New River	This study
P. ceratophyllum	14	Watauga River	This study
Heteranthera dubia	14	New River	This study
Nitella flexilis	4	Watauga River	This study
Ranunculus penicillatus var. *calcareus*	33	Bere Stream, England	Dawson. 1980
Emergent			
Justicia americana	10-14	New River	This study
J. americana	27-63	Chowan River, North Carolina	Twilley. 1976
Typha latifolia	18-24	New River	This study
T. latifolia	180	Par Pond. South Carolina	Boyd. 1970
T. latifolia	69	Claytor Lake, Virginia	Webster and Simmons. 1978
Rorippa nasturtium-aquaticum	3	Bere Stream, England	Dawson. 1980

*Time interval for 50% loss (T½).

Cummins (1974), macrophyte processing rates we observed are an order of magnitude faster in most cases (Table 6). The fast category of terrestrial plants has processing rates comparable to the slow category of emergent hydrophytes. Rapid growth and turnover of aquatic plant material would indicate that macrophytes are important in slowing nutrient export in the essentially one-way flow of lotic systems (Nelson and Scott, 1962).

Few benthic invertebrates were collected from litter bags in either river. In the New River more macroinvertebrates were collected in the control bags (plastic strips) than in all other bags combined. Snails were apparently feeding on periphyton that colonized the plastic and the bags. A total of thirty-four invertebrates were found—27 Chironomidae, 4 mayflies, and 3 snails. In the Watauga River snails, caddis flies, and midges were found in low numbers (less than 5) associated with litter bags (in and on). Apparently, most of the biological processing of macrophyte litter in these rivers can be attributed to microbial activity. The same observation was made for allochthonous organic materials (leaves) in the New River by Paul et al. (1978).

Table 6. Processing Rates of Allochthonous Plant Materials and Aquatic Macrophytes

	$K, g\ g^{-1}d^{-1}$	$t_{1/2}\ (0.693/K)$, days
Allochthonous*		
Group I, fast	>0.010	<46
Cornus amomum		
Fraxinus americana		
Group II, medium	0.005–0.0010	46–138
Carya glabra		
Salix lucida		
Decodon verticillatus		
Group III, slow	<0.005	>138
Quercus alba		
Populus tremuloides		
Autochthonous		
Submerged, fast	>0.10	7
Potomogeton crispus		
Elodea nuttallii		
Submerged, slow	<0.10	7–14
Podostemum ceratophyllum		
Heteranthera dubia		
Emergent, slow	<0.10	7–24
Justicia americana		
Typha latifolia		

*Data from Petersen and Cummings, 1974.

CONCLUSIONS

On an annual basis, both emergent and submerged macrophytes contribute notably to autotrophic production in the New and Watauga rivers. Their contribution is evidenced by the large portion of the river bottom that they cover, their rate of production, and their contribution to hydrophyte dentrital pools. Rapid and extensive growth of macrophytes tends to conserve or temporarily store nutrients and slow their essentially unidirectional flow. The relatively rapid processing rate of hydrophyte detritus indicated that the plant material had structural and nutrient characteristics that can be used readily and efficiently by the system. Although aquatic macrophytes may not serve as a major source of food while they are actively growing, the energy stored is available as detritus upon scenescence. Macrophyte detritus is processed in a relatively short period of time and, therefore, in all probability, in a short distance in a lotic system. Through this mechanism of rapid growth and turnover, macrophytes can contribute to maximizing the efficiency of lotic systems.

REFERENCES

Barr, A. J., J. H. Goodnight, J. P. Sall, and J. T. Helwig, 1976, *A User's Guide to SAS 76*, Statistical Analysis System Institute, Raleigh, NC.

Beal, E. D., 1977, *A Manual of Marsh and Aquatic Vascular Plants of North Carolina*, Technical Bulletin No. 247, Agricultural Experiment Station, Raleigh.

Boyd, C. E., 1970, Losses of Mineral Nutrients During Decomposition of *Typha latifolia*, *Arch. Hydrobiol.*, 66: 511–517.

Bray, J. D., D. B. Lawrence, and L. C. Pearson, 1959, Primary Production in Some Minnesota Terrestrial Communities for 1957, *Oikos*, 10: 38–49.

Cushing, C. E., 1976, Lotic Ecology: Current Research and New Directions, *Proceedings of the Symposium on Terrestrial and Aquatic Ecological Studies of the Northwest*, Mar. 26–27, 1976, pp. 385–386, Eastern Washington State College Press, Cheney, WA.

Cummins, K. W., 1974, Structure and Function of Stream Ecosystems, *Bio-Sciences*, 24: 631–641.

———, 1977, From Headwater Streams to Rivers, *Am. Biol. Teacher*, 39: 305–312.

———, G. L. Spengler, G. M. Ward, R. M. Speaker, R. W. Ovink, D. C. Mahan, and R. L. Mattingly, 1980, Processing of Confined and Naturally Entrained Leaf Litter in a Woodland Stream Ecosystem, *Limnol. Oceanogr.*, 25: 952–957.

Dale, H. M., and T. J. Gillespie, 1977, Diurnal Temperature Changes in Shallow Water Produced by Populations of Artificial Aquatic Macrophytes, Can. J. Bot., 56: 1099–1106.

Davis, G. J., and M. M. Brinson, 1976, *The Submersed Macrophytes of the Pamlico River Estuary, North Carolina,* Water Resources Research Institute, University of North Carolina, Raleigh.

Dawson, F. H., 1976, The Annual Production of the Aquatic Macrophyte *Ranunculus penicillatus* var. *calcareus* (R. W. Butcher) C. D. K. Cook, *Aquat. Bot.* 2: 51–73.

————, 1980, The Origin, Composition and Downstream Transport of Plant Material in a Small Chalk Stream, *Freshwater Biol.,* 10: 419–435.

de la Cruz, A. A., 1973, The Role of Tidal Marshes in the Productivity of Coastal Water, *ASB Bull.,* 20: 147–160.

Edwards, R. W., 1968, Plants as Oxygenators in Rivers, *Water Res.* 2: 243–248.

———— and M. Owens, 1960. The Effects of Plants on River Conditions. I. Summer Crops and Estimates of Net Productivity of Macrophytes in a Chalk Stream, *J. Ecol.,* 48: 151–160.

Fisher, S. G., and S. R. Carpenter, 1976, Ecosystem and Macrophyte Primary Production of the Fort River, Massachusetts, *Hydrobiologia,* 47: 175–187.

Good, R. E., D. F. Whigham, and R. L. Simpson, 1978, *Freshwater Wetlands— Ecological Processes and Management Potential,* Academic Press, Inc., New York.

Haslam, S.M., 1978, *River Plants,* Cambridge University Press, London.

Howard-Williams, C., and B. R. Davies, 1979, The Rate of Dry Matter and Nutrient Loss from Decomposing *Potamogeton pectinatus* in a Brackish South-Temperate Coastal Lake, *Freshwater Biol.,* 9: 13–21.

Hynes, H. B. N., 1970 *The Ecology of Running Waters,* The University of Toronto Press, Toronto.

Jenny, H., S. P. Gessel, and F. T. Bingham, 1949, Comparative Study of Decomposition Rates of Organic Matter in Temperate and Tropical Regions, *Soil Sci.* 68: 419–432.

Jewell, W. J., 1971, Aquatic Weed Decay: Dissolved Oxygen Utilization and Nitrogen and Phosphorus Regeneration, *J. Water Pollut. Control Fed.,* 43: 1457–1467.

Keefe, C., 1972, Marsh Production: A Summary of the Literature, *Contrib. Mar. Sci.,* 16: 163–181.

Kvet, J., J. Svobaoda, and K. Fiala, 1969, Canopy Development in Stands of *Typha latifolia L.* and *Phragmites communis* Trin. in South Moravia, *Hydrobiologia,* 10: 63–75.

Lambe, H. R., and J. R. Wohler, 1973, Studies on the Decomposition of a Duckweed (Lemnaceae) Community, *Bull. Torr. Bot. Club,* 100: 238–240.

Latter, P. M., and J. B. Cragg, 1967, The Decomposition of *Juncus squarrosus* Leaves and Microbiological Changes in the Profile of *Juncus* Moors, *J. Ecol.,* 55: 465–582.

Mason, C. F., and R. J. Bryant, 1975, Production, Nutrient Content, and Decomposition of *Phragmites communes* Trin. and *Typha angustifolia* L., *J. Ecol.,* 63: 71–95.

Mathews, C. P., and A. Kowalczewski, 1969, The Disappearance of Leaf Litter and Its Contribution to Production in the River Thames, *J. Ecol.*, 57: 543–552.

Minshall, G. W., 1978, Autotrophy in Stream Ecosystems, *BioScience* 28: 767–771.

Nelson, D. J., and D. S. Scott, 1962, Role of Detritus in the Productivity of a Rock Outcrop Community in a Piedmont Stream, *Limnol. Oceanogr.*, 7: 396–413.

Novak, J. T., A. S. Goodman, and D. L. King, 1975, Aquatic Weed Decay and Color Production, *J. Am. Waterworks Assoc.*, 67: 134–139.

Oosting, H. J., 1956, *The Study of Plant Communities*, Freeman and Company, San Francisco.

Owens, M., and R. W. Edwards, 1961. The Effects of Plants on River Conditions. II. Further Crop Studies and Estimates of Net Productivity of Macrophytes in a Chalk Stream, *J. Ecol.*, 49: 199–126.

———, and R. W. Edwards, 1962, The Effects of Plants on River Conditions. III. Crop Studies and Estimates of Net Productivity of Macrophytes in Four Streams in Southern England, *J. Ecol.*, 50: 159–162.

Paul, R. W., Jr., E. F. Benfield, and J. Cairns, Jr., 1978, Effects of Thermal Discharge on Leaf Decomposition in a River Ecosystem, *Verh. Internat. Verein. Limnol.*, 20: 1759–1766.

Peltier, W. H., and E. B. Welch, 1964, Factors Affecting Growth of Rooted Aquatics in a River, *Weed Sci.*, 17: 412–416.

Petersen, R. C., and K. W. Cummins, 1974, Leaf Processing in a Woodland Stream, *Freshwater Biol.*, 4: 343–368.

Radford, A. E., H. E. Ahles, and C. R. Bell, 1968, *Manual of the Vascular Flora of the Carolinas*, The University of North Carolina Press, Chapel Hill.

Rodgers, J., 1953, *Geologic Map of East Tennessee*, Bulletin 58, Part 2, Tennessee Division of Geology, Nashville.

Schulthorpe, C. D., 1967, *The Biology of Aquatic Vascular Plants*, Edward Arnold Publishers, Ltd., London.

Strahler, A. N., 1957, Quantitative Analysis of Watershed Geomorphology, *Trans. Am. Geophys. Union*, 38: 913–920.

Twilley, R. A., 1976, Phosphorus Cycling in *Nuphar* Communities, in M. Brinson and G. J. Davis (Eds.), *Primary Productivity and Mineral Cycling in Aquatic Macrophyte Communities of the Chowan River, North Carolina*, pp. 36–91, Water Resources Research Institute Publication, University of North Carolina, Raleigh.

Van Zon, J. D. J., 1977, Introduction to Biological Control of Aquatic Weeds, *Aquat. Weeds*, 3: 105–109.

Webster, J. R., and G. M. Simmons, Jr., 1978, Leaf Breakdown and Invertebrate Colonization on a Reservoir Bottom, *Verh. Internat. Verein. Limnol.*, 20: 1587–1596.

Westlake, D. F., 1965, Some Basic Data for Investigations of the Productivity of Aquatic Macrophytes, in C. R. Goldman (Ed.), *Primary Productivity in Aquatic Environments*, University of California Press, Berkeley.

———, 1969. Sampling Techniques and Methods for Estimating Quantity and Quality of Biomass of Macrophytes, in R. A. Vollenweider (Ed.), *A Manual on Methods for Measuring Primary Production in Aquatic Environments*, Blackwell Scientific Publications, England.

Wetzel, R. G., and P. H. Rich, 1973, Carbon in Freshwater Systems, in G. M. Woodwell and W. V. Pecan (Eds.), *Carbon in the Biosphere*, AEC Symposium Series, CONF 720510, pp. 241–262, NTIS, Springfield, VA.

Wood, R. D., and K. Imahori, 1965, A Revision of the Characeae Vol. 1, *Monograph of the Characeae*, Verlag Von V. Cramer, Weinheim.

15. BIOTIC INTERACTIONS OR ABIOTIC LIMITATIONS? A MODEL OF LOTIC COMMUNITY STRUCTURE

Barbara L. Peckarsky

Entomology Department
Cornell University
Ithaca, New York

Rocky Mountain Biological Lab
Crested Butte, Colorado

ABSTRACT

The relative roles of biological interactions and physical–chemical factors in structuring benthic stream communities have not been established. It is proposed that the relative importance of physical–chemical factors, predation, and competition depends on the physical harshness and resultant potential secondary productivity of each particular stream. If streams are placed on a gradient from harsh to benign physical conditions, harsher streams may be characterized by unfavorable ranges of diel and seasonal fluctuation of such factors as current, depth, substrate shifts, temperature, and availability of habitable space. Biological interactions may be relatively unimportant if harsh physical conditions eliminate predators and maintain prey populations at low numbers. As streams become more benign, biological interactions may increase in importance as a result of the release of physical limitations on species distributions. Predation may override competition in benign streams where predators are not excluded by physical conditions, where prey have effective defenses, or where prey cannot find spatial refuge from predators. Competition may dominate as a determinant of species distribution within prey refuges, in systems where prey defenses are effective, or within moderately harsh systems where predator populations do not reach levels high enough to exert sufficient pressure on prey populations. Data from a Wisconsin and a Colorado stream are presented in support of this hypothesis.

INTRODUCTION

I have previously reported on experiements designed to determine the role of biological factors, such as benthic density (Peckarsky, 1979b; 1981), food (Peckarsky, 1980a; Peckarsky and Dodson, 1980b), predation (Peckarsky and Dodson, 1980a), and competition (Peckarsky and Dodson, 1980b) on habitat choice by stream invertebrates. These experiments established that biological variables can have a significant influence on benthic community structure in streams. Biological factors were manipulated in cages with nearly constant physical factors, such as temperature, current velocity, and substrate, since investigators have shown that these variables may also be important determinants of distributions of invertebrates in streams (Cummins and Lauff, 1969; Minshall and Minshall, 1977; Rabeni and Minshall, 1977).

A remaining problem is to sort out the relative roles of all these factors in determining the distribution and abundance of stream invertebrates. Stream ecologists have been unsuccessful in solving this problem for a number of reasons. First, most of the evidence available is correlative. Only recently have experimental manipulations allowed us to answer questions of cause and effect. Also, manipulations have generally been simple, testing one factor at a time. More elaborate experiments are needed to test interactions among various factors. Another important restriction is that we suffer from the "my stream" syndrome; most of us concentrate on a very limited range of conditions. Because we do not incorporate a broad spatial perspective into our thinking, we limit our ability to generalize about lotic community dynamics. We also must encompass a broader temporal concept of the stream ecosystem. Processes dominating stream dynamics during one season may differ widely from those during another. It is this temporal and spatial perspective that is lacking in contemporary stream theory.

This paper has two purposes: first, to report the results of experiments conducted during several seasons in a Wisconsin and a Colorado stream to measure the effects of certain physical variables on the colonization of cages by invertebrates and, second, to propose a conceptual model that defines conditions under which physical–chemical and biological factors (predation and competition) may operate to influence invertebrate community structure in streams. The model incorporates data from all experiments conducted in the two streams and from other reports in the literature.

Physical variables tested in this study were chosen on the basis of reports in the literature of their significance in determining distributions of stream benthos and their suitability to manipulation within stainless-steel-screen cages (Peckarsky, 1979b). The specific hypotheses tested were that colonization of habitat by stream benthos differs with distance from the stream bank, current velocity, depth within the substrate, and substrate heterogeneity.

Bishop and Hynes (1969) observed that benthic invertebrates colonized traps by moving upstream within the substrate. Colonization rates were higher adjacent to the stream banks in winter and in more midstream areas in the summer. Differences were not statistically significant, however. Hayden and Clifford (1974) noted the migration of the mayfly *Leptophlebia cupida* along the stream banks during all seasons, presumably because of lower current velocity. Elliott (1971) also found upstream movement to be highest near stream banks where discharge was low and stones smaller. Cummins (1964) observed that some caddis flies inhabit stream margins during early instars and migrate to center stream in later life stages because of changes in food requirements and case-building materials.

Allen's paradox (Allen, 1951) that trout consumed more invertebrates than the production measured by biologists sparked a quest for missing benthos. The hyporheic zone has been suggested as a potential reservoir of invertebrates not detected by conventional surface-sampling techniques (Bishop, 1973; Hynes, 1974; Williams and Hynes, 1974). Williams and Hynes measured maximum benthic density at a depth of 10 to 20 cm within the substrate of the Speed River, Ontario. Poole and Stewart (1976), however, found maxima at 0 to 10 cm in the Brazos River, Texas. Bishop and Hynes (1969) observed higher colonization of invertebrates at lower levels in the substrate (9 cm deep), whereas others (Hayden and Clifford, 1974) found upstream migration to occur at the substrate surface.

The quality of the substrate has been tested directly as an important determinant of habitat selection by stream invertebrates. Substrate type (Hildrew et al., 1980; Thorup, 1966), particle size (Reice, 1980; Williams and Mundie, 1978), and heterogeneity (Hart, 1979; Williams, 1980) have been documented as causal factors of the distribution and abundance of stream benthos.

MATERIALS AND METHODS

Sites

Experiments were conducted in two third-order streams, Otter Creek, Sauk County, Wisconsin, and the East River, Gunnison County, Colorado, from 1976 to 1978. Dates of experiments, ranges of current velocity, water depth, and maximum/minimum water temperatures are shown in Table 1 for each trial. The substrate at the sites of both streams consisted of coarse cobble material interspersed with finer stones and gravel. Otter Creek substrate was generally more heterogeneous, with particle sizes ranging from fine sand to boulders 1 m in diameter. The high-elevation (3100 m) East River is a generally larger stream, but sampling sites were chosen within an area where the stream divided into two or three smaller channels with depth and width similar to Otter Creek. Otter Creek receives allochthonous input from extensive deciduous riparian vegetation, whereas the East River, sparsely bordered by willows (*Salix* spp.) and various conifers, receives less detrital input. (See Peckarsky, 1979a; 1979b; 1980a, for more complete descriptions of the streams.)

Experimental Design

Transects

Cages were buried along perpendicular stream transects on four dates in Otter Creek and two dates in the East River (Table 1). Four cages were buried at each transect at similar depth within the substrate and at approximately 0.5-m intervals from the stream bank to center stream and covered with approximately 5 to 10 cm of substrate material. Twelve cages along three transects were used in all trials, except the last trial in each stream, for which 16 cages allowed four replicate transects. Transects were at least 18 m apart.

Substrate was standardized among cages as follows: All cages received 15 stones, four <5 cm, six <7.5 cm, three <10 cm, and two <12.5 cm in largest diameter (determined by passing them through graduated hose clamps). Texture classes were qualitatively determined according to the ratio of smooth to rough faces (>1, smooth; approximately 1, intermediate; and <1, rough).

Table 1. Physical Data on the Two Streams During Trials

Site and Date*	Trial	Number of cages buried	Current,† cm/s	Depth,‡ cm	Temperature,§ °C
Otter Creek Oct. 29–Nov. 1, 1976	Transect	12	0.18–0.21	13.2	5.6/3.3
May 23–26, 1977		12	0.29–0.24	11.5	19.4/15.0
Oct. 7–11, 1977		12	0.23–0.36	17.8	11.1/7.2
May 26–30, 1978		16	0.45–0.44	18.2	18.3/15.0
East River Aug. 1–5, 1977		12	0.22–0.22	9.4	17.7/7.2
Aug. 2–5, 1978		16	0.33–0.27	13.3	15.0/4.4
Otter Creek Sept. 17–20, 1976	Depth	12	0.18–0.18	5.0	14.4/11.7
Apr. 29–May 2, 1977		12	0.41–0.37	13.9	14.4/10.0
Sept. 30–Oct. 4, 1977		12	0.26–0.26	8.8	7.8/13.3
Apr. 24–28, 1977		16	1.10–0.88	25.2	6.1/12.8
East River July 20–23, 1977		12	0.29–0.36	1.5	18.3/10.0
July 24–27, 1978		16	0.81–0.65	8.9	15.0/5.0
Otter Creek June 20–23, 1977	Substrate	12	0.18–0.21	14.0	17.8/13.3
Sept. 27–30, 1977		12	0.21–0.24	16.3	15.0/11.7
June 6–9, 1978		16	0.24–0.20	14.5	16.7/11.2
East River Aug. 14–17, 1977		12	0.20–0.23	6.3	18.9/11.1
July 6–9, 1979		16	1.64–1.32	30.6	12.8/3.3

*Burial and retrieval dates.
†Average current velocity on burial and retrieval dates.
‡Average depth to top of upper cage.
§Maximum/minimum water temperature over sampling period.

Depth

Pairs of cages were buried stacked vertically within the substrate of both streams (Figure 1). Six replicate pairs were used for all except the last trial in each stream, for which eight replicate pairs of cages were buried. The top cage was flush with the surface of the substrate at a depth of 0 to 10 cm; the lower cage was at a depth of 10 to 20 cm beneath the surface of the substrate. Substrates were standardized as described above.

Substrate

Pairs of directly adjacent cages were buried. One of each pair contained the heterogeneous substrate described; the other contained 10 10-cm stones of intermediate texture. Cages were covered with 5 to 10 cm of natural substrate.

All cages were oriented to receive active colonizers walking or swimming upstream and were buried for 3 to 4 days (see Table 1), a duration chosen to maximize numbers of trails and yet allow adequate numbers of colonizing insects for statistical comparison between experimentals and controls at each date (Peckarsky, 1979b). The short-term nature of these experiments precludes investigation of longer-term effects. Current velocity was measured at each cage with a Marsh-McBirney model 201 current meter, and water depth to the top of the upper cage was recorded.

Statistical comparisons between paired replicates were made with a Wilcoxon Sign Rank Test (depth and substrate), and Spearman Rank

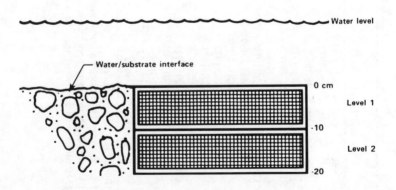

Figure 1. Schematic representation of cage orientation for depth experiment.

Correlation coefficients were calculated to determine associations between number of invertebrates per cage and distance to the closest stream bank or current velocity (transects).

RESULTS

Transects

No association was shown between total number of invertebrates per cage and distance from the stream bank for fall or spring trials in Otter Creek or for summer 1978 trials in the East River. In the summer of 1977, however, a larger number of invertebrates colonized cages in the center stream than cages closer to the banks (Figure 2a). Analysis of 22 common taxa in Otter Creek and 14 in the East River showed that no Otter Creek taxa differentially colonized cages at different distances from the stream bank on any dates (Table 2). *Cinygmula* sp. (Figure 2b) and Turbellaria (Figure 2c) preferentially colonized cages in the center of the East River during the summer of 1977.

Total invertebrates and 22 individual taxa also showed no differential colonization of cages at different current velocities for any trial in Otter Creek. During the summer of 1977, total invertebrates (Figure 3a) and *Cinygmula* sp. (Figure 3b) preferentially colonized cages at higher current velocity. Chloroperlidae spp. preferentially colonized cages at lower current velocities (Figure 3c). No such associations were shown for summer 1978 trials in the East River. Spearman Rank Correlation Coefficients were also calculated on the current velocity vs. distance to nearest stream bank for each trial. No significant associations were obtained.

Depth

The median number of invertebrates colonizing cages at the two different depth strata in both streams is shown in Figure 4. There was no significant difference between the number retrieved at 0 to 10 and 10 to 20 cm depths for any trial in Otter Creek, and there were no consistent seasonal trends (Figure 4a). However, more total numbers of invertebrates colonized the upper cage (0 to 10 cm) during both summers in the East River (Figure 4b, $P < 0.05$). Table 2 summarizes the data for 23 taxa in Otter Creek and 18 taxa in the East River. Several species showed preferences for the upper or lower strata in both streams. For example, the perlodid stonefly *Isoperla cotta*, the ephemerellid mayfly, *Ephemerella subvaria*, the limnophilid caddisfly, *Pycnopsyche* sp., and the black

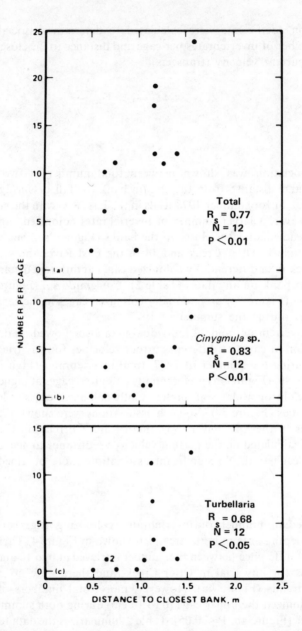

Figure 2. Numbers of invertebrates colonizing cages at different distances from the stream bank. East River, summer 1977.

Table 2. Taxa Analyzed Separately for Transect and Depth Experiments

Species	Transects†	Depth§ Number at 0-10 cm	Number at 10-20 cm
Otter Creek			
Baetis phoebus	+	19	12
Ephemerella subvaria	+	19**	6
Heptagenia hebe	+	1	15***
Stenonema fuscum	+	11	33***
Paraleptophlebia sp.	+	14	24
Paracapnia angulata	+	2	6
Amphinemura delosa	+	10	21
Acroneuria lycorias	+	2	6
Isoperla cotta	+	58*	33
Taeniopteryx nivalis	+		
Micrasema rusticum	+	26	24
Glossosoma sp.	+	6	1
Hydropsychidae slossonae	+	16	23
Lepidostoma sp.	+	8	17
Mystacides sp.	+	4	2
Ocecetis sp.		5	5
Pycnopsyche sp.	+	17*	4
Limnephilidae spp.			
Optioservus fastiditus	+	9	6
Nigronia sericornis		2	4
Sialis sp.	+	1	2
Atherix variegata	+		
Chironomidae spp.	+	45	43
Prosimulium tuberosum	+	40**	6
Antocha sp.		2	5
Tipula sp.		4	6
Minnows	+		
Total		355	348
East River			
Baetis bicaudatus	+	57	15
Ephemerella infrequens	+	6	2
Cinygmula sp.	+	138*	60
Epeorus longimanis		5	0
Rhithrogena hageni		4	2
Paraleptophlebia vaciva		0	4
Ameletus velox	+	20	10
Chloroperlidae spp.	+	29	74**
Zapada haysi	+	3	1
Kogotus modestus	+	16*	8
Pteronarcella badia	+	4	6
Arctopsyche grandis	+	1	0
Rhyacophila tucula		0	1
R. valuma		4	0
Heterlimnius sp.	+	4	2
Chironomidae spp.	+	22	15
Prosimulium sp.	+	35**	4
Hydrachnellae spp.		2	3
Turbellaria spp.	+	9	19
Total		353*	241

†A plus (+) indicates that taxa were analyzed separately in transect experiments.

§Asterisks represent conventional levels of significance. Absence of data indicates that taxa were not individually analyzed for depth experiments.

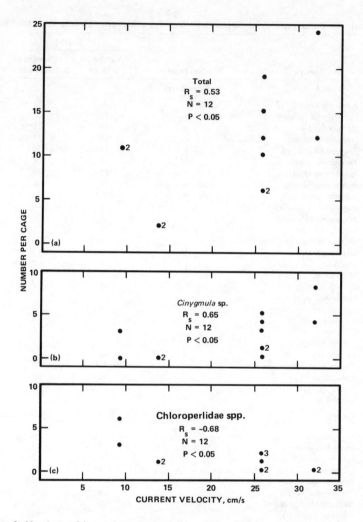

Figure 3. Numbers of invertebrates colonizing cages at different current velocities, East River, summer 1977.

fly, *Prosimulium magnum* preferred the upper cages, and the heptageniid mayflies *Heptagenia hebe* and *Stenonema fuscum* preferred the lower cages in Otter Creek. In the East River the perlodid stonefly *Kogotus modestus*, the heptageniid, *Cinygmula* sp., and *Prosimulium* sp. colonized upper cages significantly more abundantly, and Chloroperlidae spp. appeared in greater numbers in lower cages. Others, such as the baetid mayfly *B. bicaudatus* in the East River showed nonsignificant trends

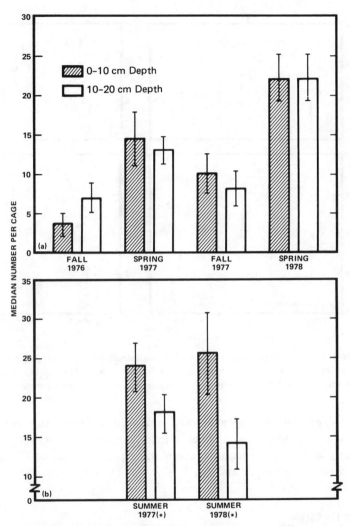

Figure 4. Number of invertebrates ($\bar{x} \pm$ SE) colonizing cages at different depths in the substrate. (a) Otter Creek. (b) East River. Asterisk (*) indicates $P < 0.05$ level of significance.

toward one stratum; a few were found in almost equal numbers in upper and lower cages (see Table 2).

Substrate Heterogeneity

As shown in Figure 5, most trials produced no significant differences in median species richness and median numbers of individuals between

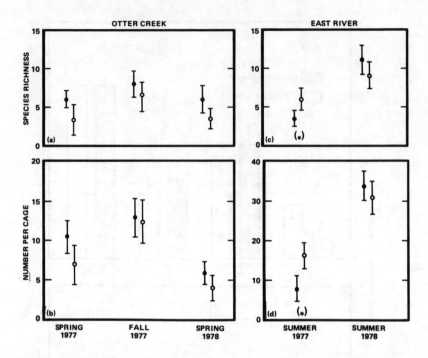

Figure 5. Species richness (a)(c) and median number of individuals (b)(d) per cage ($\tilde{x}$ ± SE) with homogeneous and heterogeneous substrates. ●, Homogeneous substrate. ○, Heterogeneous substrate. Asterisk (*) indicates $P < 0.05$ level of significance.

paired cages with homogeneous and heterogeneous substrates. However, both species richness and number of individuals were significantly higher in cages with heterogeneous substrate during the summer of 1977 in the East River (Figures 5c and 5d, $P < 0.05$).

DISCUSSION

The physical factors tested did not show any overwhelming effects on the colonization of substrate-filled cages in either stream. The only factor that significantly influenced Otter Creek benthos was depth within the substrate. Although total numbers of invertebrates were nearly equal in upper and lower cages, several species consistently appeared in one stratum or the other. *Isoperla cotta*, a somewhat stout stonefly predator; *E. subvaria*, an herbivorous mayfly; *Pycnopsyche* sp., a very large leaf shredder; and *P. magnum*, the filter-feeding black fly larva, preferred the upper cages. These insects might be restricted to upper substrate strata by their large size or by food requirements. Simuliids, for example, attach to the substrate surface and filter seston from the water column. Large leaf material and periphyton are more abundant at the substrate surface for leaf shredders and grazers, such as *Pyncnopsyche* sp. and *E. subvaria*. The dorsoventrally flattened heptageniids, *H. hebe* and *S. fuscum*, inhabit interstitial spaces quite easily and would be expected to reside on unexposed substrate as long as current and oxygen conditions were adequate (Wiley and Kohler, 1980).

Physical parameters also did not produce consistent effects on colonization of East River benthos in summer 1978 trials. The only factor having significant influence was depth within the substrate. More total numbers and numbers of several individual taxa appeared in the upper cages than in the lower cages. The large stonefly predator *K. modestus* may have been restricted from foraging at lower levels by its size. *Baetis bicaudatus* and *A. velox* tended to colonize upper cages more readily, although differences were not statistically significant. Both mayflies are excellent swimmers (Gilpin and Brusven, 1970), and baetids have been documented to remain on exposed substrate surfaces (Wiley and Kohler, 1980) and commonly enter the water column (Hughes, 1966; Peckarsky, 1980b). Simuliids, again, were found almost exclusively in the upper cages because of the nature of their filter-feeding habits. *Cinygmula* sp., a dorsoventrally flattened mayfly, might be expected to inhabit deeper strata, like the Otter Creek heptageneids. Gilpin and Brusven (1970) reported this mayfly to be most abundant in moderate to fast riffles, however, and to be absent from mud and silt substrates. Perhaps its respiratory requirements restrict it to upper substrate levels.

The only trials in which physical effects were consistently important determinants of species distributions and abundances were those of summer 1977 in the East River. Very low snowfall during the winter of 1976–1977 resulted in drought conditions in Colorado. The East River had extremely low water depth and current velocity and unusually high temperatures during the following summer (Table 1) in comparison with all other summers in my experience (Peckarsky, 1979a).

Invertebrates colonized cages in center stream and at higher current velocities than those at the periphery of the stream or at low current velocities. Benthos concentrated at the upper substrate strata in preference to 10- to 20-cm depths within the substrate. A significantly greater number of species and individuals were recovered from heterogeneous substrates as opposed to homogeneous substrates. These results suggest that suitable habitat space may have been limiting during the summer of 1977.

The high-temperature–low–flow regime could have produced respiratory stress for invertebrates adapted to cooler, more rapid current conditions. Higher densities of benthos would be expected in center stream, at high current velocity, and in upper cages under such potential stress. Organisms should choose their habitat to minimize respiratory stress (Wiley and Kohler, 1980). Presumably, during "normal flow" conditions in other seasons, the cages along the entire stream transect and range of current velocities tested provided adequate oxygen conditions for the benthos.

The results of the substrate experiments were not consistent with those of most other similar studies, except in the summer of 1977. Williams (1980) showed that a significantly higher number of taxa colonized heterogeneous than homogeneous substrates. Wene and Wickloff (1940) demonstrated that the number of insects on heterogeneous substrates increased by 26%. Hart (1979) concluded that spatial heterogeneity or substrate complexity provide greater resource availability and lead to increased numbers of species and individuals. Wise and Molles (1979), however, found greater numbers of species on small gravel and greater numbers of individuals on large gravel than on a heterogeneous mixture of the two sizes.

Reice (1980) and Hart (1979) suggested that the available surface area of substrate is an important variable determining the resultant diversity of stream invertebrate communities. I did not determine surface area of substrates used in this experiment. It is reasonable to assume that the intermediate-textured stones of homogeneous size (10 cm) offered a surface area comparable to that of the heterogeneous substrate. Alternatively, the short-term experiment might not have allowed enough time to detect substrate preferences. The inconsistency of the results for summer 1977 might, again, be due to peculiarities of the current–temperature–depth regime and the interaction with the substrates provided rather than to the substrate alone.

THE MODEL

An attempt was made to incorporate the data from these experiments and others on biological interactions conducted in both streams into a general model of lotic community structure. The basic idea is not original in ecology; it was developed by Menge (1976) in interpreting data on the organization of the marine rocky intertidal invertebrate community. The high intertidal is a physically harsh habitat tolerated by few species and is structured directly by physical factors, such as wave action and frequent desiccation. The low intertidal is relatively benign, supporting large populations of starfish predators, whose feeding constitutes the dominant influence on prey community structure. In the middle intertidal zone, conditions become sufficiently harsh to limit populations of starfishes, and thus prey populations are effectively released from regulatory control by predators. This zone constitutes a refuge from predation but allows prey populations to increase to a level where resources (space) are limited; thus competitive interactions become the dominant process organizing the invertebrate community.

Can such a model be applied to streams? Results of experiments on factors determining invertebrate distributions in the two widely differing stream systems tested here showed that the importance of invertebrate predation, competition, and physical factors was not the same in both streams (Peckarsky, 1979a; Peckarsky and Dodson, 1980b). I have attempted to compare conditions of relative harshness between the two streams to determine whether the concept developed by Menge has application to stream ecosystems.

I identified several problems in applying a harsh-to-benign gradient to stream ecosystems. Streams, unlike the marine rocky intertidal, do not exist in a spatial or temporal continuum from harsh to benign physical conditions. Hypotheses become very difficult to test experimentally, since we cannot readily identify conditions that are harsh to stream invertebrates. The following is a working definition of a harsh habitat; it will be subject to subsequent modification, but it provides potential schemes for ranking habitats in terms of a continuum of harsh-to-benign conditions for comparative purposes. "Harsh" is a set of physical–chemical conditions that impose physiological problems for many stream invertebrates. These might include manmade perturbations, such as acid mine drainage, channelization, and stream regulation, to which many species have not evolved adaptive mechanisms. Natural harshness may be imposed by seasonal and diel fluctuations that are unpredictable or hazardous to stream species or by extreme lack of seasonal fluctuations, as in a spring ecosystem, which presents problems to insects whose life cycles depend on proximal cues involving temperature fluctuations. Therefore a benign stream is not necessarily one lacking in disturbance or fluctuations but one that could potentially support a highly productive consumer community. Alternatively, streams may be placed relatively on a two-dimensional scale, such as oligotrophic ↔ eutrophic, tolerable ↔ intolerable, moderate ↔ erratic or extreme.

Comparative physical–chemical and biological data measured during 3 years of study of the East River and Otter Creek are presented qualitatively in Table 3. They support the contention that the East River can be ranked as a harsher stream than Otter Creek, by the previous definition.

Table 3. Comparisons between the East River, Colorado, and Otter Creek, Wisconsin

Parameter	East River, Colorado	Otter Creek, Wisconsin
Elevation	3100 m	300 m
Substrate	More homogeneous	More heterogeneous
Current velocity	Higher absolute, more seasonal and diel fluctuation	Lower absolute, less seasonal and diel fluctuation
Temperature	More diel fluctuation	Less diel fluctuation
	Similar seasonal fluctuation	
Depth, width	More seasonal and diel fluctuation	Generally smaller and shallower, less fluctuation
Habitable space	More temporal fluctuation	Less temporal fluctuation
Abiotic fluctuations	Greater	Fewer
Environmental predictability (?)	Less	Greater
Food, primary productivity	Low allochthonous input, autotrophic, more oligotrophic	High allochthonous input, heterotrophic, more eutrophic
Biotic diversity	Lower	Higher
Density of invertebrates (?)	Higher	Lower
Invertebrate predators	Fewer species, less dense	More species, more dense
Vertebrate predators	Four trout species	Many—trout, minnows, amphibians, snakes

This ranking is based on relative biological productivity, as well as physical–chemical factors (Sander, 1968). Question marks appear when ranking is uncertain. (More complete descriptions appear in Peckarsky, 1979a; 1979b). Similar comparisons have been made between temperate and tropical streams (Stout and Vandermeer, 1975; Fox, 1977).

Otter Creek provides suitable habitat for a wide range of invertebrates. Predation by large stoneflies has been shown to exert a significant influence on benthic distributions in this stream (Peckarsky, 1980b; Peckarsky and Dodson, 1980a; 1980b). In an interactive experiment to test the relative effects of predation and competition, access to cages by predators eliminated or reduced to insignificant the effect of the presence of competitors on colonization by mayflies. Only in cages offering a refuge from predation (by restrictive mesh size) was a competitive effect measured (Peckarsky and Dodson, 1980b). In the present study the physical factors tested had little influence on habitat choice by benthos. The results of these single-factor and interactive experiments are consistent with the hypothesis that invertebrate predation is the most important factor structuring communities of stream benthos in the benign, temperate, woodland stream.

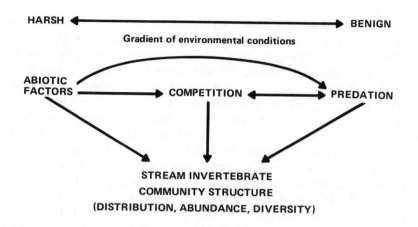

Figure 6. Conceptual model of stream community structure. (See text for explanation.)

The results of identical experiments in the harsher, high-altitude stream were different. Experiments on predation effects produced variable results in the East River (Peckarsky and Dodson, 1980a). During the summer of 1977, the unusually low-discharge–high-temperature regime could be considered harsh, producing stressful conditions for the benthos. Effects of predation were not significant during this summer. This was also the only summer in which physical factors were significant determinants of benthic distributions. The interactive experiment testing the effects of predation and competition showed that competition had the predominant influence over colonization of cages by prey (Peckarsky and Dodson, 1980b). Access to cages by predators did not override the effects of the presence of competitors in determining habitat choice by prey. These results are consistent with the hypothesis that in "normal" seasons the East River supports lower populations of invertebrate predators, which, thus, exert lower predation pressure on prey. Conditions are favorable enough, however, to support prey populations dense enough to compete for potentially limited habitable space or, perhaps, a limited allochthonous food source (Peckarsky, 1980a). Competition was the dominant biological factor structuring the benthic communities in the East River. Under the unusually stressful drought conditions of 1977, physical factors were also significant determinants of benthic distributions.

A model of lotic community structure incorporating the results of these experiments is given in Figure 6. In summary, when physical conditions are benign enough to support a large invertebrate predator population, effects of predation may maintain prey populations at levels low enough to minimize competition for limited habitable space (or, in some cases, food). When some factor (such as differential physical–chemical harshness, differential vertebrate predation, prey defenses, or the existence of refuges from predators) reduces the effectiveness of invertebrate predators, competition for limited habitable space or food may result. When physical–chemical conditions become so harsh that physiological tolerance by prey populations is difficult, biological interactions are probably unimportant in determining stream community structure, and chance dispersal (Harrison, 1980), and the physical factors themselves may directly structure the insect community. These conditions might be present in highly stressed streams, such as heavy-metal- or nutrient-polluted streams, streams with acid stress, or intermittent streams.

Stout (1981) showed that abiotic factors control populations of insects in extremely harsh streams in Costa Rica. Matthews and Hill (1980) suggested that habitat partitioning among species of stream fishes is evident only when environmental conditions are relatively mild in a stream in central Oklahoma. Closely related species converged to similar habitats when physical–chemical conditions were rigorous. Matthews and Hill documented the importance of physical–chemical factors in regulating distributions of stream fishes under unstable conditions. Kraemer (1979) suggested that harsh conditions in altered rivers eliminated competitors of the introduced Asiatic clam (*Corbicula*), but, in more benign, unaltered streams, competition was prevalent between *Corbicula* and the native Unionidae.

These studies provide preliminary evidence of the generality of this model, but complete experiments must be designed to test interactive effects of physical–chemical and biological parameters on community structure of stream invertebrates. Rigorous tests of this model will require such hypothesis testing under a wide range of controlled environmental conditions.

ACKNOWLEDGMENTS

I thank Stan Dodson, Rosemary Mackay, Andy Sheldon, John Neess, James Ward, Manuel Molles, Seth Reice, Chuck Hawkins, the Cornell Ecology Group, Bill Matthews, and Bill Matter for critical comments and valuable discussions of these ideas. Stan Dodson and Dick Gange designed and built the cages; Beth French edited the manuscript; and Steve Horn and Cheryl Hughes prepared the illustrations. Leanne Mumpy and Steve Horn provided laboratory and field assistance. Jo Ann Hayes was an expert stone measurer and found more 10-cm stones than anyone else. The experiments were supported by an NSF Doctoral

Dissertation Grant, a National Academy of Sciences Grant, and University of Wisconsin Graduate Research grants. Research was conducted in partial fulfillment of a Ph.D. degree at the University of Wisconsin, Madison.

REFERENCES

Allen, K. R., 1951, The Horokiwi Stream, *N.Z. Mar. Dep. Fish. Bull.*, 10.

Bishop, J. E., 1973, Observations on the Vertical Distribution of the Benthos in a Malaysian Stream, *Freshwater Biol.*, 3: 147–156.

———, and H. B. N. Hynes, 1969, Upstream Movements of Benthic Invertebrates in Speed River, Ontario, *J. Fish. Res. Board Can.*, 26: 279–298.

Cummins, K. W., 1964, Factors Limiting the Microdistribution of Larvae of Caddisflies, *Pycnopsyche lepida* (Hagen) and *P. guttifer* (Walker) in a Michigan Stream, *Ecol. Monogr.*, 34: 271–295.

———, and G. H. Lauff, 1969, The Influence of Substrate Particle Size on the Microdistribution of Stream Macrobenthos, *Hydrobiologia*, 34: 145–181.

Elliott, J. M., 1971, Upstream Movements of Benthic Invertebrates in a Lake District Stream, *J. Anim. Ecol.*, 40: 235–252.

Fox, L. R., 1977, Species Richness in Streams: An Alternative Mechanism, *Am. Nat.*, 111: 1017–1021.

Gilpin, B. R., and M. A. Brusven, 1970, Food Habits and Ecology of Mayflies of the St. Marie's River in Idaho, *Melandaria*, 4: 20–40.

Harrison, R. G., 1980, Dispersal Polymorphisms in Insects, *Annu. Rev. Ecol. Systemat.*, 11: 95–118.

Hart, D. D., 1979, Diversity in Stream Insects: Regulation by Rock Size and Microspatial Complexity, *Verh. Internat. Verein. Limnol.*, 20: 1376–1381.

Hayden, W., and H. F. Clifford, 1974, Seasonal Movements of the Mayfly *Leptophlebia cupida* (Say) in a Brown-Water Stream of Alberta, Canada, *Am. Mid. Nat.*, 91: 90–102.

Hildrew, A. G., C. R. Townsend, and J. Henderson, 1980, Interactions Between Larval Size, Microdistribution and Substrate in the Stoneflies of an Iron-Rich Stream, *Oikos*, 35: 387–396.

Hughes, D. A., 1966, On the Dorsal Light Response of a Mayfly Nymph, *Anim. Behav.*, 14: 13–16.

Hynes, H. B. N., 1974, Further Studies on the Distribution of Stream Animals Within the Substrate, *Limnol. Oceanogr.*, 19: 92–99.

Kraemer, L. R., 1979, *Corbicula* (Bivalvia: Sphaeriacea) vs. Indigenous Mussels (Bivalvia: Unionacea) in U.S. rivers: A Hard Case for Interspecific Competition? *Am. Zool.*, 19: 1085–1096.

Matthews, W. J., and L. G. Hill, 1980, Habitat Partitioning in the Fish Community of a Southwestern River, *Southwest. Nat.*, 25: 51–66.

Menge, B. A., 1976, Organizaiton of the New England Rocky Interdial Community: Role of Predation, Competition, and Environmental Heterogeneity, *Ecol. Monogr.*, 46: 355–393.

Minshall, G. W., and J. N. Minshall, 1977, Microdistribution of Benthic Invertebrates in a Rocky Mountain (U.S.A.) Stream, *Hydrobiologia*, 55: 231–249.

Peckarsky, B. L., 1979a, Experimental Manipulations Involving the Determinants of the Spatial Distribution of Benthic Invertebrates Within the Substrate of Stony Streams, Ph.D. Thesis, University of Wisconsin, Madison.

———, 1979b, Biological Interactions as Determinants of Distributions of Benthic Invertebrates Within the Substrate of Stony Streams, *Limnol. Oceanogr.*, 24: 59–68.

———, 1980a, Influence of Detritus upon Colonization of Stream Invertebrates, *Can. J. Fish. Aquat. Sci.*, 37: 957–963.

———, 1980b, Predator-Prey Interactions Between Stoneflies and Mayflies: Behavioral Observations, *Ecology,* 61: 932–943.

———, 1981, Reply to Comment by Sell, *Limnol. Oceanogr.*, 26: 982–987.

———, and S. I. Dodson, 1980a, Do Stonefly Predators Influence Benthic Distributions in Streams? *Ecology*, 61: 1275–1282.

———, and S. I. Dodson, 1980b, An Experimental Analysis of Biological Factors Contributing to Stream Community Structure, *Ecology*, 61: 1283–1290.

Poole, W. L., and K. W. Stewart, 1976, The Vertical Distribution of Macrobenthos Within the Substratum of the Brazos River, Texas, *Hydrobiologia*, 50: 151–160.

Rabeni, C. F., and G. W. Minshall, 1977, Factors Affecting Microdistribution of Stream Benthic Insects, *Oikos*, 29: 33–43.

Reice, S. R., 1980, The Role of Substratum in Benthic Macroinvertebrate Microdistribution and Litter Decomposition in a Woodland Stream, *Ecology*, 61: 580–590.

Sanders, H. L., 1968, Marine Benthic Diversity: A Comparative Study, *Am. Nat.*, 102: 243–282.

Stout, J., 1981, How Abiotic Factors Affect the Distribution of Two Species of Tropical Predaceous Aquatic Bugs (Family: Naucoridae), *Ecology*, 62: 1170–1178.

———, and J. Vandermeer, 1975, Comparisons of Species Richness for Stream-Inhabiting Insects in Tropical and Mid-Latitude Streams, *Am. Nat.*, 109: 263–280.

Thorup, J., 1966, *Substrate Type and Its Value as a Basis for the Delimitation of Bottom Fauna Communities in Running Waters*, Special Publication No. 4. Pymatuning Laboratory of Ecology, University of Pittsburgh.

Wene, G., and E. L. Wickloff, 1940, Modification of Stream Bottom and Its Effect on the Insect Fauna. *Can. Entom.*, 72: 131–135.

Wiley, M. J., and S. L. Kohler, 1980, Positioning Change of Mayfly Nymphs Due to Behavioral Regulation of Oxygen Consumption, *Can. J. Zool.*, 58: 618–622.

Williams, D. D., 1980, Some Relationships Between Stream Benthos and Sub-strate Heterogeneity, *Limnol. Oceanogr.*, 25: 166–171.

———, and H. B. N. Hynes, 1974, The Occurrence of Benthos Deep in the Sub-stratum of a Stream, *Freshwater Biol.*, 4: 233–256.

———, and J. H. Mundie, 1978, Substrate Size Selection by Stream Inverte-brates and the Influence of Sand, *Limnol. Oceanogr.*, 23: 1030–1033.

Wise, D. H., and M. C. Molles, Jr., 1979, Colonization of Artificial Substrates by Stream Insects: Influence of Substrate Size and Diversity, *Hydrobiologia*, 65: 69–74.

16. PREDATION AND SUBSTRATUM: FACTORS IN LOTIC COMMUNITY STRUCTURE

Seth R. Reice

Department of Zoology
University of North Carolina
Chapel Hill, North Carolina

ABSTRACT

Substratum particle size has a greater impact on the microdistribution of lotic macroinvertebrates than does feeding by large predators. In a natural stream riffle (New Hope Creek, North Carolina) uniform patches of denuded cobbles or pebbles were exposed in cages for 3 months of colonization. The substrates were enclosed in wire-mesh baskets (25 by 25 by 10 cm; mesh size, 6.35 mm) with and without a mesh cover. The cobbles averaged 12.08 ± 0.34 cm (long axis), and the pebbles averaged 4.00 ± 0.07 cm. The mesh cover was designed to restrict access by fish and salamanders but to permit entry by invertebrate predators and migration by all macroinvertebrates. Each basket had three 5-g leaf packs of *Cornus florida* leaves attached to the rock surface. Three replicate baskets of each treatment combination were sampled biweekly. Species abundances were analyzed by substrate particle size, leaves vs. mineral substrate, and open vs. exclusion cages.

Major differences in species abundances were detected between the two particle sizes. Many common species showed preferences for one substrate size. Nearly all species specialized on either leaves or mineral substrates. No species' abundance was significantly affected by predator exclusion. This suggests that lotic macroinvertebrate microdistributions in New Hope Creek are more influenced by substrata than by large (mostly vertebrate) predators.

Analysis of pairwise associations among species showed that patterns of association were affected by the predator exclusion, however. Different species pairs were more significantly associated in exclusion cages than in open ones. Therefore large predators appear to affect interspecific interactions more than simple single-species abundances.

INTRODUCTION

Substrate particle size is a major determinant of the distribution and abundance of lotic macroinvertebrates: Cummins and Lauff (1969), Higler (1975), Allan (1975), Minshall and Minshall (1977), Rabeni and Minshall (1977), Hart (1978), Wise and Molles (1979), Shelly (1979), and Reice (1980) analyzed the role of substrate in detail. Reice (1981) showed that substrate type also affects the web of interspecific interactions in streams.

The role of predation on benthic community structure has not been widely tested in streams. Allan (1975) showed that the absence of trout was correlated with increased lotic invertebrate densities, but Allan (1982) showed no significant effect after experimentally removing trout. In other aquatic systems [i.e., ponds (Hall et al., 1970; Gilinsky, 1980) soft-bottom marine communities (Virnstein, 1977), salt marshes (Vince et al., 1976; Kneib, 1980), and the rocky intertidal zone (Paine, 1966; Dayton, 1971; Menge and Sutherland, 1976)] predation has been demonstrated to be a key factor in community structure. Peckarsky and Dodson (1980a) and Obendorfer et al. (1980) conducted some of the first invertebrate predator manipulations in streams.

Peckarsky and Dodson (1980a; 1980b) manipulated invertebrate predator and prey densities inside buried cages. The number of invertebrate predators colonizing cages with either high or zero prey density did not differ (Peckarsky and Dodson, 1980b). When a large stone fly predator (*Acroneuria lycorias*) was confined within a cage, prey immigration was less than that into cages without *Acroneuria*. Prey organisms avoided the predator even when it was not free to attack them (Peckarsky and Dodson, 1980a). Obendorfer et al. (1980) confined invertebrate predators within small cages with precolonized leaf packs enclosed and monitored leaf pack community structure and decomposition rate. One predator, *Megarcys signata* (Plecoptera), significantly reduced the total number of organisms present. The other, *Rhyacophila* sp. (Tricoptera), did not reduce the total number of organisms. Leaf pack decomposition was slowed when predators were added. The slowest decomposition occurred with *Megarcys*, which consumed the shredder *Zapada*. The reduction of population size of a major shredder could slow decomposition. Only *Zapada* densities showed a response to the predator manipulations. The work on the effects of predation on lotic communities has just begun.

To pursue the relative importance of substrate and predation, I manipulated substrate type (2 sizes of mineral substrates and leaf packs) and large predators to test their effects on the structure and function of a riffle community. Fish and salamanders were excluded from patches of pebbles and cobbles. The development of the community on these exclusion patches was compared with that where all predators had free access. In addition, I studied the community structure and decomposition of leaf packs on all patches to test a different type of substrate and determine response in terms of the functioning of the system. On the basis of the available data, substrate size was expected to have a significant effect. Since there was no data base on fish or salamander predation, however, no direction of response was predicted.

The experiment was designed to partition and evaluate the effects of substrate type and vertebrate predation in a lotic ecosystem.

MATERIALS AND METHODS

The experiment was set up in New Hope Creek (Orange County, North Carolina), a third-order stream. The site was a single riffle, 20 m long and 8 m wide, with a mean depth of 50 cm. The natural substrate was predominantly cobbles, with a mixture of pebbles, gravel, and sand and an occasional boulder. The study site was located in a gated section of the Duke Forest.

The experimental units were replicated substrate patches in galvanized hardware-cloth baskets (6.35-mm mesh). The 25-by-25-by-10-cm baskets were filled with either cobbles or pebbles. Dimensions of the substrates are given in Table 1. Both substrates were granite and were free of animals

Table 1. Substrate Dimensions*

	Long axis,† cm	Short axis,† cm	Surface area per basket (cm^2)
Cobbles	12.08 ± 0.34	4.72 ± 0.21	4,285
Pebbles	4.00 ± 0.07	1.58 ± 0.04	10,858

* N = 100.

† Average ± standard error.

when the experiment began. The cobbles had been collected from the streambed, scrubbed, and air dried for 1 year. The pebbles were crushed stone incubated in the stream for 2 months and then removed, scrubbed, and air dried for 1 year [these are two of the substrates used in previous work (Reice, 1980; 1981).] The baskets were filled with a single substrate type, placed in the stream bed, and allowed to be colonized.

All baskets had three leaf packs of 5 g dry mass of *Cornus florida* (flowering dogwood) leaves. The leaves were collected in autumn after the abscission layer had formed but before they fell to the ground. The leaf packs were made by briefly soaking (10 min.) the dried leaves (50° C for 48 h) and stapling them with a plastic I Bar fastener (Buttoneer, Dennison Manufacturing Co.). They were then sewn with 89-N breaking-strength (20-lb test) nylon fishing line. The leaf packs were held snug against the upper surface of the substrate, and the line was tied to the bottom of the basket. The leaf pack microhabitat is selected differentially from the mineral substrates by the lotic macrobenthic invertebrates (Reice, 1980).

Predation was manipulated by covering half the baskets with a cover of the same size mesh (6.35 mm). This allowed all New Hope Creek macroinvertebrates to enter, including the large invertebrate predators [*Acroneuria abnormis* (Plecoptera); *Argia* sp., *Boyeria viposa* and *Enallagma; Libellula* sp.; *Macromia* sp. (Odonata); *Corydalis* sp.; and *Nigronia serricornis* (Megatoptera)]. The cage was designed to exclude fish (darters and dace) and salamanders—i.e., large, vertebrate predators. Juvenile fish or salamanders were rarely found in exclusion cages (three individuals). Larger adults were found fairly commonly in the open cages. The covered cages effectively excluded adult salamanders and fish.

A complete sample set was two predation treatments (FISH, NO FISH), two substrate treatments (COBBLES, PEBBLES), and three replicates, i.e., 2 × 2 × 3, or 12 sample baskets. Each basket had three leaf packs, for a total of 36 leaf packs per set. All baskets were placed in New Hope Creek on October 18, 1976. Five complete sets were collected every two weeks until December 27, 1976. More samples were planned, but the experiment was terminated by a flood. Baskets were lifted off the stream bed and transferred into a fine-mesh receiver box placed immediately downstream of the sample basket. Animals dislodged from the baskets were washed onto the receiver screen (0.75 mm) to minimize animal losses. Leaf packs were placed in labeled, individual, sealed plastic bags. Substrates were transferred to buckets. The empty baskets and sample receivers were searched for animals, and all animals were preserved in 70% ethyl alcohol. In the laboratory animals were washed from the substrates with a water jet into a Number 60 Tyler Series sieve (250-μm opening) and transferred to a white enamel pan. Then the macroinvertebrates were removed. The leaf packs were gently washed, and all animals were collected. The leaves were squeezed dry, dried in a forced-air oven (50° for 48 h), and weighed. Animals were keyed to species whenever possible.

Sixty-six taxa, excluding Chironomidae, were identified. Analyses of Chironomidae from the New Hope Creek watershed in a later study indicated the presence of at least 21 genera. Of the taxa identified to genus or better, 29 were rare (less than 10 total individuals). The analyses were limited to the common taxa, excluding the pooled Chironomidae. The 17 most abundant taxa were analyzed for co-occurrence patterns.

The experimental design was a three-way analysis of variance (predation × substrate × date) performed separately on the most common species abundances, leaf pack biomass, and other parameters on leaf pack and substrate samples. Animal numbers (n) were transformed by $\sqrt{n + 0.5}$ to homogenize variance and normalize the clumped distributions. Animal distributions were analyzed both by abundance (number of individuals per sample) and density (number of individuals per square meter of substrate surface area). Surface areas of substrates are listed in Table 1. The surface area per leaf pack was estimated to be 2040 cm^2, or approximately 6120 cm^2 per basket. For a discussion of determination of surface area and use of this measure of density, see Reice (1980). Each sample was examined for the presence or absence of each species to determine patterns of co-occurrence. Then each sample was classified for each species pair according to the grid of Figure 1. Only presence–absence data are used, and abundances are ignored. The distribution of samples in the grid was tested by the chi square 2 × 2 contingency test for deviation from randomness (Pielou, 1977):

$$\text{chi square} = \frac{N(ad - bc)^2}{mnrs} \tag{1}$$

This is appropriate for species that occupy discrete habitat units. A significant chi square test shows the presence of an interaction between the species but gives no indication of the strength or nature of the relationship. It shows that the two species did not occur independently of another; i.e., that given each species distribution, the pair co-occurred more or less often than one would predict from a mutually random distribution. To test whether species co-occur more frequently (positive association) or less frequently (negative association) then expected, I calculated the V statistic:

$$V = \frac{ad - bc}{+(mnrs)^{1/2}} \tag{2}$$

The V statistic assumes values from +1 for complete positive association ($b = 0$, $c = 0$) and −1 for complete negative association ($a = 0$, $d = 0$). Chance occurrence of positive and negative associations should be equal. The value of V is zero when there is no association; V is essentially a non-parametric correlation coefficient. The product moment correlation coefficient (r) cannot be used with accuracy on populations that are patchily distributed. A single sample with high densities of both species will produce very high r values, even if both species are usually rare and do not co-occur. The V values must be interpreted with caution. For a pair of species with a significant interaction ($P(X^2) < 0.05$), a high positive V value shows that the species co-occur more frequently than we should expect. It does not explain why they do. A strong negative V value indicates more than expected disassociation. This can be interpreted as evidence for competition, but it may simply show that the species have alternative substrate preferences. Knowledge of species life-history characteristics, distributions, and functional role is necessary to interpret these data. Analysis of co-occurrence patterns also can provide hypotheses about interspecific relationships for subsequent experimental tests.

RESULTS

Substrate Effects

Most common taxa (14 of 17) had a significantly higher abundance and density in either leaf packs or mineral substrates (Table 2). The results parallel my earlier findings (Reice 1980) of widespread preferences for leaves or stones in the lotic fauna. Table 2 shows density data (number of individuals per square meter of surface area); this more fairly compares substrate samples of such varied surface area. Many of the smaller taxa (Oligochaeta, Nematoda, Ostracoda, Copepoda, and Cladocera), which were not keyed to genus, were very dense in leaf packs and virtually absent in mineral substrates. This may reflect a tendency to lose small individuals during the washing of cobbles and pebbles more than a true preference for leaves.

Many of the leaf-pack specialists are adapted to feed on leaves. *Allocapnia* and *Taeniopteryx* are shredders (Merritt and Cummins, 1978) and *Physa* scrapes (rasps) the leaf surfaces. The larger predators (*Argia, Enallagma,* and *Acroneuria abnormis*) may be more free to maneuver in the larger spaces between stones than in the tightly packed leaves. They may, in fact, be more abundant in natural (loose) aggregations of leaves. *Simulium* uses the stones as a holdfast site for filter feeding.

	Species B present	Species B absent	
Species A present	a	b	m = a + b
Species A absent	c	d	n = c + d
	r = a + c	s = b + d	N = a + b + c + d

Figure 1. 2 × 2 grid of species co-occurrences to calculate chi-square and V statistics.

Table 2. Preferences of Common Taxa for Leaf Packs or Mineral Substrates by Density†‡

Order	Prefer leaf packs	Prefer mineral substrata
Ephemeroptera	*Ephemerella doris****	*Isonychia* sp.** *Stenachron interpunctatum**** *Stenonema annexum**** *Stenonema rubrum****
Plecoptera	*Allocapnia* sp.*** *Taeniopteryx maura****	*Acroneuria abnormis****
Tricoptera		*Polycentropus* sp.*
Odonata		*Argia* sp.*** *Enallagma* sp.***
Diptera	Empididae*	*Simulium* sp.***
Gastropoda	*Physa* sp.*	

† Analysis of variance of $\sqrt{(n + 0.5)}$ m^{-2}.
‡ * = $P(F) \leqslant 0.05$, ** = $P(F) \leqslant 0.01$; and *** = $P(F) \leqslant 0.001$.

To compare the preferences between mineral substrates, I used the density measure again. This compensates for the fact that a basket of cobbles has less than half the surface area of a basket of pebbles. By density criteria, only *Simulium* (of the mineral substrate specialists) distinguished between cobbles and pebbles, and it preferred cobbles. By simple counts, however, *Stenacron, Polycentropus, Enallagma,* and *Argia* were significantly more abundant in pebbles. There were nearly fourfold more Chironomids (total numbers for all cages) in pebbles (8814) then in cobbles (2371). Two leaf-pack specialists were significantly [P(F < 0.05)] more abundant in pebbles than in cobbles (*Ephemerella doris*, 2 : 1, and *Allocapnia*, 2.85 : 1). In summary, species preferences for cobbles vs. pebbles were demonstrated. The strongest effect was that the majority of species showed clear selection for leaf packs or mineral substrates.

Predation Effects

The exclusion of fish and salamanders had little effect on the abundance of all species. In many cases the two treatments appeared to be more like replicates than a manipulation. In mineral substrates the numbers of animals in exclusion and open baskets were within 10% of each other for 28 of 66 taxa and in leaf packs, for 18 of 66 taxa. In the ANOVA of the square-root transformed data (abundance or density) for mineral substrates, there were no significant differences as a result of the predation manipulation. All leaf-pack specialists (Table 2) were more abundant with large predators excluded, but the differences were small and insignificant. With predators present, the mean number of individuals per basket was 227.97. With predators excluded, the number per basket was 235.14. Species richness and taxonomic diversity also did not respond to the manipulation. There were no significant differences in the abundance or density of nearly all individual taxa. The variation from basket to basket within the three replicate samples in a given sample set typically exceeded the differences between those of the alternative predation treatments. Differences between colonization of leaf packs and mineral substrates and between colonization of cobbles and pebbles were far more numerous and significant than differences between predation treatments. Conventional measures of population size and community structure reveal no effects of fish and salamander exclusion.

Decomposition of Leaf Packs

One variable that did respond to the predation manipulation was leaf-pack biomass (Figure 2). After leaching, there was little change in

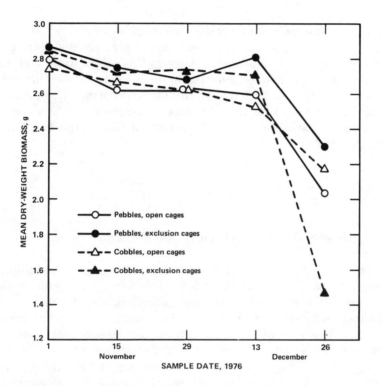

Figure 2. Change in mean dry-weight biomass of leaf packs through time, by treatment. Variances surpassed for clarity of presentation.

biomass until December 13, 1976. Until then, the main effect was that decomposition (weight loss) was greater on pebbles than on cobbles, with slight differences between predator exclusion and open baskets. By December 26, weight loss on cobbles with large predators excluded accelerated, resulting in a significant predation effect $[P(F \leqslant 0.01)]$, and a time × predation interaction $[P(F \leqslant 0.001)]$. Although significant, the difference is rather small.

Of the original leaf pack weight, 70.8% was lost on cobbles without fish. When fish had access to the leaf packs on cobbles, 56.6% of the original weight was lost. Fish exclusion did not affect decomposition on pebbles. A potential explanation for the increased decomposition rate on cobbles without large predators lies in the populations of the shredders *Taeniopteryx maura* and *Allocapnia* sp. On December 13, there were averages of 5.22 *Taeniopteryx* and 3.44 *Allocapnia* per leaf pack on cobbles in exclusion cages. The feeding of the shredders was not detected until the next sample. This was the highest density of *Allocapnia* for any date and treatment combination. Only open pebbles had more *Taeniopteryx* on that or any other date ($\bar{x}$ = 5.56 per leaf pack). Note that open pebbles cages had the second most rapid weight loss over the period from December 13 to 26. It is likely that the action of these leaf-shredding stone flies accounted for the substantial decomposition in this interval. Their concentration in this treatment combination on this date has no ready explanation.

PATCHINESS OF ANIMAL DISTRIBUTION

The common macroinvertebrates were distributed in a distinctly clumped or patchy manner ($S^2/\bar{x} > 1$). This is not unusual for benthic organisms. It is noteworthy, however, when you consider that (1) the pebble substrate was uniform, (2) all cobble patches were nearly identical (3) the configuration of the basket sampler homogenized much patch-to-patch variation within a substrate type, and (4) the patches were all set in a single riffle. Despite all these homogenizing factors, the clumped distributions still emerged. Certain baskets seemed extraordinarily attractive to the fauna. Analyzing the distribution of nine of the dominant species, I found that one basket sample (cobbles with large predators excluded, December 27, replicate 1) had the highest observed population of seven of the nine species. Another sample (pebbles, with predators present, December 13, replicate 1) had the highest abundance for three species (one tied) and the second highest for a fourth species. There was nothing about these samples to explain the peak abundances for so many taxa simultaneously, but their occurrences suggest that there may indeed be strong positive associations between species. The microhabitat of these patches may be uniquely favorable, despite all the attempts at making them similar. This may be the result of eddy currents which carried the animals there or enriched the local food supply. All the organisms found are mobile colonists in these previously bare substrate patches. Therefore it is likely that, if the densities were too high or competition was too severe, the organisms could emigrate (see Peckarsky, 1979). They coexist at extremely high densities. Total individuals range up to 799 in a $1/16$-m^2 basket of pebbles, or nearly 13,000 /m^2. I suspect that interspecific interactions are indeed important at such high densities. To examine this question, let us look at the results of the tests of interspecific association.

Table 3. Frequency, Sign, and Strength of Association (V) by Microhabitat

Microhabitat	N	%	Positive	Negative	$\overline{X}_v$	S
All samples	46	(33.8)	42	4	0.254	0.071
Mineral Substrates Only						
	16	(11.8)	16	0	0.359	0.080
All cobbles	14	(10.3)	13	1	0.451	0.088
All pebbles	7	(5.1)	6	1	0.461	0.076
Predator exclusions only	18	(13.2)	18	0	0.437	0.070
Cobbles, exclusion cages*	5	(3.7)	5	0	0.690	0.193
Pebbles, exclusion cages*	6	(4.4)	5	1	0.630	0.069
Open cages only	11	(8.1)	10	1	0.457	0.092
Cobbles, open cages*	8	(5.9)	7	1	0.617	0.110
Pebbles, open cages*	3	(2.2)	2	1	0.560	0.045
Leaf Packs Only						
	11	(8.1)	10	1	0.382	0.159
Cobbles	10	(7.4)	9	1	0.487	0.132
Pebbles	6	(4.4)	4	2	0.496	0.112
All exclusion cages	6	(4.4)	6	0	0.570	0.161
Cobbles, exclusion cages*	3	(2.2)	3	0	0.763	0.103
Pebbles, exclusion cages*	4	(2.9)	3	1	0.789	0.204
All open cages	4	(2.9)	3	1	0.556	0.300
Cobbles, open cages*	5	(3.7)	5	0	0.616	0.076
Pebbles, open cages*	6	(4.4)	4	2	0.682	0.170

*Classes of data where the strongest associations occurred.

Analysis of Interspecific Associations

Ranges of V values for all pairs of significantly co-occurring species were calculated [$P(X^2) < 0.05$]. The V values measure the strength and direction of interspecific association for each pair of taxa. The mean and variance of V is shown for nested subsets of the data in Table 3. Matrices of V-values for significantly co-occurring pairs are shown in Figures 3 through 6 for cobbles and pebbles, with predators excluded and present. Interspecific interactions in leaf packs are discussed but not illustrated. The seventeen most common species were tested, for a total of 136 possible pairs. The larger the fraction of the data included in the analysis (Table 3), the more significant co-occurrences there are. The mean V

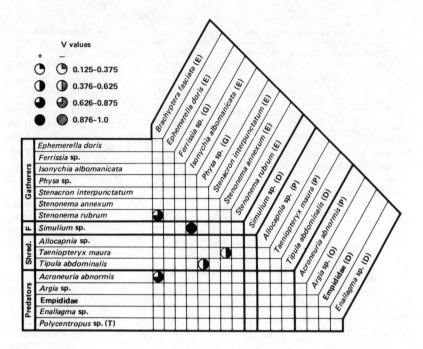

Figure 3. Species association matrix for cobbles substrates with fish and salamanders excluded.

value declines as the size of the fraction increases, however. This is because the pairs that are significant in one treatment combination are not significant in another. Since the strong interactions, for example, in cobbles with predators excluded are not reinforced in other treatments, the strength of the relationship declines as more data are pooled. Therefore the most instructive data came from single substrates in either the exclusion or exposed condition [these are indicated by an asterisk (*) in Table 3].

A very small fraction of the potential interspecific associations were significant. In the classes of data where the strongest associations occurred (*, Table 3), the number of significant pairs was approximately 5% in all cases. Therefore the huge number of nonsignificant interspecific associations means that these species were distributed randomly with respect to each other. Since a 0.05 level of significance was used, we would expect 5% of the cases to be significant by chance alone.

Significant negative associations are even more rare. If these associations are chance events and are not biologically meaningful, 50% of them should be negative, but negative associations represent only 6 of the 40

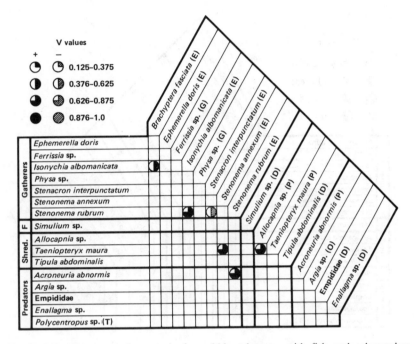

Figure 4. Species association matrix for pebble substrates with fish and salamanders excluded.

significantly associated pairs (* cases), or only 15%. In any treatment (i.e., leaf packs on pebbles, open cages), this represented 0, 1, or 2 negative associations out of 136 species pairs, for a maximum of 1.5% of the possible interactions. This scarcity of dissassociation occurred despite the fact that most common species specialized on cobbles or pebbles or in leaves or mineral substrates. Given their propensities, the species did not appear to alter their distributions to avoid one another.

Even though there was no direct response of individual species abundances, the substrate specific pattern of interspecific interactions did respond to the exclusion of fish and salamanders. Examination of the pairs that are significant in exclusion vs. open cages on cobbles (Figure 3 vs. Figure 5) and on pebbles (Figure 4 vs. Figure 6) reveals that within a substrate the co-occurrences on open vs. predator exclusion patches are mutually exclusive sets. Predator exclusion and open cages for a given substrate size shared no significantly associated species pairs. The predation manipulation resulted in unique responses by the community. It is also possible that all the associations occurred by chance since they represent <5% of the total. Still, several individual interactions are worth examining to decipher the nature of the responses. *Acroneuria abnormis*

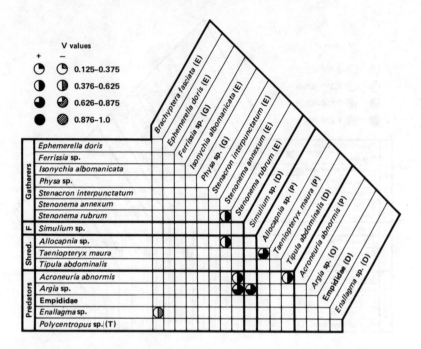

Figure 5. Species association matrix for cobble substrates in open baskets.

and *Stenonema rubrum* were positively associated. This is a predator–prey relationship (*Acroneuria* eats *Stenonema rubrum*, O. E. Walton and S. R. Reice, personal observation). This resulted in a positive association on open cobble and covered pebble patches. The association persisted despite the manipulations. [Note that the probability of any species pairs being significantly associated by chance alone is $1/136$ = 0.00735 for any treatment combination. The probability of the same pairs being associated in two conditions simultaneously is $(0.00735)^2$ = 0.000054, or 1 out of 18,519 cases.] *Allocapnia* and *Taeniopteryx*, both shredders, followed the same pattern. If they were not associated in substrates, they were associated in leaf packs (exclusion cobbles and open pebbles). There is a curious relationship between *Enallagma* (a damselfly predator) and *Brachyptera fasciata (Plecoptera)*, a collector–gatherer. These species were positively associated in pebbles and negatively associated in cobbles (open patches). When the vertebrate predators were excluded, the species showed no relationship. It is possible that in the more open matrix of cobbles, *Brachyptera fasciata* moves to avoid its predator. Overall one-third of the significant relationships on all four treatment combinations pooled were predator–prey pairs. By chance alone, we would predict $60/136$ = 44% to be predator–prey pairs.

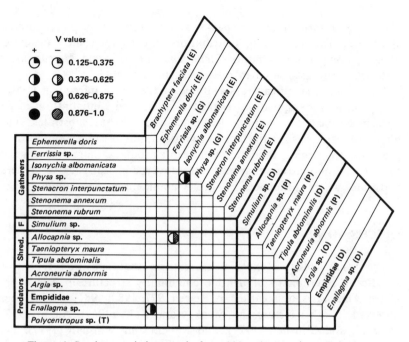

Figure 6. Species association matrix for pebble substrates in open baskets.

There are only two other negative associations in mineral substrates. *Allocapnia* and *Ferrissia* disassociate in open pebble baskets. The negative association between two morphologically and functionally similar species, *Stenacron interpunctation* and *Stenonema rubrum*, is the only clear example of potential competition found. It is possible that this is competitive exclusion, but it occurred only in pebbles when the cage was closed to large predators. If the closed cage is a refuge, this suggests that competition becomes important when predation pressure is reduced.

In leaf packs the patterns of association were similar. There were fewer significant relationships in leaf packs than in mineral substrates (*, Table 3). Again exclusion and open cages of a given substrate shared no significant interspecific associations. *Allocapnia* and *Taeniopteryx*, two stone fly shredders, showed positive association in two of the treatment combinations, as noted. *Allocapnia* and *Ephemerella doris* also showed positive association two of four times (P = 0.000054). Five of the 18 cases (28%) were positive co-occurrences among species that select leaves preferentially (Table 2). Two cases of negative association appear to be potential competition among mayflies (*Ephemerella doris* vs. *Stenonema annexum* on open pebbles and *E. doris* vs. *S. rubrum* on pebbles in exclusion cages). This reflects their conflicting substrate preferences.

Ephemerella doris prefer leaf packs, and *Stenonema* sp. prefer mineral substrates. This pattern of distribution may be their mechanism for avoiding competition in nature. *Stenonema annexum* and *Allocapnia* are also negatively associated in open baskets of pebbles.

In summary, significant associations of taxa were infrequent (~5%). The predation manipulation resulted in a changed configuration of co-occurrence matrices in both leaf-pack and mineral substrates. This alteration came despite the lack of significant numerical responses to the predation treatments. Predation treatments produced more differences in the association matrices than did substrate type or leaf-pack vs. mineral substrate, where differences in species abundances were common.

DISCUSSION

The results of this experiment reaffirm that substrate type is a significant determinant of benthic community structure in New Hope Creek. This was shown earlier by Reice (1977; 1980; 1981) and has been observed in other streams by Mackay and Kalff (1969), Allan (1975), Minshall and Minshall (1977), and Rabeni and Minshall (1977). The direct effects of substrate size and shape have been elegantly demonstrated by Hart (1978). Laboratory studies have also enriched our understanding of the role of substrate size (Cummins and Lauff, 1969; Higler, 1975; and Shelley, 1979). Many species selectively colonize and inhabit particular substrates. The patchiness of distribution of substrates is a major factor in the distribution of the lotic fauna.

Preferences for leaf packs or mineral substrates by the common taxa were shown (as in Reice, 1980). Examination of functional feeding groups aided in interpretation of the species distributions (Cummins and Klug, 1979; Merritt and Cummins, 1978). For example, two shredders, *Allocapnia* and *Taeniopteryx*, preferred leaf packs, but *Tipula abdominalis*, another shredder, was twice as abundant in mineral substrates as in leaf packs (ns). This may have been because the leaf packs were too tightly packed to permit the large *Tipula* larvae to enter.

Analysis of the temporal trends in distribution of the shredders provided a viable explanation of the rate of decomposition of leaf packs. Simple analysis of species richness or total numbers of individuals had shown no helpful patterns. New insight into stream processes is a valuable by-product of the growing use of functional groups in the study of lotic communities.

Some collector–gatherers preferred leaf packs (*Ephemerella doris*, *Physa* sp.), whereas others preferred mineral substrates (*Isonychia al-*

bomanicata, Stenacron interpunctatum, Stenonema spp.). We need more knowledge to explain why they partition the habitat in this manner. Invertebrate predators were more common in mineral than leafy substrates. This suggests that my leaf packs may serve as refuges from predators, as well as an abundant source of food for detritivores (see Reice, 1978). Further study of the interplay between substrate preference and functional groups should prove profitable.

A major goal of this experiment was to test the role of large vertebrate predators in the benthic community. The major findings are that predator exclusion had no direct effects on species richness and diversity, total number of individuals, and individual species population size. The "predation" effects on species abundance and density were miniscule in comparison with the effects of substrate type. A potentially significant portion of the community that was not manipulated here was the large complex of invertebrate predators. They may well have a greater effect on prey distribution and abundance than the vertebrate predators. In New Hope Creek, invertebrate predators per square meter outnumber vertebrate predators by more than 65 : 1 (predaceous Chironomidae and Hydracarina are excluded). The relative prey consumption of vertebrate and invertebrate predators has not been evaluated. All baskets had substantial populations of many invertebrate predators. Fish and salamander densities (approximately $1/m^2$) may be too low, particularly in this season (fall–winter), to have much direct effect on benthic macroinvertebrate densities. However, Allan (1982) also showed no effect of drastic (75%) reductions in fish density on the benthic macroinvertebrate community.

It is difficult to manipulate lotic invertebrate predators successfully in situ. Peckarsky and Dodson (1980a; 1980b), manipulating invertebrate predator and prey densities within submerged cages, found that prey organisms responded to the presence or absence of *Acroneuria*, but predators did not respond to elevated prey densities. The importance of predator–prey interactions in streams is still a wide-open question. If proper experiments can be devised, I suspect that invertebrate predation will be found to be a critical determinant of community structure in lotic benthic systems.

The analysis of the co-occurrence matrices allows us to examine the nature and extent of interspecific interactions among the lotic macroinvertebrates. Significant interspecific interactions were rare, consituting about 5% of the total number of "pairwise" interactions tested. This is what we would expect by chance alone at a 5% significance level. In Table 3, columns N and %, we see that the percentages for the starred lines are near this 5% figure. If this was simply an exercise in probability, these results could be dismissed, but two lines of reasoning suggest that the results hold genuine biological meaning. First, negative associations are too rare to be accounted for by chance (15% observed, 50% expected by chance). Second, several species pairs are significantly associated in more than one treatment. For example, *Taeniopteryx* and *Allocapnia* showed significant positive association in four cases—leaf packs on cobbles (exclusion cages), leaf packs on pebbles (open cages), cobble substrate (open cages), and pebble substrate (exclusion cages). This association is clearly not a chance event ($P = 2.92 \times 10^{-9}$). This is a repeated finding of positive association between two potential competitors, both leaf-shredding Plecoptera of about the same size. The species may not actually be in competition as long as leaves are plentiful. Since leaf packs were never fully consumed in the course of this study and detritus was abundant in New Hope Creek, there was little reason for or evidence of competitive exclusion.

Many of the positive associations (and one negative association) were predator–prey pairs (22% in leaf packs and 36% in mineral substrates). Only the disassociation of *Ephemerella doris* and *Stenonema annexum* in leaf packs in open baskets of pebbles fits the classic picture of competitive exclusion between similar species. Yet it occurred only in one treatment combination and only in leaf packs. In general, evidence for competition was scarce. Considering the extreme degree of spatial heterogeneity in streams, the importance of disturbances in the form of floods, and the general availability of food resources year round, competition may not be a significant determinant of community organization in streams (see Reice, 1981).

An intriguing problem is raised by the association analysis: Why did pebble (or cobble) baskets with and without a mesh cover have different constellations of interacting species? It can be argued that this is the effect of excluding fish and salamanders; i.e., the presence or absence of fish and salamanders permit particular taxa to coexist. Note that large predator exclusion did not radically (or even marginally) affect the abundance of any species or the species diversity in the patches. This is in contrast to the rocky intertidal community of the Pacific Northwest (Paine, 1966; Dayton, 1971).

Another possibility is that the presence of the mesh covers had other effects more important than simply excluding large predators. The caging could produce subtle changes in the microhabitats to favor some species pairs in one condition and other pairs in different conditions. Rabeni and Minshall (1977) pinpointed the importance of accumulated fine particulate detritus in gravel; it makes gravel a more attractive microhabitat (particularly for collectors). This was not measured here, but it is reasonable to think that small reductions in current velocity as a result of the presence of the mesh covers could increase the deposition of detritus in the exclusion cages. Velocity differences alone could affect the colonizaiton patterns. A velocity–substrate interaction could result in convergence between communities on covered large substrates and open smaller substrates. No evidence for that was found here. The pairs shared in common between treatment combinations were typically between open cobbles and covered pebbles. Another possible mechanism is that the closed cage heightened species interactions, particularly between invertebrate predators and their prey. Clearly the manipulation affected the web of interspecific associations on the same substrate. The specific mode of action that produced this effect is still unknown. This work suggests that higher order effects on interspecific relationships were more responsive to the manipulation than were single species population sizes. New questions about stream communities can be addressed by analyzing patterns of interspecific associations.

This research reemphasizes the importance of substrate type as a determinant of lotic community structure and function. It does not eliminate vertebrate predation as a significant factor, but it does direct our attention to the dynamics of the abundant and diverse invertebrate predator fauna. Study of this group should provide more understanding of the dynamics of the lotic community.

ACKNOWLEDGMENTS

This research was supported by National Science Foundation grants DEB-7680443 and BMA74-15332. I want to thank John Alderman, Philip Service, Ellen Gilinsky, and O. Eugene Walton, Jr., for taxonomic and field assistance. I am grateful to Cindy Thompson and Sharon Wagoner for their technical support. Two reviewers, Bobbi Peckarsky and David Allan, made valuable suggestions for improving the manuscript.

REFERENCES

Allan, J. D., 1975, The Distributional Ecology and Diversity of Benthic Insects in Cement Creek, Colorado, *Ecology*, 56: 1040–1053.
———, 1982, The Effect of Reduction in Trout Density on the Invertebrate Community of a Mountain Stream, *Ecology* 63: 1444–1455.

Cummins, K. W., and M. J. Klug, 1979, Feeding Ecology of Stream Inverte-brates, *Annu. Rev. Ecol. Systemat.*, 10: 147–172.

———, and G. H. Lauff, 1969, The Influence of Substrate Particle Size on the Microdistribution of Stream Macrobenthos, *Hydrobiologia*, 34: 145–181.

Dayton, P. K., 1971, Competition, Disturbance and Community Organization: The Provision and Subsequent Utilization of Space in a Rocky Intertidal Community, *Ecol. Mongr.*, 41: 351–389.

Gilinsky, E., 1980, The Role of Predation and Spatial Heterogeneity in Deter-mining Community Structure: The Experimental Manipulation of a Pond System, Ph.D. Thesis, University of North Carolina, Chapel Hill.

Hall, D. J., W. E. Cooper, and E. E. Werner, 1970, An Experimental Approach to the Production Dynamics and Structure of Freshwater Animal Communities, *Limnol. Oceanogr.*, 15: 839–928.

Hart, D. D., 1978, Diversity in Stream Insects: Regulation by Rock Size and Microspatial Complexity, *Verh. Internat. Verein. Limnol.*, 20: 1376–1381.

Higler, L. W. G., 1975, Reaction of Some Caddis Larvae (Trichoptera) to Different Types of Substrate in an Experimental Stream, *Freshwater Biol.*, 5: 151–157.

Kneib, R. T., 1980, The Responses of a Soft-Sediment Intertidal Community to Experimental Manipulations of the Population Size Structure and Density of a Predator, Fundulus heteroclitus (L.), Ph.D. Thesis, University of North Carolina, Chapel Hill.

Mackay, R. J., and J. Kalff, 1969, Seasonal Variation in Standing Crop and Species Diversity of Insect Communities in a Small Quebec Stream, *Ecology*, 50: 101–109.

Menge, B. A., and J. P. Sutherland, 1976, Species Diversity Gradients: Synthesis of the Roles of Predation, Competition, and Temporal Heterogeneity, *Am. Nat.*, 110: 351–369.

Merritt, R. W., and K. W. Cummins (Eds.), 1978, *An Introduction to the Aquatic Insects of North America*, Kendall/Hunt Publishing Co., Dubuque, IA.

Minshall, G. W., and J. N. Minshall, 1977, Microdistribution of Benthic Invertebrates in a Rocky Mountain (U.S.A.) Stream, *Hydrobiologia*, 55: 231–249.

Obendorfer, R. Y., J. V. McArthur, and J. R. Barnes, 1980, The Role of Predators in the Maintenance of Leaf Pack Community Structure, abstract, North American Benthological Society Meeting, Savannah, GA.

Paine, R. T., 1966, Food Web Complexity and Species Diversity, *Am. Nat.*, 100: 65–75.

Peckarsky, B. L., 1979, Biological Interactions as Determinants of Distributions of Benthic Invertebrates Within the Substrate of Stony Streams, *Limnol. Oceanogr.*, 24: 59–68.

———, and S. I. Dodson, 1980a, Do Stonefly Predators Influence Prey Distributions in Streams? *Ecology*, 61(6): 1275–1282.

———, and S. I. Dodson, 1980b, An Experimental Analysis of the Effects of Biological Interactions in Structuring Stream Insect Communities, *Ecology*, 61(6): 1283–1290.

Pielou, E. C., 1977, *Mathematical Ecology*. John Wiley & Sons, Inc. New York.

Rabeni, C. F., and G. W. Minshall, 1977, Factors Affecting Microdistribution of Stream Benthic Insects, *Oikos*, 29: 33–43.

Reice, S. R., 1977, The Role of Animal Associations, Current Velocity and Sediments in Lotic Litter Decomposition, *Oikos*, 29: 357–365.

_____, 1978, The Role of Detritivore Selectivity in Species-Specific Litter Decomposition in a Woodland Stream, *Verh. Internat. Verein. Limnol*, 20: 1396–1400.

_____, 1980, The Role of Substratum in Benthic Macroinvertebrate Micro-distribution and Litter Decomposition in a Woodland Stream, *Ecology*, 61: 580–590.

_____, 1981, Interspecific Associations in a Woodland Stream, *Can. J. Fish. Aquat. Sci.* 38: 1271–1280.

Shelly, T. E., 1979, The Effect of Rock Size Upon the Distribution of Species of Orthocladiinae (Chironomidae: Diptera) and *Baetis intercalaris* McDunnough (Bactidae: Ephemeroptera), *Ecol. Entomol.*, 4: 95–100.

Virnstein, R. W., 1977, The Importance of Predation by Crabs and Fishes on Benthic Infauna in Chesapeake Bay, *Ecology*, 58: 1199–1217.

Vince, S., I. Valiela, N. Backus, and J. M. Teal, 1976, Predation by the Salt Marsh Killifish *Fundulus heteroclitus* in Relation to Prey Size and Habitat Structure: Consequences for Prey Distribution and Abundance, *J. Exp. Mar. Biol. Ecol.*, 23: 255–266.

Wise, D. H., and M. C. Molles, Jr., 1979, Colonization of Artificial Substrates by Stream Insects: Influence of Substrate Size and Diversity, *Hydrobiologia*, 65: 69–74.

17. THE INTERMEDIATE-DISTURBANCE HYPOTHESIS: AN EXPLANATION FOR BIOTIC DIVERSITY PATTERNS IN LOTIC ECOSYSTEMS

James V. Ward

Department of Zoology and Entomology
Colorado State University
Fort Collins, Colorado

Jack A. Stanford

University of Montana Biological Station
Bigfork, Montana

ABSTRACT

The intermediate-disturbance hypothesis predicts that biotic diversity will be greatest in communities subjected to moderate levels of disturbance. It is consistent with diversity patterns observed in natural and altered lotic ecosystems. Species diversity is suppressed in stream habitats exposed to disturbances that are severe (organic loading or acid mine drainage) or frequent (diel flow fluctuations). In addition, habitats with enhanced environmental constancy (spring sources or streams below storage reservoirs) exhibit suppressed diversity even if adverse conditions (e.g., oxygen deficits) are not apparent. It is postulated that "undisturbed" lotic systems are in fact "disturbed" and that the high biotic diversity of natural streams is a function of moderate perturbation. Diversity is enhanced by the spatio-temporal heterogeneity resulting from intermediate disturbance, which maintains the community in a nonequilibrium state. This theory may account for much of the diversity variance within stream systems and between different types of lotic habitats.

INTRODUCTION

The intermediate-disturbance hypothesis (Connell, 1978) predicts that a greater biotic diversity will be maintained in communities subjected to intermediate levels of disturbance than in those undergoing either greater or lesser perturbation. As stated by Osman (1977), "There appears . . . to be an optimal frequency of disturbance at which diversity is maximized." Support for the intermediate-disturbance hypothesis can be derived from terrestrial (e.g., Harper, 1969), marine (e.g., Paine, 1971), or lentic (e.g., Dodson, 1970) systems. There is a relatively rich literature relating the concept to intertidal and subtidal communities (Paine, 1966; 1969; 1971; Osman, 1977; Dayton, 1971; Menge, 1979).

It is postulated here that the intermediate-disturbance hypothesis largely explains patterns of biotic diversity observed in natural and altered lotic ecosystems. After a brief review of the salient features of the hypothesis, this paper explores the relationships between disturbance level and the diversity (species richness) of stream communities. Diversity is viewed primarily in an ecological time frame, i.e., "the maintenance of diversity, as opposed to the generation of diversity" (Huston, 1979), unless otherwise stated.

THE INTERMEDIATE-DISTURBANCE HYPOTHESIS

The intermediate-disturbance hypothesis is diagrammed in Figure 1. Simply stated, a certain level of "disturbance" results in a higher biotic diversity than greater or lesser perturbation. Disturbance level may be roughly equated with environmental heterogeneity if the latter is viewed broadly in a spatio-temporal context. The initial disturbance can be biotic (e.g., predation or disease) or abiotic (e.g., fire, storms, or wave action). The severity of the disturbance can be considered from the standpoint of intensity [e.g., intensity of predation (Paine, 1966)], frequency [e.g., frequency of fire (Taylor, 1973)], or both.

Connell (1978) presented evidence that the high diversity of trees in tropical rain forests and corals on tropical reefs is a function of disturbances, such as storms, which have an intermediate frequency of occurrence. Connell postulated that these systems exist in a "non-equilibrium state which, if not disturbed further, will progress toward a low-diversity equilibrium community." Examining burned areas in Yellowstone National Park, Taylor (1973) found that areas burned 25 years previously had the greatest diversity of plant, bird, and mammal species. Areas burned with greater or lesser frequency exhibited lower

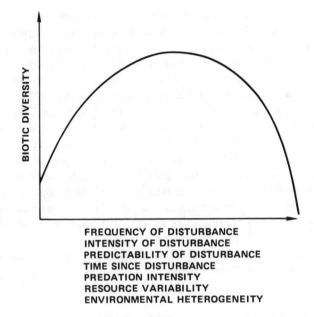

Figure 1. Theoretical relationship between biotic diversity and various measures of "disturbance" (modified from Connell, 1978).

biotic diversity. Loucks (1970) suggested that the natural tendency for random environmental perturbation in temperate forests maintains peak diversity in the vegetation community. Intermediate levels of disturbance or physiological stress have been shown to result in the greatest diversity of plant species in pastures (Grime, 1973). Osman (1977) reported a low diversity of epifaunal marine species on large rocks (high stability) and small rocks (low stability) but a high diversity on intermediate-sized rocks of moderate stability. Moderate predation pressure may maintain a high level of diversity of prey species, as Paine (1966) showed by comparing intertidal areas containing starfishes with those from which these predators had been removed. Many additional examples could be drawn from studies of terestrial and especially marine systems (see Fox, 1979, and Huston, 1979, for additional references).

What accounts for the enhanced diversity associated with moderate disturbance? Disturbances that are too frequent (e.g., annual burning or low stability intertidal rocks) or too intense (e.g., toxic inputs or severe predation) are thought to suppress biotic diversity by causing local extinction of certain species or increasing dominance. Environmental homogeneity may likewise result in local extinction of populations and shifts in dominance as competitive exclusion becomes operative under equilibrium conditions (see Huston, 1979). For example, the most competitive species monopolized the most stable intertidal rocks to the virtual exclusion of other epifaunal groups (Osman, 1977).

An intermediate level of disturbance is thought to maintain nonequilibrium conditions that allow the coexistence of a diverse assemblage of species; the equilibrium requirement of competitive exclusion is not fulfilled. For this to occur, however, it is necessary that the extent of disturbance exhibit less variance than the rate of recovery (Fox, 1979), which will differ from one community to another and on a spatiotemporal scale within communities (Caswell, 1978). Intermediate disturbance may maintain communities at an intermediate stage of succession with a higher diversity than earlier or later stages; this in some communities is partly a function of enhanced opportunities for potential invaders (Levin and Paine, 1974).

EVIDENCE FROM LOTIC ECOSYSTEMS

It has been suggested that the intermediate-disturbance hypothesis may apply to a variety of systems (Levin and Paine, 1974). Although the concept has not previously been specifically related to lotic ecosystems, there is evidence supporting the applicability of the hypothesis to running waters. Patrick (1970), for example, considered competition among benthic plants to be relatively unimportant in determining the community structure of natural streams. She contended that unpredictable fluctuations in nutrients maintain diversity by preventing monopolization of resources by one or a few species. Biotic diversity is maintained partly by species replacement, as changing environmental conditions favor different assemblages of species. Ward (1976a; 1976b) postulated that the diversity of many natural streams is maintained by their nonequilibrium conditions and that anthropogenic alterations that enhance environmental constancy may increase coactive patterns and cause a shift toward an equilibrium community, accompanied by a reduction in diversity. Species diversity of stream organisms has also been associated with the extent of seasonal (Ide, 1935) and diel (Vannote et al., 1980) thermal variation, which may be partly responsible for the maintenance of lifecycle diversity (Ward, 1976a). Moderate predation pressure has been implicated in the maintenance of biotic diversity in streams (Patrick, 1970), and the "keystone species" concept (Paine, 1969) has been postulated as influencing diversity in stream systems (Fox, 1977; Ward and Stanford, 1979). Gray (1980) presented evidence suggesting that desert stream macroinvertebrate communities may respond to flood events in a manner consistent with the intermediate-disturbance hypothesis.

The species richness of selected stream macroinvertebrate communities is plotted against "extent of disturbance" in Figure 2. Examples were drawn from lotic systems ranging from small brooks to mid-sized rivers. Data points for natural streams (i.e., those relatively unaffected by man's activities) are plotted in the center of the figure; the (+) indicates a "normal" level of disturbance. It is not implied that all natural streams exhibit the same degree of environmental variance; habitats with lesser (−) and greater (++) than normal disturbance are likewise somewhat arbitrarily positioned on the horizontal axis. Because of variations in collecting efforts, sampling techniques, and levels of taxonomic resolution, species diversity values from different studies are of only rough comparative value. However, habitats indicated with solid circles have reference plots with corresponding numbers to indicate different sites from the same study. For example, number 6 in the center (130 species) is

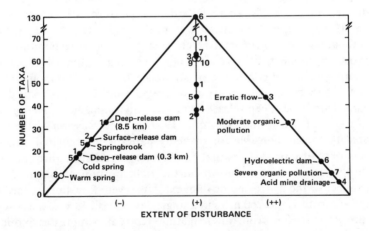

Figure 2. Taxonomic diversity of selected stream macroinvertebrate communities and extent of disturbance. Lotic habitats with minimal anthropogenic impact and "normal" environmental variability are plotted in the center (+) to indicate a moderate extent of disturbance. Habitats exhibiting lesser (−) and greater (++) than normal disturbance are also shown (see text for further explanation). Arabic numerals indicate citations: 1, Ward, 1974; 1976b. 2, Ward and Short, 1978. 3, Ward and Short, 1978. 4, Herricks and Cairns 1974. 5, Gray and Ward 1979; Martinson and Ward, unpublished. 6, Stanford, unpublished. 7, Gaufin and Tarzwell, 1956. 8, Ward, unpublished. 9, Ames, 1977. 10, Ward, 1975. 11, Ames, 1977.

the Middle Fork of the Flathead River, Montana, which exhibits considerable natural environmental heterogeneity, whereas number 6 on the right (15 species) is the South Fork of the Flathead River below a hydroelectric dam, where diel flow fluctuations are extreme. The reference plots enable comparisons of species richness between differentially disturbed sections of the same stream (or stream system), thus reducing other differences that influence diversity, such as stream type and zoogeographical factors. In some instances species richness does not fully elucidate the differences between disturbance levels because of the more even distribution of taxa in the natural systems. For example, although 25 taxa occurred below the surface-release dam, in contrast to 36 at a station above the reservoir (Figure 2), Shannon index values (1.7 vs. 4.1) better contrast the diversity differences. It was not possible in all cases to calculate diversity indices from the data available, however.

Several lotic habitats with enhanced environmental constancy are exemplified in Figure 2 (see figure legend for citations). Habitats were selected in which the low biotic diversity is not attributable to low oxygen or toxic substances. Extreme physico-chemical constancy characterizes the warm (25°C) and cold (8°C) spring sources. Explanations for reduced diversity in springs (relative to environmental homogeneity) have been addressed in detail elsewhere (Ward and Dufford, 1979). The spring brook (Figure 2), a short distance from the cold spring source, exhibits slightly greater environmental heterogeneity and a corresponding slight increase in number of species. The remaining examples of habitats exhibiting enhanced constancy are streams below storage impoundments. A notable increase in diversity is exhibited from 0.3 to 8.5 km downstream from the deep-release dam as environmental heterogeneity increases. See Ward and Stanford (1979) for a discussion of the interrelated factors responsible for diversity patterns in regulated streams.

A variety of lotic systems that exhibit excessive disturbance caused by anthropogenic factors is shown on the right portion of Figure 2. The stream below the hydroelectric dam exhibits severe flow fluctuations that may vary from 7.5 to 260 m^3/s in a single day. Erratic flow of a less severe nature downstream from an irrigation reservoir suppresses biotic diversity to a much lesser degree. A significant, but incomplete, recovery of macroinvertebrate diversity is shown from the septic zone to the recovery zone of a stream stressed by organic pollution. On the basis of species diversity, the stream receiving acid mine drainage indicates the most severe disturbance (Figure 2).

DISCUSSION AND CONCLUSIONS

It is our contention that "undisturbed" lotic systems are in fact "disturbed" and that the high biotic diversity of certain natural streams is a function of moderate perturbation. We propose that the diversity of natural lotic systems is maximized by the spatio-temporal heterogeneity resulting from moderate disturbance, which maintains the community in a nonequilibrium state. We feel that the concept of resource variability itself behaving as an abstract resource (Levins, 1979) is applicable to lotic ecosystems, especially as related to species packing.

It must be emphasized that biotic diversity varies greatly from one natural stream to another and longitudinally in the same stream system (Vannote et al., 1980). This variation in diversity in natural streams is generally consistent with the intermediate-disturbance hypothesis, however. The middle reaches of the steam continuum, the region of greatest environmental heterogeneity, also exhibit the highest biotic diversity (Vannote et al., 1980). Headwater reaches and the lower portions of rivers have lower diversity values associated with more constant environmental conditions (but see Horwitz, 1978). The unsuspected magnitude and unpredictability of environmental variance in tropical streams (Stout and Vandermeer, 1975) may, at least partly, account for the higher diversity of tropical than temperate stream insects, just as Connell (1978) used environmental inconstancy to explain the high diversity of trees and corals in tropical regions. To confirm or refute the relationship between disturbance level and biotic diversity, we must examine specific lotic ecosystems displaying variability in the frequency and severity of disturbance. In addition, a more precise and objective method of determining the extent of disturbance is needed.

Community structure is shaped by a myriad of physical, chemical, and biological variables acting synergistically. A single ecological measure of disturbance, such as the efficiency of energy flow or the rate of species replacement, may serve to integrate the biologically significant environmental variables in running waters. Until controlled experimental treatments are explicitly designed to quantify spatio-temporal heterogeneity, however, it is perhaps premature to attempt a rigorous mathematical description of disturbance in lotic ecosystems.

ACKNOWLEDGMENTS

Appreciation is extended to R. W. Pennak, University of Colorado, and V. H. Resh, University of California, Berkeley, for reading the manuscript. Discussions with R. T. Paine, University of Washington, and M. L. Rosenzweig, University of Arizona, were most helpful. Additional intellectual stimulation was provided by the students in the Graduate Limnology Seminar, Colorado State University, fall 1979. This manuscript was prepared while J. V. Ward was supported by the Colorado Experiment Station.

REFERENCES

Ames, E. L., 1977, Aquatic Insects of Two Western Slope Rivers, Colorado, M. S. Thesis, Colorado State University, Fort Collins.

Caswell, H., 1978, Predator-Mediated Coexistence: A Nonequilibrium Model, *Am. Nat.,* 112: 127-154.

Connell, J. H., 1978, Diversity in Tropical Rain Forests and Coral Reefs, *Science,* 199: 1302-1310.

Dayton, P. K., 1971, Competition, Disturbance, and Community Organization: The Provision and Subsequent Utilization of Space in a Rocky-Intertidal Community, *Ecol. Monogr.* 41: 351-389.

Dodson, S. I., 1970, Complementary Feeding Niches Sustained by Size-Selective Predation, *Limnol. Oceanogr.,* 15: 131-137.

Fox, L. R., 1977, Species Richness in Streams—An Alternative Mechanism, *Am. Nat.,* 111: 1017-1021.

Fox, J. F., 1979, Intermediate-Disturbance Hypothesis, *Science,* 204: 1344-1345.

Gaufin, A. R., and C. M. Tarzwell, 1956, Aquatic Macroinvertebrate Communities as Indicators of Pollution in Lytle Creek, *Sewage Industr. Wastes,* 28: 906-924.

Gray, L. J., 1980, Recolonization Pathways and Community Development of Desert Stream Macroinvertebrates, Ph.D. Thesis, Arizona State University, Tempe.

——, and J. V. Ward, 1979, Food Habits of Stream Benthos at Sites of Differing Food Availability, *Am. Midl. Nat.,* 102: 157-167.

Grime, J. P., 1973, Control of Species Density in Herbaceous Vegetation, *J. Environ. Manag.,* 1: 151-167.

Harper, J. L., 1969, The Role of Predation in Vegetational Diversity, *Brookhaven Symp. Biol.,* 22: 48-62.

Herricks, E. E., and J. Cairns, Jr., 1974, *Rehabilitation of Streams Receiving Acid Mine Drainage,* Virginia Water Resources Research Center, Bulletin 66, Virginia Polytechnic Institute and State University, Blacksburg.

Horwitz, R. J., 1978, Temporal Variability Patterns and the Distributional Patterns of Stream Fishes, *Ecol. Monogr.*, 48: 307-321.

Huston, M., 1979, A General Hypothesis of Species Diversity, *Am. Nat.*, 113: 81-101.

Ide, F. P., 1935, The Effect of Temperature on the Distribution of the Mayfly Fauna of a Stream, *Publs. Ont. Fish. Res. Lab.*, 50: 1-76.

Levin, S. A., and R. T. Paine, 1974, Disturbance, Patch Formation, and Community Structure, *Proc. Nat. Acad. Sci.*, 71: 2744-2747.

Levins, R., 1979, Coexistence in a Variable Environment, *Am. Nat.*, 114: 765-783.

Loucks, O. L., 1970, Evolution of Diversity, Efficiency, and Community Stability, *Am. Zool.*, 10: 17-25.

Menge, B. A., 1979, Coexistence Between the Seastars *Asterias vulgaris* and *A. forbesi* in a Heterogeneous Environment: A Non-Equilibrium Explanation, *Oecologia*, 41: 245-272.

Osman, R. W., 1977, The Establishment and Development of a Marine Epifaunal Community, *Ecol. Monogr.*, 47: 37-63.

Paine, R. T., 1966, Food Web Complexity and Species Diversity, *Am. Nat.*, 100: 65-75.

———, 1969, The *Pisaster-Teguia* Interaction: Prey Patches, Predator Food Preferences, and Intertidal Community Structure, *Ecology*, 50: 950-961.

———, 1971, A Short-Term Experimental Investigation of Resource Partitioning in a New Zealand Rocky Intertidal Habitat, *Ecology*, 52: 1096-1106.

Patrick, R., 1970. Benthic Stream Communities, *Am. Sci.*, 58: 546-549.

Stout, J. and J. Vandermeer, 1975, Comparison of Species Richness for Stream-Inhabiting Insects in Tropical and Mid-Latitude Streams, *Am. Nat.*, 109: 263-280.

Taylor, D. L., 1973, Some Ecological Implications of Forest Fire Control in Yellowstone National Park, Wyoming, *Ecology*, 54: 1394-1396.

Vannote, R. L., G. W. Minshall, K. W. Cummins, J. R. Sedell, and C. E. Cushing, 1980, The River Continuum Concept, *Can J. Fish. Aquat. Sci.*, 37: 130-137.

Ward, J. V., 1974, A Temperature-Stressed Stream Ecosystem Below a Hypolimnial Release Mountain Reservoir, *Arch. Hydrobiol.*, 74: 247-275.

———, 1975, Bottom Fauna-Substrate Relationships in a Northern Colorado Trout Stream: 1945 and 1974, *Ecology*, 56: 1429-1434.

———, 1976a, Effects of Thermal Constancy and Seasonal Temperature Displacement on Community Structure of Stream Macroinvertebrates, in G. W. Esch and R. W. McFarlane (Eds.), *Thermal Ecology II*, ERDA Symposium Series, CONF-750425, pp. 302-307, NTIS.

———, 1976b, Comparative Limnology of Differentially Regulated Sections of a Colorado Mountain River, *Arch. Hydrobiol.*, 78: 319-342.

———, and R. G. Dufford, 1979, Longitudinal and Seasonal Distribution of Macroinvertebrates and Epilithic Algae in a Colorado Springbrook-Pond System, *Arch. Hydrobiol.*, 86: 284-321.

———, and R. A. Short, 1978, Macroinvertebrate Community Structure of Four Special Lotic Habitats in Colorado, U.S.A., *Verh. Internat. Verein. Limnol.,* 20: 1382-1387.

———, and J. A. Stanford, 1979, Ecological Factors Controlling Stream Zoobenthos with Emphasis on Thermal Modification of Regulated Streams, in J. V. Ward and J. A. Stanford (Eds.), *The Ecology of Regulated Streams,* pp. 35-55, Plenum Press, New York.

18. INVERTEBRATE DRIFT AND PARTICULATE ORGANIC MATERIAL TRANSPORT IN THE SAVANNAH RIVER BELOW LAKE HARTWELL DURING A PEAK POWER GENERATION CYCLE

William J. Matter

School of Renewable Natural Resources
University of Arizona
Tucson, Arizona

Patrick L. Hudson

U. S. Fish and Wildlife Service
Southeast Reservoir Investigations
Clemson, South Carolina

Gary E. Saul

Army Engineers Waterways Experiment Station
Vicksburg, Mississippi

ABSTRACT

The Savannah River below Lake Hartwell, Georgia–South Carolina, receives hypolimnetic water discharged from the reservoir for peak-power generation. Invertebrates and particulate organic material (POM) in the water column were collected during a 24-h release cycle at sites 1.0, 4.5, and 12.5 km downstream from the dam. Water was released during a 6-h period, reaching a maximum generation discharge of 688 m^3/s. River discharge was less than 10 m^3/s during nongeneration periods.

Highest POM concentrations were associated with the initial downstream surge of water at the start of power generation; values were 200 to

357

400 times greater than those during nongeneration periods. The POM rapidly decreased to less than one-tenth the original surge levels. Much of the POM originated in the tailwater, and concentrations increased at successive downstream sites.

Of the drifting invertebrates, 80 to 93% originated in the reservoir; the rest, primarily Oligochaeta, Diptera, and Ephemeroptera, were from the tailwater. Densities of benthic invertebrates were highest during passage of the initial release surge, whereas densities of invertebrates originating in the reservoir peaked 2 to 3 h after the initial surge at each station, during maximum release. Densities of drifting benthic organisms decreased rapidly after the initial surge despite increasing discharge, were not uniform throughout the water column, and increased with increasing distance downstream. Evidence of "behavioral" drift suggested that benthic production may still be in excess, despite repeated "catastrophic" losses during releases.

INTRODUCTION

Reservoirs may be viewed as distinctly different in form and function from either lakes or streams because of the dramatic physical, chemical, and biological changes that occur after impoundment (Baxter, 1977). Tailwaters also represent unique combinations of physical and chemical conditions since they are strongly influenced by processes within reservoirs and the timing and magnitude of releases. Reservoir releases during electric power generation often subject tailwaters to large flow variations over a short period of time, with concomitant changes in depth, water temperature and quality, and particulate matter load.

Attempts to demonstrate relationships between the transport of particulate organic material (POM) and hydraulic characteristics of natural and regulated streams have not been completely successful (Sedell et al., 1978; Bilby and Likens, 1979; Webster et al., 1979). Similarly, the degree to which reservoir releases act to scour and transport benthic invertebrates and their particulate food sources out of tailwaters is poorly understood. In this report we describe the magnitude, composition, and periodicity of invertebrate drift and POM transport in the Lake Hartwell tailwater during a single peak-power generation water release cycle.

STUDY SITE

Lake Hartwell is a 22,640-ha reservoir, with a volume of 3.15×10^9 m³, situated on the Savannah River between Georgia and South Carolina. The lake has a maximum depth of 53 m and exhibits thermal stratification during the summer (Dudley and Golden, 1974); the top of the thermocline is at about 7.5 m. Four 66,000-kW generators draw reservoir water through 7.3-m diameter penstocks located at a center-line depth of 30.0 m at full power pool elevation of 201 m msl. Water is released primarily for electrical generation during peak power demand (about 1200 to 2100 h, Monday through Friday). The maximum generation release capacity is about 708 m³/s. Nongeneration flows within the study area include dam and dike leakage plus inflow from small streams (approximately 3 to 10 m³/s). Hypolimnetic water released into the tail water seldom exceeds 20°C.

MATERIALS AND METHODS

We used drift nets to estimate the transport of benthic invertebrates and POM during different flow conditions. The rectangular nets (15 by 15 cm at the mouth and 145 cm long) were constructed of 450-μm-mesh nylon net secured to a brass rod frame. Nets were held in the water column (about 8 cm off the bottom) by steel rods anchored in a concrete block (Figure 1). During high flow an additional net of identical design was held just below the water surface by a boat-mounted or handheld frame at each station and left in position over the same time period as the lower net. Nets were set within 5 m of the river bank so that they would be accessible during high flows. Samples were taken 1.0, 4.5, and 12.5 km downstream from the dam (stations 1, 2, and 3, respectively), between 0900 h on July 12 and 0930 h on July 13, 1979. Water released between 1255 and 1904 h on July 12 rapidly increased river discharge to about 690 m³/s and raised the water level about 1.7 m. Results of preliminary sampling led us to consolidate drift data into four distinct periods, each having unique flow conditions: the period of low flow before power generation began (pregeneration), the passage of the initial release surge (surge), the period of high flow caused by water release but not including the surge (high-flow), and the period of falling water levels after generation ended (postgeneration). For the 24-h period sampled, the duration of the pregeneration, high-flow, and postgeneration periods

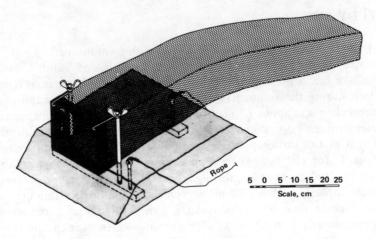

Figure 1. Bottom-anchored drift net used in Lake Hartwell tailwater. Eyebolts and steel rods are threaded into metal plates and then sunk into concrete. The net is attached by eyed hooks to steel rods and secured by wing nuts.

were 380, 350, and 690 min. The initial surge was defined as the first 20 min of water release. Nighttime drift during the postgeneration period was sampled to separate behavioral from flow-induced drift.

Flow rates were measured at depths adjacent to drift nets (with a Price AA current meter), since they could not be measured within the net mouths. The duration of the ten net sets at each station was adjusted (from 15 to 60 min) to minimize net clogging. Collected material was preserved in 10% formalin. Invertebrates were sorted, identified, enumerated, dried for 24 h at 60°C, and weighed.

The POM (other than invertebrates) collected in drift nets was filtered onto a pre-ashed, tared glass-fiber filter, and dried for 24 h at 60°C to determine dry weight. Filters were then ignited (550°C for 15 min) and reweighed to determine ash and ash-free dry weight (AFDW). Additional POM samples were collected at each station during pregeneration, surge and high-flow by filtering 20 l of water through an 80-μm mesh screen. The AFDW of the residue and 1 l of filtrate was measured.

Because reservoir releases are expressed as volume per unit of time, we expressed drift data as "drift density", i.e., the number of organisms per unit volume of water passing a given point. An estimate of total reservoir and benthic invertebrate drift through the entire tailwater cross section at each of the stations over 24 h was calculated as:

$$D = \sum_{i=1}^{4} d_i v_i$$

where: D = total drift

     i = one of four flow conditions, (i.e., pregeneration, surge, high-flow, and post-generation)

     d_i = mean density or weight of invertebrates (number/m^3 or mg/m^3) during the ith flow conditions,

     v_i = total volume of water passing through the cross section during the ith period

RESULTS

From 79 to 93% of the drifting invertebrates (number and biomass) collected at the sampling stations originated in the reservoir and were carried into the tailwater with the released water (Table 1). The density of reservoir invertebrates was always higher in the surface net than in the bottom net.

Aquatic Oligochaeta made up about 50% of the lotic benthic invertebrates, followed by Diptera (about 30%, predominantly Chironomidae, subfamily Orthocladiinae). Ephemeroptera, predominantly *Pseudocloeon*, were relatively rare at station 1 but were common at the other sites, especially at station 2 (Table 1). A variety of other benthic invertebrate taxa were represented by only a few individuals collected in only one or two time periods and seldom recorded from all stations. Higher densities of Ephemeroptera were recorded in the surface net than in the bottom net, but other benthic invertebrates were evenly distributed.

An analysis of variance showed that total drift density was not significantly different between stations but was significantly different

Table 1. Percentage of Total Number of Drifting Invertebrates and Total Number of Drifting Benthic Invertebrates in Major Taxonomic Groups at Three Stations in the Lake Hartwell Tailwater

	Total invertebrates			Benthic invertebrates		
	Station			Station		
Group	1	2	3	1	2	3
Reservoir forms*	92.9	92.8	79.3			
Tailwater forms						
Oligochaeta	4.6	3.4	10.4	64.8	47.7	50.2
Diptera	2.2	2.3	7.5	30.2	31.4	36.5
Ephemeroptera	0.2	1.3	1.9	3.1	17.4	9.0
Miscellaneous invertebrates†	0.1	0.2	0.9	2.0	3.5	4.3

*Invertebrates primarily of reservoir origin (i.e., zooplankton and *Chaoborus*).

†Benthic invertebrates originating in the tail water but collected in only small numbers; taxa included are Trichoptera, Coleoptera, Odonota, Amphipoda, Isopoda, Hydracarina, and Nematoda.

(P < 0.001) over the four time periods. Drift density was significantly different (P < 0.001) between reservoir forms and benthic forms, and the interaction of invertebrate group with time period was also significant (P < 0.001). The significant interaction resulted from the occurrence of maximum densities of benthic invertebrates with passage of the initial release surge and rapid decline to near pregeneration levels, whereas the density of reservoir invertebrates peaked during maximum release, 2 to 3 h after the initial surge at each station (Table 2, Figure 2). Drift density of benthic forms alone was significantly different (P < 0.02) among stations; no significant differences were found for total invertebrates, however, because of the similar numerical dominance of reservoir zooplankton at the three stations. Benthic drift densities were highest at station 2 and lowest at station 1. Ephemeroptera exhibited a definite increase in drift density during the night (Figure 3) but was the only taxon to do so.

The relative abundance of major taxa in the benthic drift changed during the four periods (Table 2). Oligochaeta made up very little of the drift during pregeneration and postgeneration periods but represented 50 to 65% of the drift during the surge and subsequent high-flow period. Ephemeroptera made up 45 to 85% of the benthic drift during pregeneration and postgeneration but represented only a small proportion of the drift during the surge and high-flow periods. Diptera generally represented 30 to 40% of the drift in all time periods but tended to be relatively less abundant in postgeneration samples.

Total estimated 24-h drift of reservoir forms ranged from 932×10^6 to 1592×10^6 individuals and from about 9.7 to 10.4 kg (Table 3). Since significant differences in density of reservoir forms between stations were not observed, the range in values is caused by both random error and differences in total discharge at the three sites. The reported values are underestimates of the actual levels because of the relatively large mesh size used, but measurements of the relative abundance of reservoir forms among stations and time periods probably reflect the actual differences in drift.

Total estimated 24-h drift of benthic forms ranged from 28×10^6 to 168×10^6 individuals (stations 1 and 3, respectively) and from 2.0 to 2.7 kg (Table 3). The increase in benthic invertebrate drift downstream was a product of both increasing drift density and stream volume. The distance downstream that benthic organisms entering the water column at stations 1 and 2 are carried is unknown.

Changes in POM over time were similar to that observed for benthic drift; the maximum density was associated with the passage of the release surge (Table 4). Almost 20% of the transport of POM > 450 μm occurred during the passage of the initial release surge. Most of the remainder was

Table 2. Drift Density for Major Invertebrate Groups at Three Stations
in the Lake Hartwell Tailwater*

Station and group	Time			
	Pregeneration	Surge	High-Flow	Postgeneration
Station 1				
Reservoir forms†	17.5	32.1	89.5	69.1
Oligochaeta	0	42.0	1.0	0.2
Diptera	<0.1	19.5	0.5	<0.1
Ephemeroptera	0	0.7	0	0.5
Miscellaneous invertebrates‡	0	1.0	<0.1	<0.1
Total	17.6	95.3	91.0	70.0
Total benthos§	<0.1	63.2	1.6	0.9
Station 2				
Reservoir forms	3.4	29.0	139.3	54.2
Oligochaeta	<0.1	46.5	1.4	0.1
Diptera	0.2	24.6	1.3	0.4
Ephemeroptera	0.3	2.7	<0.1	3.8
Miscellaneous invertebrates	0.1	2.7	<0.1	0.1
Total	4.0	105.5	142.0	58.6
Total benthos	0.7	76.6	2.8	4.5
Station 3				
Reservoir forms	16.1	25.0	149.6	39.7
Oligochaeta	0.2	26.0	9.2	0.2
Diptera	0.3	19.8	4.7	0.9
Ephemeroptera	0.2	5.0	1.0	0.9
Miscellaneous invertebrates	0	3.0	0.3	<0.1
Total	16.9	78.8	164.8	41.7
Total benthos	0.8	53.9	15.1	2.0

*Drift density in numbers per cubic meter, July 12–13, 1979, at different times during a hydropower water release cycle.

†Invertebrates of reservoir origin (i.e., zooplankton and *Chaoborus*).

‡Benthic invertebrates from the tailwater, collected in small numbers, including Trichoptera, Coleoptera, Odanata, Amphipoda, Isopoda, Hydracarina, and Nematoda.

§Excludes reservoir forms.

carried in the high flow immediately following the surge. The POM tended to increase with distance downstream under all flow conditions (Table 4). The POM >450 μm was about two to three times greater at station 3 than at the upstream stations. Data from 20-*l* grab samples indicate that particles less than 80 μm (and, therefore, not collected in drift nets) consistently make up the largest part of the POM transport;

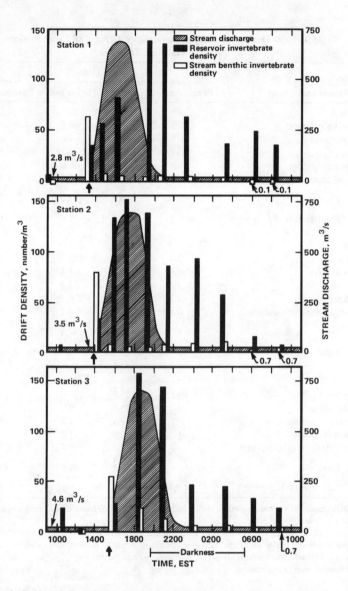

Figure 2. Drift density and approximate stream discharge at three sites in the Lake Hartwell tailwater over a peak-power generation cycle. The arrow indicates the time of passage of the initial release surge.

however, a 20-*l* sample may be inadequate for estimating large-sized POM (Table 4).

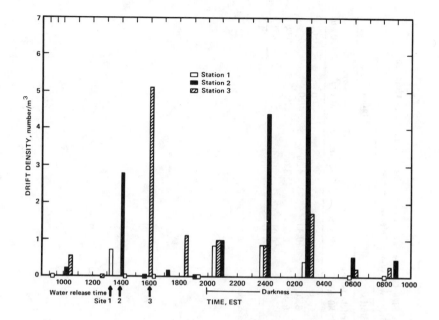

Figure 3. Drift density of Ephemeroptera at three sites in the Lake Hartwell tailwater over a peak-power generation cycle. Arrows indicate passage of the initial release surge at each site.

Table 3. Estimates of the Total Number and Weight of Reservoir-Originated and Benthic Invertebrates Drifting Through a Cross Section of the Lake Hartwell Tailwater Over a 24-h Peak-Power Water-Release Cycle

Period	Station					
	1		**2**		**3**	
	Reservoir forms	Benthos	Reservoir forms	Benthos	Reservoir forms	Benthos
			Estimated Number, X 10^6			
Pregeneration	1.1	<0.1	0.3	<0.1	2.0	0.1
Surge	6.0	11.8	0.3	14.3	4.7	10.1
High-flow	915.5	16.4	1,308.6	26.9	1,567.5	157.0
Postgeneration	9.6	0.1	9.1	0.8	18.1	0.9
Total	932.3	28.2	1,323.4	42.1	1,592.3	168.1
			Estimated Weight,* g, AFDW			
Pregeneration	12	3	25	12		
Surge	108	351	28	271		
High-flow	10,148	2,312	9,514	1,681		
Postgeneration	113	25	94	72		
Total	10,381	2,691	9,661	2,036		

*Biomass not available for station 3.

Table 4. Mean Density* of Particulate Organic Material in Different Size Classes† at Three Stations in the Lake Hartwell Tailwater at Different Times During a Hydropower Water-Release Cycle

Flow conditions	Station 1			Station 2			Station 3		
	Particle (size)			Particle (size)			Particle (size)		
	≤80 μm	>80 μm	>450 μm	≤80 μm	>80 μm	>450 μm	≤80 μm	>80 μm	>450 μm
Pregeneration	1.50	0	<0.01	1.50	0	<0.01		0	0.01
Surge	1.40	0.10	0.28	4.00	3.70	0.30	4.30	4.20	0.58
High-flow	2.00	0.60	0.02	2.50	0.90	0.02	3.00	0.60	0.06
Postgeneration			<0.01			<0.01			<0.01

*Grams per cubic meter, AFDW.

†Data for sizes ≤ or >80 m are based on a 20-*l* grab sample and for >450 μm are based on drift net samples.

DISCUSSION

Although the data presented are primarily from a single 24-h cycle, the same trends have been measured in subsequent sampling. The density of reservoir invertebrates in the drift appears to depend directly on the quantity of water released; i.e., disproportionately greater amounts of zooplankton were entrained as release volumes increased. The high density of zooplankton found near the water surface may have been caused by air bubbles trapped under the carapaces of entrained organisms. Measures of plankton abundance, when made at the water surface, may overestimate true densities. Hudson and Lorenzen (1981) suggested that organisms entrained in releases be viewed as gains to the tailwater system rather than as losses from the reservoir. Predation on reservoir invertebrates by the tailwater fauna may be severely limited during high-flow periods, however, especially if concentrations of prey are near the surface.

Densities of drifting benthic invertebrates were always highest during the passage of the initial release surge; this suggests that the combination of scour forces and the probability of exposure to these forces was particularly great over that short time period. Elliott (1967) and Anderson and Lehmkuhl (1968) reported increases in drift rate with increasing discharge but found virtually no change in drift density. Releases to the Lake Hartwell tailwater may cause more acute disturbances than witnessed in most previous studies. Brooker and Hemsworth (1978) did observe an increase in drift density during a 2-day release from a reservoir in Wales. They also reported a sudden increase in the number of drifting organisms with the passage of the initial release surge. Predation on benthic organisms in the drift is likely to occur during the passage of the surge but not during the high-flow period.

The decline in the density of benthic drift after the initial surge in the Lake Hartwell tail water is probably not the product of reduced scour force since discharge continued to increase after the surge. Instead the decline may be attributed to downward movement of some invertebrates into the substrate (Hynes, 1974; Poole and Stewart, 1976) and other behavioral adaptations.

Anderson and Lehmkuhl (1968) found that Ephemeroptera retained diel periodicity in drift abundance despite catastrophic drift losses related to seasonal spates. Brooker and Hemsworth (1978) and Armitage (1977) found much the same situation in streams subject to water releases from reservoirs. Waters (1961) and Dimond (1967) maintained that behavioral drift is largely a density related response that regulates population levels according to available resources. As such, behavioral drift should not be observed where continual catastrophic losses reduce benthic populations to levels well below the resource base. Only Ephemeroptera (primarily *Pseudocloeon*) exhibited a definite nocturnal increase in behavioral drift (density and rate) in the Lake Hartwell tailwater. This same pattern occurred on weekends when water was not released. The occurrence of behavioral drift may indicate that production of *Pseudocloeon* exceeds local resources despite losses during water releases. However, to suggest that production of a single taxon exceeds local resources does not imply that existing or alternative release regimes do not have an impact on the overall diversity and production of invertebrates in the tailwater.

The fluctuations in water depth and velocity accompanying peak power water releases might be seen as regularly occurring "floods" because, not unlike natural floods, they cause flushing of POM from tailwater reaches (Ward, 1976). Large water releases occur far more frequently than natural floods, however. Reservoirs are particle traps and retain up to 90% of the POM carried in from the watershed (Lind, 1971; Armitage, 1977; Simons, 1979). Because reservoirs release particulate material in the form of limnetic plankton, the tailwaters below peak power generators do not receive the same type of allochthonous POM recharge from upstream areas as observed in natural streams.

Substantial amounts of POM were transported out of the Lake Hartwell tailwater. The distribution of large-sized POM in the water column may be too patchy to be adequately measured by 20-*l* samples. Qualitative observation indicated that much of the POM collected from the water column consisted of fragmented periphytic algal mats and aquatic macrophytes found in the tailwater and, thus, is autochthonus. The increasing density of POM moving downstream from the dam suggests that autochthonous and allochthonous sources contributed POM along the length of the tailwater. The virtual absence of large POM shredders and the predominance of periphyton scrapers in the benthos indicate that POM inputs may not represent the most important energy sources within the stream reach and time period sampled. This POM may be used further downstream or in the next reservoir.

ACKNOWLEDGMENTS

We gratefully acknowledge the field assistance of Jim Oliver, Philip Moore, William Painter, and Marc Zimmerman. Laboratory space and materials were supplied by James Clugston, Chief, Southeast Reservoir Investigations, U. S. Fish and Wildlife Service. This study was supported by the Office, Chief of Engineers, through the U. S. Army Engineer

Waterways Experiment Station as part of the Environmental and Water Quality Studies Program.

REFERENCES

Anderson, N. H., and D. M. Lehmkuhl, 1968, Catastrophic Drift of Insects in a Woodland Stream, *Ecology*, 49: 198-206.

Armitage, P. D., 1977, Invertebrate Drift in the Regulated River Tees, and an Unregulated Tributary Maize Beck, below Cow Green Dam, *Freshwater Biol.*, 7: 167-183.

Baxter, R. M., 1977, Environmental Effects of Dams and Impoundments, *Annu. Rev. Ecol. Systemat.* 8: 255-283.

Bilby, R. E., and G. E. Likens, 1979, Effect of Hydrologic Fluctuations on the Transport of Fine Particulate Organic Carbon in a Small Stream, *Limnol. Oceanogr.*, 24: 69-75.

Brooker, M. P., and R. J. Hemsworth, 1978, The Effect of the Release of an Artificial Discharge of Water on Invertebrate Drift in the River Wye, Wales, *Hydrobiologia*, 59: 155-163.

Dimond, J. B., 1967, Evidence That Drift of Stream Benthos Is Density Related, *Ecology*, 48: 855-857.

Dudley, R. G., and R. T. Golden, 1974, Effect of a Hypolimnion Discharge on Growth of Bluegill (*Lepomis macrochirus*) in the Savannah River, Georgia, Completion report USDI/OWRR Proj. No. B-0570GA, University of Georgia, Athens.

Elliott, J. M., 1967, Invertebrate Drift in a Dartmoor Stream, *Arch. Hydrobiol.*, 63: 202-237.

Hudson, P. L., and W. E. Lorenzen, 1981, Manipulation of Reservoir Discharge to Enhance Tailwater Fisheries, in R. M. North, L. B. Dworsk, and D. J. Allee (Eds.), *Unified River Basin Management Symposium*, pp. 568-579, American Water Resources Association, Minneapolis.

Hynes, H. B. N., 1974, Further Studies on the Distribution of Stream Animals Within the Substratum, *Limnol. Oceanogr.*, 19: 92-99.

Lind, O. T., 1971, The Organic Matter Budget of a Central Texas Reservoir, in G. E. Hall (Ed.), *Reservoir Fisheries and Limnology*, Special Publication 8, pp. 193-202, American Fisheries Society, Washington, D.C.

Poole, W. C., and K. W. Stewart, 1976, The Vertical Distribution of Macrobenthos Within the Substratum of the Brazos River, Texas, *Hydrobiologia*, 50:151-160.

Sedell, J. R., R. J. Naiman, K. W. Cummins, G. W. Minshall, and R. L. Vannote, 1978, Transport of Particulate Organic Material in Streams as a Function of Physical Processes, *Verh. Internat. Verein. Limnol.*, 20: 1366-1375.

Simons, D. B., 1979, Effects of Stream Regulation on Channel Morphology, in J. V. Ward and J. A. Stanford (Eds.), *The Ecology of Regulated Streams*, pp. 95-111, Plenum Press, New York.

Ward, J. V., 1976, Effects of Flow Patterns Below Large Dams on Stream Benthos: A Review, in J. F. Orsborn and C. H. Allman (Eds.), *Instream Flow Needs*, Vol 2, pp. 235-253, American Fisheries Society, Washington, D.C.

Waters, T. F., 1961, Standing Crop and Drift of Stream Bottom Organisms, *Ecology*, 42: 532-537.

Webster, J. R., E. F. Benfield, and J. Cairns, Jr., 1979, Model Predictions of Effects of Impoundment on Particulate Organic Material Transport in a River System, in J. V. Ward and J. A. Stanford (Eds.), *The Ecology of Regulated Streams*, pp. 339-364, Plenum Press, New York.

19. FOOD CONSUMPTION BY TROUT AND STONEFLIES IN A ROCKY MOUNTAIN STREAM, WITH COMPARISON TO PREY STANDING CROP

J. David Allan

Department of Zoology
University of Maryland
College Park, Maryland

ABSTRACT

"Allen's paradox" stems from the finding that trout consumed 40 to 150 times the standing stock of invertebrates in the Horokiwi stream (Allen, 1951). Current estimates of invertebrate turnover rates indicate that such a relationship is unlikely. Even allowing for certain corrections in Allen's estimates, the paradox appears to remain.

A 5-yr study of predator–prey relationships in a Colorado trout stream provided data to re-examine the potential role of top predators in limiting the abundance of stream benthos. The predators investigated were trout and setipalpian stoneflies. At a 3350-m altitude site, trout were absent, and the principal predator was *Megarcys signata* (Perlodidae). At a 3050-m site, brook trout (*Salvelinus fontinalis*), *M. signata,* and *Kogotus modestus* (Perlodidae) were the main predators, whereas at a 2740-m site brown trout (*Salmo trutta*), *Claassenia sabulosa* (Perlidae), and *K. modestus* filled this role.

Feeding rates for each predator, based on gut analysis and water temperature, together with seasonal standing crop of predators, were used to estimate annual prey consumption (dry weight). Values for the stoneflies were 3.03 g m^{-2} yr^{-1} at the 3350-site, 3.30 g m^{-2} yr^{-1} at the 3050-m site, and 2.68 g m^{-2} yr^{-1} at the 2740-m site. Trout consumed 5.5 g m^{-2} yr^{-1} at the 3050-m site and 5.29 g m^{-2} yr^{-1} at the 2740-m site. Thus total prey consumption ranged roughly from 3 and 9 g m^{-2} yr^{-1} dry weight, and trout, where present, consumed about twice as much as stoneflies.

The standing crop biomass of total invertebrates was on the order of 1 g/m^2, based on Surber samples and weighings. The standing crop biomass of Ephemeroptera and Diptera was estimated quite precisely since they were the principal prey items of all macropredators. Values typically ranged from 0.3 to 0.9 g/m^2. The ratio of amount consumed to amount available (based on total fauna) was estimated to be 2.5:1 (3350 m), 8.4:1 (3050 m), and 8.7:1 (2740 m); higher values were obtained if only Ephemeroptera and Diptera were considered as prey.

Although grazing appeared to equal or exceed production (assuming the ratio of production to biomass is between 3:1 and 8:1), the discrepancy was much less than with other similar estimates. Furthermore, invertebrate biomass must have been underestimated (the hyporheic fauna was not sampled), whereas estimated prey consumption was near maximum. The Allen paradox of prey consumption greatly exceeding prey available is not fully resolved but appears of a magnitude within reach of a solution from improved accuracy in estimation techniques.

INTRODUCTION

"Allen's paradox" is a term coined by Hynes (1970) and based on the finding of K. R. Allen in a classic study (Allen, 1951) that trout consumed 40 to 150 times the standing stock of invertebrates in the Horokiwi stream. Current estimates of invertebrate turnover rates (Waters, 1977) indicate that such a relationship is unlikely. Even allowing for certain corrections (Gerking, 1962) in Allen's estimates, the paradox appears to remain. There have been few attempts since to estimate total predation by fish.

The potential impact of invertebrate predators in running water communities has rarely been assessed. Invertebrate residents of stony streams, such as setipalpian stoneflies, larvae of the Megaloptera, net-spinning and rhyacophilid caddis flies and the occasional mayfly nymph, act as predators for at least part of their life cycles. A wide variety of invertebrates may serve as prey, but the predominant prey are Chironomidae and Ephemeroptera (Allan, 1982a). For setipalpian stoneflies, often the principal invertebrate predator, there is typically a shift of diet with development. Chironomidae are preyed upon almost exclusively by the earliest carnivorous stages, but the Ephemeroptera (often *Baetis*) and trichopterans become more important to larger individuals, which develop a generally broader diet. The kind of prey ingested is a function of predator body size and the availabilty of prey in the vulnerable size range for a given size of predator.

In laboratory streams the perlid stonefly *Hesperoperla pacifica* was estimated to consume more food than did sculpins or trout, despite the fact that its biomass was less than that of the fish (Davis and Warren, 1965; Brocksen et al., 1968). This conclusion was based on laboratory estimates of food conversion efficiencies, which were lowest in the stoneflies, and were not obtained by direct estimates of food available and food eaten. Nonetheless, the potential importance of invertebrate predation was suggested by the demonstration that increasing biomass of stone flies plus sculpins combined was inversely correlated with benthic standing crop. Siegfried and Knight (1976) estimated prey consumption rates directly in the closely related *Calineuria californica*. By combining these figures with estimates of prey and predator density, they were able to predict total prey consumption over various intervals, with the result that the impact of *C. californica* predation appeared considerable.

This chapter reports a study of predation in a Colorado trout stream, which included estimates of prey choice and predation rates by setipalpian stoneflies (Allan, 1982a) and trout (Allan, 1981), and an experimental removal of trout from a section of stream (Allan, 1982b). Data from these papers are combined here to estimate total annual prey consumption by trout and stonefly predators and to attempt to estimate their impact on the prey community.

STUDY SITE

Cement Creek is a third-order stony-bottom stream originating in snow melt at 3600 m and joining the East River at 2600 m. It is part of the Gunnison River drainage, Gunnison County, Colorado. Allan (1975) described the stream in detail.

The only vertebrate predators, apart from the occasional dipper (*Cinclus mexicanus*), are trout. The cutthroat (*Salmo clarkii*) is found only at high elevations, principally from 3200 to 3320 m. The brook trout (*Salvelinus fontinalis*) occurs from 2900 to 3320 m and is the common species at higher elevations. The brown trout (*Salmo trutta*) occurs below 3050 m and is common only below 2900 m. Rainbow trout (*Salmo gairdneri*) are stocked for and rapidly removed by fishermen.

The principal macroinvertebrate predators are setipalpian stoneflies, of which three species are common. *Megarcys signata* (Perlodidae) occurs at the highest elevations; it is joined by *Kogotus modestus* (Perlodidae) at middle elevations, and *Claassenia sabulosa* (Perlidae) and a few *K. modestus* are found at the lowest elevations. Additional predaceous macroinvertebrates (not included in the analysis) are *Rhyacophila* caddis larvae, which occur at all elevations; *Ephemerella doddsi* (Ephemeroptera), which is common at high elevations; *Arctopsyche grandis* (Trichoptera: Hydropsychidae), which occur at low evelations; and very occasional setipalpian stoneflies *Isoperla* spp and *Hesperoperla pacifica* (Perlidae), which are at mid to low elevations.

This paper concentrates on three sampling sites. Site 1, at 3350 m, contained no trout. The only common stonefly and the principal predator was *M. signata*. Site 2, at 3050 m, supported brook trout, *M. signata,* and *K. modestus*. Site 3, at 2740 m, supported brown trout, *C. sabulosa,* and *K. modestus.*

METHODS

Only a broad outline of methods is given here; details are available in related papers (Allan, 1981; 1982a; 1982b). Temperature was measured weekly with maximum–minimum thermometers at each site. Trout were censused by electroshocking to estimate standing crop and production. Food consumption was estimated from stomach analysis conducted at eight 3-h intervals over 24 h on five occasions from early June to the end of September. Invertebrate densities were estimated from Surber samples (0.093-m^2 area, 0.3-mm-mesh net) collected at each site over 18 months. Typically six samples were collected at each site on each date and were pooled; however, at site 2 on numerous dates I collected 12 replicates, which were not pooled. Stonefly densities were estimated from Surber collections. Additional stonefly collections were made by kick sampling (0.5-mm net) for dissection of foreguts to estimate prey consumption.

RESULTS

Consumption of Prey by Setipalpian Stone flies

Consumption of prey was estimated in the same manner for each stonefly species, according to data from Allan (1982a). The amount of food contained in a full foregut was estimated to be roughly 10% of the predator's dry weight. This value was obtained by measuring head widths of prey found in foreguts and using regressions of dry weight on head width to compute dry weight of foregut contents and of the predators. The average weight of an individual predator on a given date (mg dry wt) was then multiplied by 10% to compute an average full meal. Gut clearance time was estimated to be 1 day for *M. signata* at summer temperatures on the basis of direct weighings of foregut contents of individuals collected at midnight, held without food in the stream, and frozen after 4, 8, 12, and 16 h. I assumed longer gut clearance times, 4 days during winter when water temperatures were 0° and 2 days during the brief warming and cooling periods of spring and fall (Figure 1). Total prey consumption over an interval (mg/m²) was calculated from:

$$\frac{\text{dry wt of a full foregut (mg/ind)}}{\text{gut clearance time (d)}} \times \text{predator density (individuals/m}^2) \qquad (1)$$

$$\times \ \text{duration of interval (d)}$$

Megarcys signata, Site 1

Total annual prey consumption was estimated to be 3.03 g m⁻² yr⁻¹ (Table 1). All feeding was assumed to stop on June 15 to provide a 1-yr

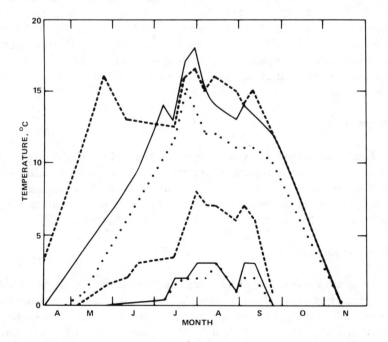

Figure 1. Maximum and minimum weekly water temperatures at site 1 (· · · ·), site 2 (———), and site 3 (– – – –) in 1978.

Table 1. Estimated Prey Consumption by *Megarcys signata* at Site 1

Date (1977)	Standing crop of M. signata		Estimated dry weight of full foregut, mg/individual	Gut clearance time, days	Biomass of prey consumed, mg m^{-2}-interval^{-1}
	mg/m^2	number/m^2			
June 16	4.1	30.6	0.013	1	5.9
June 28	5.7	37.5	0.015	1	9.9
July 7	16.3	52.1	0.031	1	30.3
July 18	33.8	62.7	0.054	1	75.7
August 1	74.3	41.3	0.180	1	211.2
August 25	101.7	17.9	0.568	1	152.4
September 12	67.6	9.0	0.751	1	515.7
October 12	276.2	19.7	1.402	2	520.8
November 29	157.8	12.6	1.252	4	333.0
March 17	228.3	10.8	2.114	2*	1173.9
June 27	223.2	9.0	2.480		
Total					3028.8 mg m^{-2} yr^{-1}

*A 1-day gut clearance time was used from June 1–15, after which feeding was assumed to cease.

total; this is supported by observations that *M. signata* ceased feeding at about this time, just before emergence. Estimates of standing crop of *M. signata* and contents of a full foregut appear quite reliable; the major weakness in these calculations lies in estimating gut clearance time, especially at low temperatures. If feeding is assumed to cease at 0°C, then the estimate of total annual consumption is reduced to 2.18 g m^{-2} yr^{-1}, 75% of the previous value.

Megarcys signata, Site 2

Total annual prey consumption was estimated to be 2.83 g m^{-2} yr^{-1} (Table 2). A certain amount of interpolation of predator standing crop was necessary, but the result is consistent with that observed at site 1. If, again, feeding is assumed to cease at 0°C, total annual consumption is estimated at 1.60 g m^{-2} yr^{-1}, 57% of the previous value.

Kogotus modestus, Site 2

Total annual consumption was estimated to be 0.47 g m^{-2} yr^{-1} (Table 3). If feeding is assumed to cease at 0°C, total annual consumption becomes 0.39 g m^{-2} yr^{-1}, 84% of the previous value. The estimates for *K. modestus* are much lower than for *M. signata* because *K. modestus* is smaller and maintains a substantially smaller biomass despite its greater numbers.

Table 2. Estimated Prey Consumption by *Megarcys signata* at Site 2

Date (1977)	Standing crop of *M. signata*		Estimated dry weight of full foregut, mg/individual	Gut clearance time, days	Biomass of prey consumed, mg m⁻²-interval⁻¹
	mg/m²	number/m²			
June 28	0	0	0.015		
				1	0.6
July 7	1.7	5.4	0.031		
				1	4.7
July 18	6.8	12.6	0.054		
				1	38.6
August 1	48.4	26.9	0.180		
				1	151.7
August 25	78.0	13.7	0.568		
				1	82.4
September 12	13.5	1.8	0.751		
				1	396.8
October 12	251.0	17.9	1.402		
				2–4*	1102.6
January 6	402.4	33.6	1.200		
				4	511.2
March 16	190.3	9.0	2.114		
				2	361.6
June 1		+	2.480		
				1	176.2
July 7	44.6	1.8	2.500		
Total					2826.4 mg m⁻² yr⁻¹

*A 2-day gut clearance time used from October 12 to November 29, and 4 day from November 29 to January 6.

+Weight interpolated, and all feeding assumed to cease on June 15.

Table 3. Estimated Prey Consumption by *Kogotus modestus* at Site 2

Date (1977)	Standing crop of *K. modestus*		Estimated dry weight of full foregut, mg/individual	Gut clearance time, days	Biomass of prey consumed, mg m⁻²-interval⁻¹
	mg/m²	number/m²			
August 1	0	0	0.005		
				1	4.4
August 25	3.7	68.2	0.005		
				1	13.7
September 12	10.0	44.9	0.022		
				1	28.9
October 12	8.3	25.1	0.033		
				2	34.3
November 29		*	0.074		
				4	27.2
January 6	33.6	59.2	0.057		
				4	45.3
March 17	18.9	39.5	0.048		
				2	117.7
June 1		*	0.175		
				1	131.6
July 7	39.8	13.4	0.296		
				1	23.9
July 14	28.9	10.8	0.268		
				1	39.1
July 31	13.3	3.6	0.370		
Total					466.1 mg m⁻² yr⁻¹

*Estimated by interpolation.

Kogatus modestus, Site 3

Kogotus modestus was quite rare at site 3, and, as a result, total annual prey consumption was very low, 0.11 g m⁻² yr⁻¹ (Table 4). If feeding is

Table 4. Estimated Prey Consumption by *Kogotus modestus* at Site 3

Date (1977)	Standing crop of K. modestus		Estimated dry weight of full foregut, mg/individual	Gut clearance time, days	Biomass of prey consumed, mg m^{-2}-interval^{-1}
	mg/m^2	number/m^2			
August 25	0	0	0.005		
September 12	3.2	14.4	0.022	1	1.8
October 12	6.5	19.7	0.033	1	14.1
November 29	8.0	10.8	0.074	2	19.6
January 7	7.2	12.6	0.057	4	7.5
March 16	2.6	5.4	0.048	4	8.2
April 20	4.5	2.7	0.168	4	3.7
June 1		*	0.175	1	19.3
July 7	5.3	1.8	0.296	1	20.0
August 24	0	0	0.370	1	14.4
Total					108.6 mg m^{-2} yr^{-1}

*Estimated by interpolation.

assumed to cease at 0° C, total annual consumption becomes 0.09 g m^{-2} yr^{-1}, 82% of the previous value.

Claassenia sabulosa, Site 3

Because *C. sabulosa* is semivoltine, it was not possible to estimate the average weight of an individual on any given date, as was done for the univoltine perlodids (Tables 1 through 4). Instead I used actual standing crop (mg/m^2) and took 10% of that value to be the collective feeding for that group of individuals regardless of individual size. In principle this is no different from the other calculations, except that an average weight per individual and average density are not shown.

Total annual prey consumption for *C. sabulosa* was estimated to be 2.57 g m^{-2} yr^{-1}. If feeding is assumed to cease at 0° C (a shorter time at site 3 because of its warmer temperature regime), total annual consumption becomes 2.06 g m^{-2} yr^{-1}, 80% of the previous value (Table 5).

Consumption of Prey by Trout

Site 1

On the basis of several censuses in which 500-ft sections of stream were electroshocked, it was determined that no trout were present at site 1.

Table 5. Estimated Prey Consumption by *Claassenia sabulosa* at Site 3

Date (1977)	Standing crop of C. sabulosa, mg/m^2	Estimated dry weight of full foregut, mg/m^2	Gut clearance time, days	Biomass of prey consumed, mg m^{-2}-interval^{-1}
July 18	248.9	24.9	1	592.4
August 25	62.9	6.3	1	152.0
September 12	106.0	10.6	1	198.3
October 12	26.2	2.6	2	208.5
November 29	155.1	15.5	4	331.9
January 7	15.1	1.5	4	92.1
March 17	11.6	1.2	4	83.2
April 20	178.6	17.9	1	913.0
July 24	13.6	1.4		
Total				2571.4 mg m^{-2} yr^{-1}

Table 6. Standing Crop of All Trout (>96% Brook Trout) in Vicinity of Site 2 Based on Electroshocking, 1975

Age class	Mean wet weight, g	Number/m^2	Biomass, g/m^2
0$^+$	7.9	0.057	0.46
1$^+$	25.1	0.025	0.66
2$^+$	55.6	0.025	1.46
3$^+$	91.7	0.011	1.07
4$^+$	153.1	0.008	1.21
Total		0.126	4.86

Site 2

Brook trout made up over 96% of the trout present, the remainder being cutthroat and brown trout. Trout standing crop was estimated to be 647.1 in a 1220-m section of stream (5182-m^2 stream area) extensively electroshocked in 1975 as part of a trout removal experiment (Allan, 1982b). The standing crop in numbers and biomass is given for each age class in Table 6. Trout standing crop was estimated to be 4.86 g/m^2 wet weight. The breakdown by age classes strictly applies to mid-July, but for calculations here is assumed to be constant throughout the year.

Allan (1981) provides values for estimated daily ration from stomach analysis of brook trout collected in early June, July, August, September,

Table 7. Total Prey Consumption (mg dry wt/m^2) for Various Time Intervals and Age Classes of Trout, Site 2

Age class	May 1 to June 14	June 15 to July 31	Aug. 1 to Aug. 31	Sept. 1 to Sept. 20	Sept. 20 to Oct. 31	Summer total	Nov. 1 to Apr. 30	Year Total
0$^+$	162.9	182.5	92.6	19.5	26.3	483.7	115.6	559.3
1$^+$	149.1	694.7	83.2	31.5	63.5	1022.0	275.1	1297.1
2$^+$	149.1	694.7	83.2	31.5	63.5	1022.0	275.1	1297.1
3$^+$	91.2	807.0	36.1	11.2	79.2	1024.7	337.9	1362.6
4$^+$	66.3	547.1	24.5	7.6	53.7	699.2	245.7	944.9
Total	618.6	2926.0	319.6	101.3	286.2	4251.6	1249.4	5501.0

and late September. Table 2 of that paper gives estimates for trout weighing <25 g, which were applied to the 0$^+$ age class; for trout weighing 25 to 50 g, which were applied to the 1$^+$ and 2$^+$ age classes; and for trout weighing >50 g, which were applied to the 3$^+$ and 4$^+$ age classes. Total prey consumption (mg dry wt/m^2) for various intervals was estimated for each age class:

$$\text{Estimated daily ration (mg/individual)} \times \text{standing crop (individual/m}^2)$$

$$\times \text{ duration of interval (d)} \qquad (2)$$

These calculations (Table 7) estimate total annual prey consumption to be 4.25 g m^{-2} yr^{-1}, assuming no feeding between November 1 and April 30 when water temperatures were 0°C (Figure 1) and the stream was covered by snow and ice. However, if trout feed throughout the winter at individual consumption rates based on the late September estimate, the total is increased by 1.26 g m^{-2} yr^{-1}, to 5.51. In comparison to prey consumption by stone flies, 77% of the value obtained by including winter feeding is contributed when water temperatures are above 0°C.

I used average values for the age distribution of trout because adequate data were lacking to specify an exact age distibution for each time interval. The values used were representative of the summer period when prey consumption was greatest, however, so accuracy should not be seriously affected. The values for estimated daily ration for large trout in the July collections were greater than Elliott (1975) estimated for brown trout of comparable size in the laboratory. Limitations of these estimates are discussed in Allan (1981); the high values were caused by a few trout eating large earthworms. It is possible that the estimates of Table 7 are too high.

I attempted another estimate using Elliott's (1975) equations relating maximum daily ration for brown trout to trout size and water temperature. Trout size and abundance data again were from Table 6, and water temperature was taken from Figure 1. Obviously this is only approximate since brook trout may differ in their temperature constants. The values obtained were 2.8 g m^{-2} yr^{-1} exlcuding winter feeding and 3.13 g m^{-2} yr^{-1} including winter feeding.

Site 3

The standing crop of trout, all of which were brown trout, is given in Table 8. Results are similar to Table 6. For a given length of stream trout densities were greater at site 3 than at site 2, but, since the stream was wider, numbers per square meter did not differ greatly. Trout were separated into age groups by a size frequency histogram of 104 trout collected July 11, 1977. Computations are based on the average weight of each age class, however. The indicated age classes are only approximations, as was true for brook trout.

Since no stomach analysis was done on brown trout at site 3, daily consumption rates could not be estimated directly. Instead, Elliott's (1975) equation for size of maximum daily ration (mg dry wt/d) was used. Trout size and abundance were taken from Table 8, and water temperature was from Figure 1. The results (Table 9) place total annual prey consumption at 4.99 g m^{-2} yr^{-1} if trout cease to feed at 0° C. To take possible winter feeding into account, I used a feeding rate associated with 1° C; this raised the estimate to 5.29 g m^{-2} yr^{-1}. In comparison with prey consumption by stoneflies, 94.3% of the value obtained including winter feeding occurs when water temperature is above 0° C.

Combined Trout and Stonefly Predation

The predation figures can now be combined to estimate total predation pressure at each site. Since estimates were most speculative for the winter months, because of lack of information on feeding and gut clearance rates at 0°C, I compare, first, the data when water temperatures were >0° C and, second, the estimate including values for winter feeding (Figure 2). The overall conclusions are the same.

Total predation pressure from trout and stoneflies ranges from a low of 2.18 to a high of 8.81 g m^{-2} yr^{-1} dry wt. The effect of trout, where present, was about twice that of the predaceous stoneflies. At the highest site (site 1) where only stoneflies were present, the highest single estimate of stonefly predation was obtained.

Table 8. Standing Crop of All Trout (Brown Trout Only) in the Vicinity of Site 3, Based on Electroshocking July 1977 and July 1979

Age class	Mean wet weight, g	Numbers/m^2	Biomass, g/m^2
0^+	10.2	0.029	0.29
1^+	35.3	0.025	0.87
2^+	61.4	0.023	1.40
3^+	102.8	0.013	1.37
4^+	160.8	0.010	1.53
Total		0.100	5.46

Table 9. Total Prey Consumption (mg dry wt/m^2) for Various time Intervals and Age Classes of Trout, Site 3

Age class	Apr. 1–30	May 1–31	June 1–30	July 1 to Aug. 31	Sept. 1–30	Oct. 1–31	Summer total	Nov. 1 to Mar. 31	Year total
0^+	9.5	58.7	67.4	196.1	80.0	14.9	426.6	25.5	452.1
1^+	21.1	129.8	149.1	433.7	176.9	33.1	943.7	56.7	1000.4
2^+	29.6	181.8	208.7	607.3	247.7	46.4	1321.5	79.5	1401.0
3^+	24.8	151.9	174.5	507.6	207.0	38.8	1104.6	66.5	1171.1
4^+	26.8	164.1	188.5	548.3	223.6	42.0	1193.3	72.0	1265.3
Total	111.8	686.3	788.2	2293.0	935.2	175.2	4989.7	300.2	5289.8

DISCUSSION

The importance of setipalpian stoneflies, or any other invertebrate predator in running water, has rarely been assessed. The only figure for comparison of total annual prey consumption is that of Siegfried and Knight (1976) for *Calineuria californica*. Their figure of 3.7 g m^{-2} yr^{-1} is not greatly higher than my estimates. They make no mention of values used for gut clearance times, and water temperatures were warmer on the average in the stream they studied. Both of these factors could account for some difference. Heiman and Knight (1975), in a laboratory study of *C. californica*, found daily feeding rates to be on the order of 6 to 8% of body weight at summer temperatures and 1 to 2% during winter. These values are similar to but slightly smaller than my estimates of 10%/d during the summer and 2.5%/d during winter.

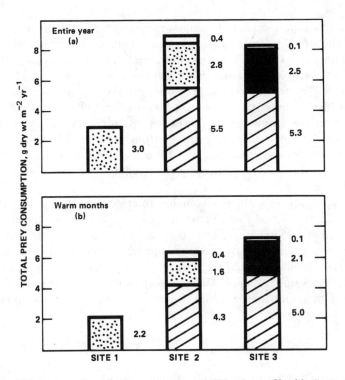

Figure 2. Estimated total prey consumption by trout (▨) and stoneflies *M. signata,* ▦; *K. modestus,* ☐; and *C. sabulosa,* ■ at three sites in Cement Creek. The warm months (b), when water was warmer than 0°C, were May 1 to Oct. 31 at sites 1 and 2 and Apr. 1 to Oct. 31 at site 3.

The results of my study indicate that, where both trout and stoneflies occur, trout consume about twice as much as stoneflies (estimates ranged from 1.7 to 2.3). Nonetheless, the amount eaten by invertebrate predators appears to be substantial and would be all the more so if additional invertebrate predators were included. Micropredators, especially among the Chironomidae, were not considered in this study. The predation rates of rhyacophilid caddis flies, which occurred at all sites in Cement Creek, were not measured. At sites 1 and 2, I believe that stoneflies were the major invertebrate predators, and thus my estimates should represent nearly the total amount of predation. At site 3, however, the large, net-spinning caddis *Arctopsyche grandis* was common, as well as a phila-potomid and the rhyacophilid species. Thus the contribution to predation by Trichoptera might substantially increase the estimate of total inverte-brate predation, to perhaps the same order of magnitude as trout predation.

In light of these results, it is not surprising that Brocksen et al. (1968) found the stonefly *Hesperoperla pacifica* to compete effectively with trout in a laboratory stream. Whether invertebrate predators actually depress trout production in nature is undemonstrated, however. The most obvious difference in their feeding is the role of the Chironomidae, which are of major importance to stoneflies (Allan, 1982a) and of lesser importance to trout (Allan, 1981), except for very young trout. In addition, the mode of feeding differs; trout presumably feed principally from the drift, and stoneflies feed from the benthos. Finally, whether trout or stoneflies are limited by the availability of food is an open question.

For comparison with the estimated food consumed at each site, the mean standing crops (mg dry wt/m^2) for *Baetis,* total Ephemeroptera, Chironomidae, Simuliidae, and total fauna are presented in Table 10. These were determined from Surber sample estimates of densities, head width measurements, and regressions of dry weight on head capsule width. The value for total fauna is most subject to error since weight–head-width regressions were incomplete or were lacking for some rare items.

Table 10. Average (April–October) Standing Crop Biomass of Invertebrates (mg dry wt/m^2) at Three Sites in Cement Creek and Ratio of Prey Consumed (from Figure 2) to Prey Available*

Invertebrate	Site 1	Site 2	Site 3
Baetis bicaudatus	136.1	45.1	111.8
	(1.0–408.2)	(1.4–284.4)	(44.8–186.0)
Ephemeroptera	825.6	354.2	231.1
	(211.2–1595.5)	(34.6–915.6)	(94.3–351.5)
Chironomidae	41.5	35.0	35.3
	(7.5–98.4)	(2.1–113.7)	(2.1–96.3)
Simuliidae	1.6	3.4	22.0
	(0.0–5.5)	(0.0–9.0)	(0.0–95.5)
Total fauna	1191.9	743.3	821.7
	(480.6–2114.1)	(272.9–1120.6)	(357.7–1961.8)
Amount consumed/ amount available			
Main prey	2.5	15.9	24.8
Total fauna	1.8	8.4	8.7

*Biomass range is given in parentheses.

Standing crop biomass (g dry wt/m^2) of all invertebrates in Cement Creek, averaged over the 7 months from April to October, was 1.12 at site 1, 0.74 at site 2, and 0.82 at site 3 for a grand average of 0.92 g dry wt/m^2. Maximum values were 1.1 to 2.1 g/m^2 (Table 10). This time interval was chosen to represent the period when grazing presumably was most intense.

These values are comparable to other estimates, e.g., for the Horokiwi Stream (0.4 to 1.2, Allen, 1951), for Berry Creek, Oregon (0.4 to 0.8 in an unenriched section, Warren et al., 1964), for Afon Hirnant, Wales (0.8, Hynes and Coleman, 1968), for River Endrick, Scotland (1.6, Maitland, 1964), and for the Lusk River, Alberta (0.8 to 1.1, Radford and Hartland-Rowe, 1971). Somewhat higher values were obtained by other investigators; e.g., Horton (1961) reported values of 1.5 to 3.5 g/m^2 in the Walla Brook, United Kingdom; MacKay and Kalff (1969), 2.2 g/m^2 in West Creek, Quebec, and Coffman et al. (1971), 4.8 g/m^2 in Linesville Creek, Pennsylvania. In a comparison of streams in the Rocky Mountains, Pennak (1977) estimated an average biomass of 1 g/m^2, with a range of 0.15 to 3.7. Although techniques varied among workers (corrections were needed to compare estimates obtained by weighing specimens wet or dried, with or without storage in preservative, or by using volume to estimate wet weight), there is general agreement of a standing crop biomass (g dry wt/m^2) of $\approx$1 g in Rocky Mountain trout streams, up to 2 to 3 g in woodland streams, and an occasional estimate approaching 5 g.

The amount of prey consumed is compared to the amount available (April to October mean biomass) by using data from Figure 2 and Table 10. This ratio (see Table 10) cannot exceed the turnover ratio (production/biomass, P/B) and will equal the turnover ratio only if all mortality losses of the invertebrate community are from predation, an unlikely event. A very low ratio of prey consumed to prey available implies that predators were not responsible for a major amount of invertebrate mortality. A very high value may indicate an error in estimation.

Waters (1977) reviewed published estimates of production and turnover ratios (P/B) for secondary consumers in freshwater. For invertebrates in streams, he concluded that P/B values of 4 to 7 were typical for most univoltine and bivoltine species and 1 to 3 was usual for longer-lived species. Multivoltine Diptera and some Ephemeroptera may reach P/B values of 10. More recently, Neves (1979) estimated annual turnover rates of some chironomids to be over 50, and Hall et al. (1980) estimated an annual P/B of 26 in the bivoltine mayfly (*Tricorythodes atratus*) in a warm-water Minnesota stream. Waters (1979) acknowledged that turnover rates for some fast-growing chironomids, simuliids, and Ephemeroptera can be higher than previously thought. Benke (1976) suggested that biomass of prey consumed as a ratio of prey standing crop provides an indirect estimate of turnover rate which typically are higher in value than most direct estimates. He argued from such estimates that P/B ratios may be in excess of 30.

The annual turnover rate for the invertebrate community of a high mountain stream is not known exactly. The cold temperature regime (Figure 1) would suggest low values. I have estimated annual production, standing crop biomass, and P/B for some insects in Cement Creek (Allan, unpublished). Three estimated annual P/B values for perlodid stone flies fell between 2.1 and 2.7. Seven estimates for mayfly species ranged from 4.4 to 9.9. The two highest values were for the bivoltine *Baetis bicaudatus* at site 1 (annual P/B, 9.6) and the univoltine *Ephermerella coloradensis* at site 2 (P/B, 9.9). No chironomid turnover rates were estimated; however, since estimates of average biomass were quite low (Table 10), it is doubtful that this matters greatly. In sum, although certain recent estimates of turnover ratios indicate the possibility of high values, the weight of evidence leans toward an estimate between roughly 3 and 8 for the invertebrate community of Cement Creek.

At site 1, where only stonefly predation was important, predators removed 2.5 times the main prey biomass and 1.8 times the total faunal biomass. These values are well within estimated turnover ratios and are strikingly lower than estimates for sites 2 and 3. The implication is that food was plentiful, but the actual availability of prey to the predator is unknown. Certainly the stoneflies did not consume an impossible amount of prey. At sites 2 and 3, in contrast, total estimated predation removed between 8 and 9 times the average biomass of the total invertebrate fauna and 16 to 25 times the biomass of preferred prey. Quite possibly the estimate of the main prey at site 3 is low, because it does not include trichopterans. Caddis larvae were a much greater part of the standing crop biomass at this site, and, since the diet of brown trout at site 3 was not investigated, it is possible that caddis larvae were important prey.

No correction has been made in these calculations for the role of terrestrial food in trout diet. During the spring and summer, when trout feeding was greatest at site 2 (Allan, 1981), terrestrial items constituted only 2.4% by numbers of total prey consumption, whereas in late summer, when trout feeding rates declined, the figure was 10.5%. This would require only a very small correction, which was judged unnecessary in light of other sources of error.

My results show that total prey consumption by predators is in excess of production, or at least nearly equals total production, at the two sites where trout were present. Allen's paradox is demonstrated once again.

What is noteworthy here is the fact that the excess of consumption over production is much less than in comparable studies. Allen (1951) estimated prey consumption to be 40 to 150 times the standing crop of invertebrates, a figure criticized as too high because of overestimation of trout feeding rates (Gerking, 1962) and underestimation of prey densities (Macan, 1958). Using Allen's original estimates of trout and invertebrate biomass and Elliott's (1975) estimates of the maximum ration a brown trout can consume in a day as a function of its size and temperature, I recalculated the ratio of prey consumed to invertebrate standing crop. For all sections of the Horokiwi combined, the maximum value is about 20. This does not take into account the criticism that invertebrates were underestimated because of too wide a mesh. Horton (1961), using methods similar to Allen's, estimated annual prey consumption to be 9 to 26 times mean standing crop, but she almost certainly overestimated trout feeding rates by at least a factor of two. Finally, in a careful study of cutthroat trout in a laboratory stream, Warren et al. (1964) estimated the ratio of total consumption to mean standing crop to be 11.5 and 4.0 in two unenriched stream sections and 2.3 and 2.8 in two sections enriched with sucrose. Since trout derived 29 to 58% of their diet from terrestrial items, these figures appear well within conservative estimates for P/B.

The tentative conclusion I draw from these comparisons is that Allen's paradox, although not fully resolved, is of a smaller magnitude than it once appeared to be. In fact, if numbers of invertebrates are underestimated by anything near the 5 or 10 to 1 (by numbers) factor suggested by investigations of fauna deep in the substrate (Hynes and Coleman, 1968; Hynes, et al., 1976), prey might be quite abundant relative to consumption. We clearly need more information on the hyporheic fauna, particularly in terms of biomass, to evaluate prey standing crop. It is also likely that my figures for prey consumption represent maximum rations. The estimate of food consumption by brown trout at site 3 assumed that these predators could fill their guts as soon as they were empty. Clearly, further research on these points is called for, with the likely result that the ratio of prey consumed to prey available will continue to fall to within reasonably conservative estimates of P/B values.

There is some reason to believe that these predators are in fact not limiting the density of their prey. Allan (1982b) experimentally reduced trout densities in a 1220-m section of Cement Creek to between 10 and 25% of initial standing crop biomass for 4 years. Invertebrate density did not increase; this suggests no significant effect of trout in reducing prey abundance. Cooper et al. (1962), examining growth rate of brook trout in stream sections where fish densities were lowered by rotenone, found only a very small increase in comparison with normal sections; this again suggests that food availability was similar at high and low trout densities. Heiman and Knight (1975), comparing the growth rate of *Calineuria californica* nymphs in the laboratory fed preferred food, to nymphs in the field, found equivalent growth rates. They concluded from this that food in the field must be plentiful or a reduced growth rate relative to that in the laboratory would result. For *Megarcys signata* at site 1, the only case where my data were adequate to estimate growth rate, high constant growth was observed from June to October. Admittedly, however, this species has a monopoly on the macropredator role at this site.

A number of assumptions, each of which is a potential source of error, underlie the estimates of this study. These may be separated into error associated with estimating (1) prey standing crop, (2) prey turnover rates, (3) predator density, (4) stonefly feeding rates, (5) trout feeding rates. These assumptions are discussed further in Allan (1981; 1982a; 1982b). In brief, I assume turnover ratios (P/B) to fall between 3 and 8 for the total community, on the basis of the work of Waters (1977) and my own estimates for certain species in Cement Creek. Although much higher P/B values have recently been published (Waters, 1979), I doubt their applicability to this high-elevation stream. I believe that estimates of total prey consumed in this study are fairly accurate, but probably err on the high side because of the assumption that stoneflies are able to fill their guts fully and the use of an equation for brown trout at site 3 which predicts maximum ration. Standing crops of trout and stoneflies probably are estimated much better than those of the invertebrate community, because only the latter is likely to penetrate deep into the substrate. Although my estimates of prey standing crop are consistent with most published estimates, it seems quite plausible that all such estimates need to be adjusted upward to account for the hyporheic fauna. In conclusion, the Allen paradox of prey consumption greatly exceeding prey availability is not fully resolved, but appears to be of a magnitude within reach of a solution from improved accuracy in estimation techniques.

ACKNOWLEDGMENTS

I thank T. T. Macan and C. M. Drake for helpful discussions and for reading the manuscript, and the Freshwater Biological Association for hospitality during my sabbatical visit. The Colorado Division of Wildlife was most cooperative in allowing this research to be carried out, and Rick Sherman and Walt Burkhardt helped immeasurably. I am grateful to numerous field assistants and friends who helped with various stages of this research program. Support was provided by the National Science Foundation (DEB-75-03396 and DEB-77-11131).

REFERENCES

Allan, J. D., 1975, The Distributional Ecology and Diversity of Benthic Insects in Cement Creek, Colorado, *Ecology* 55: 1040-1053.

———, 1981, Determinants of Diet of Brook Trout (*Salvelinus fontinalis*) in a Mountain Stream, *Can. J. Fish. Aquat. Sci.*, 38: 184-192.

———, 1982a, Feeding Habits and Prey Consumption of Three Setipalpian Stoneflies (Plecoptera) in a Mountain Stream, *Ecology*, 63: 26-34.

———, 1982b, The Effects of Reduction in Trout Density on the Invertebrate Community of a Mountain Stream, *Ecology*, 63: 1444-1455.

Allen, K. R., 1951, The Horokiwi Stream: A Study of a Trout Population, *N. Z. Mar. Dep. Fish. Bull.*, 10.

Benke, A. C., 1976, Dragonfly Production and Prey Turnover, *Ecology*, 57: 915-927.

Brocksen, R. W., G. E. Davis, and C. E. Warren, 1968, Competition, Food Consumption, and Production of Sculpins and Trout in Laboratory Stream Communities, *J. Wildl. Mgt.*, 32: 51-75.

Coffman, W. P., K. W. Cummins, and J. C. Wuycheck, 1971, Energy flow in a Woodland Stream Ecosystem. I. Tissue Support Trophic Structure of the Autumnal Community, *Arch. Hydrobiol.*, 68: 232-276.

Cooper, E. L., J. A. Boccardy, and J. K. Andersen, 1962, Growth Rate of Brook Trout at Different Population Densities in a Small Infertile Stream, *Prog. Fish-Cult.*, 24: 74-80.

Davis, G. E., and C. E. Warren, 1965, Trophic Relations of a Sculpin in Laboratory Stream Communities, *J. Wildl. Mgt.*, 29: 846-871.

Elliott, J. M., 1975, Number of Meals in a Day, Maximum Weight of Food Consumed in a Day and Maximum Rate of Feeding for Brown Trout, *Salmo trutta* L., *Freshwater Biol.*, 5: 287-303.

Gerking, S. D., 1962, Production and Food Utlization in a Population of Bluegill Sunfish, *Ecol. Monogr.*, 32: 31-78.

Hall, R. J., T. F. Waters, and E. F. Cook, 1980, The Role of Drift Dispersal in Production Ecology of a Stream Mayfly, *Ecology:* 61: 37-43.

Heiman, D. R., and A. W. Knight, 1975, The Influence of Temperature on the Bioenergetics of the Carnivorous Stonefly Nymph, *Acroneuria californica* (Banks) (Plecoptera, Perlidae), *Ecology*, 56: 105-116.

Horton, P. A., 1961, The Bionomics of Brown Trout in a Dartmoor Stream, *J. Anim. Ecol.*, 30: 311-338.

Hynes, H. B. N., 1970, *The Ecology of Running Waters*, Liverpool University Press, Liverpool, England.

———, and M. J. Coleman, 1968. A Simple Method for Assessing the Annual Production of Stream Benthos, *Limnol. Oceanogr.*, 13: 569-73.

———, D. D. Williams, and N. E. Williams, 1976, Distribution of the Benthos Within the Substratum of a Welsh Mountain Stream, *Oikos*, 27: 307-310.

Macan, T. T., 1958, Causes and Effects of Short Emergence Periods in Insects, *Verh. Internat. Verein. Limnol.*, 13: 845-849.

MacKay, R. J., and J. Kalff, 1969, Seasonal Variation in Standing Crop and Species Diversity of Insect Communities in a Small Quebec Stream, *Ecology*, 50: 101-109.

Maitland, P. S., 1964, Quantitative Studies on the Invertebrate Fauna of Sandy and Stony Substrates in the River Endrick, Scotland, *Proceedings of the Royal Society of Edinburgh*, Section B, Vol. 68, Part IV (No 20), pp. 277-301.

Neves, R. J., 1979, Secondary Production of Epilithic Fauna in a Woodland Stream, *Am. Midl. Nat.*, 102: 209-224.

Pennak, R. W., 1977, Trophic Variables in Rocky Mountain Trout Streams, *Arch. Hydrobiol.*, 80: 253-285.

Radford, D. S., and R. Hartland-Rowe, 1971, Subsurface and Surface Sampling of Benthic Invertebrates in Two Streams, *Limnol. Oceanogr.*, 16: 114-120.

Siegfried, C. A., and A. W. Knight, 1976, Trophic Relations of *Acroneuria* (*Calineuria*) *californica* (Plecoptera: Perlidae) in a Sierra Foothill Stream, *Environ. Entomol.*, 5: 575-581.

Warren, C. E., J. H. Wales, G. E. Davis, and P. Doudoroff, 1964, Trout Production in an Experimental Stream Enriched with Sucrose, *J. Wildl. Mgt.*, 28: 617-660.

Waters, T. F., 1977, Secondary Production in Inland Waters, in A. Macfadyen (Ed.), *Advances in Ecological Research*, Vol. 10, pp. 91-164.

———, 1979, Benthic Life Histories: Summary and Future Needs, *Can. J. Fish. Aquat. Sci.*, 36: 342-345.

20. DO STREAM FISHES FORAGE OPTIMALLY IN NATURE?

A. John Gatz, Jr.

Department of Zoology
Ohio Wesleyan University
Delaware, Ohio

ABSTRACT

This study tests predictions of optimal foraging theory relating to differences between "pursuers" (sit-and-wait predators) and "searchers" (widely foraging predators). Theory predicts a more specialized diet and generalized habitat for pursuers than for searchers, negative skew to prey size distribution in pursuers only, larger relative prey size and more sexual dimorphism in body size in pursuers than in searchers, and greater character difference ratios in both trophic apparatus and body size in pursuers than in searchers.

Stream fishes (26 species) were positioned on the pursuer–searcher continuum according to the amount of red muscle in their axial musculature. Only fishes with intermediate or large amounts of red muscle actively cruise in search of prey; fishes with little or no red muscle closely approximate the classic sit-and-wait predatory type. Only the last prediction regarding relative magnitude of body and mouth size ratios for sympatric species was supported by the data. Problems in experimental design or in the theory itself are, of course, possible reasons why so few predictions were supported. Alternatively, selective forces for other functions could have overridden those for optimal foraging in all cases but the one. Particular selective forces that could lead to compromise of the predictions of optimal foraging theory, and hence to the discrepancies observed, include low levels of food abundance in streams, competitive interactions within predatory groups, and the effects of sexual selection.

INTRODUCTION

A recurrent theme in the development of optimal foraging theory (*sensu* Pyke et al., 1977) has been that differences should be expected between foragers with antithetical predatory styles. I investigate here the degree to which theoretically predicted differences in food, habitat, and morphology between different predatory types are realized in a group of stream fishes.

At one extreme of predatory style is a predator that expends negligible energy in search of prey but substantial amounts in pursuit. Such a predator is known variously as (1) a "pursuer" (MacArthur and Pianka, 1966), (2) a "sit-and-wait" predator (Pianka, 1966, (3) a "type I" predator (Schoener, 1969), and (4) a "passive searcher" (Eckhardt, 1979). At the opposite extreme, a predator that expends negligible energy in pursuit but substantial amounts in search of prey is referred to by these respective investigators as (1) a "searcher", (2) a "widely foraging species", (3) a "type II" predator, and (4) an "active searcher." I use the terms "pursuer" and "searcher" in this paper. A series of six differences to be expected between predators of these two types is summarized in Table 1.

Heretofore, placement of a particular species in either the pursuer or searcher category has usually depended on subjective decisions based on natural historical observations. Thus Pianka (1966) made placements after having observed foraging behavior in 12 species of desert lizards, and Schoener (1969) placed a number of species from each of several classes of vertebrates, and some invertebrates, in each category based on similar information. Conspicuously absent from these and other lists (e.g., MacArthur and Pianka, 1966) are fishes.

Table 1. Predicted Differences Between Searchers and Pursuers*

Pursuers	Searchers
1. Specialized prey size and type	1. Generalize in prey size and type
2. Overlapped habitats	2. Habitat specialize
3. Negative skew to prey size distribution	3. Normal or positive skew to prey size distribution
4. Greater relative prey size	4. Smaller relative prey size
5. More sexually dimorphic in size	5. Less sexually dimorphic in size
6. Coarse-grained differences in sizes	6. Fine-grained differences in sizes

*Sources of predictions are: 1, MacArthur and Pianka (1966); 1, 2, and 6, MacArthur and Levins (1964); and 1, 3, 4, and 5, Schoener (1969).

Fishes can be assigned to categories of predatory style objectively based on the physiological characteristics of their musculature. Boddoke et al. (1959) (see also Gatz, 1979) showed that mode of life—including manner of predation—is related to the amount of red muscle in the axial musculature of fishes. Morphological and biochemical differences between red and white fibers (see Gatz, 1973, for references) produce functional differences that enable fishes with high percentages of red muscle to cruise continuously and aerobically, whereas fishes with only or nearly only white fibers exhibit principally anaerobic bursts of swimming. Field observations of the species dealt with here verify that fishes with large amounts of red muscle are searchers whereas fishes lacking or nearly lacking red muscle are pursuers. Thus the percentage of red muscle in the caudal peduncle of a fish provides both a convenient and an objective way of quantifying fish predatory behavior.

MATERIALS AND METHODS

I collected fishes by seine on six or seven occasions at each of four to six stations in three streams in three different drainages in the Piedmont region of North Carolina and preserved them in 10% Formalin immediately after capture. All species of Esocidae (pickerels), Cyprinidae (minnows), Centrarchidae (sunfishes), and Etheostomatinae (darters) collected more than once over my 2-yr total collecting period were included in the analysis. Red muscle content of ten individuals of each species was determined by making freezing microtome sections of the entire caudal peduncle, staining for fats, and using a planimeter to determine areas of both red and white fibers on drawings of the sections traced from the image of a microprojector. No seasonal variations in red muscle content were identified.

Gut content analyses were performed on the same ten individuals and additional "replacement" individuals for any of the original fish that had empty digestive tracts. Specimens of average adult size in the streams were selected to evenly represent collections made in all seasons. Thus the food habits represent the full annual diet and not that of only a single season. Prey items were counted, measured with a millimeter rule (longest dimension), and placed into one of 12 prey-type categories—fish, copepods, ostracods, aquatic insects, terrestrial insects, crayfishes, isopods and amphipods, molluscs, non-insect terrestrial invertebrates, diatoms, filamentous algae, and vascular plants. Relative prey size was calculated as mean prey size for the species divided by mean standard length for the species.

Several of the predictions (Table 1) required nonstandard measurements. For prediction 2, a measure of habitat specialization was needed. I used the number of stations at which each species occurred as a percentage of the total number possible given the species' presence or absence in a specific drainage. Thus a species that occurred at four of six stations in the one stream in which it was found would be considered to be no more or less habitat specialized (67%) than another species that occurred in all three streams but was present at only ten of the 15 stations sampled. For prediction 3, I analyzed skew in prey-size distribution by calculating g_1 as an estimate of γ_1 and then performed a t-test of the hypothesis $g_1 = 0$ (Sokal and Rohlf, 1969). For predictions 4 and 6, measures of fish size and mouth size were necessary. I used standard length for fish size and the average of mouth width and height for mouth size. The interior lateral dimension of the mouth when fully opened was measured with vernier calipers for mouth width, and an analogous dorso-ventral dimension was used for mouth height.

RESULTS

Percentage of red muscle content for the 26 species in this study are given in Table 2, along with standard lengths, mouth sizes, and other species mean values for features used in subsequent analyses. Notice that the fishes, when separated by families, also form three non-overlapping predator-type categories according to their red muscle contents (0 to 2%, 3 to 5%, and 6 to 12%).

Prediction 1

Pursuers and searchers did not differ significantly in specialization as to prey type as was predicted (Tables 1 and 3). Furthermore, the intermediate predator, sunfishes, tended to be more specialized than either of the extreme predatory types. These same results were found no matter what index of food-type specialization was used, e.g., number of prey types; Simpson's diversity (D) of prey types; or equitability (E) of prey types, where E = H/Hmax from the Shannon–Wiener function (see Krebs, 1978).

Similarly, prediction 1 relative to food-size specialization also was not supported. Specialization was measured by using the coefficient of variation of prey sizes (Schoener, 1969). The overall pattern of results was insignificant but opposite from the theoretical expectation (Table 3). A reason for the observed pattern may be the strong correlation between predator size and variability of prey size (r = 0.756, 24 df, $P < 0.001$); the sunfishes were larger than the minnows, and the pickerels were larger still.

Table 2. Species Means (n = 10) for Various Features

Species	Red muscle, %	Standard length, mm	Mouth size, mm	Habitats %	Number of foods	Diversity of foods	Equitability of foods	Coefficient variation in food sizes	Prey size, mm
Pursuers									
Esox americanus	2.2	120.6	15.5	100	5	0.35	0.48	40	23.9
Esox niger	1.8	201.7	24.8	18	5	0.60	0.73	355	45.7
Etheostoma flabellare	0.0	47.8	4.6	78	2	0.10	0.29	20	6.3
Etheostoma fusiforme	0.0	33.2	2.4	33	2	0.32	0.72	3	1.6
Etheostoma olmstedi	0.0	46.2	3.5	100	1	0	0	2	3.2
Intermediate									
Centrarchus macropterus	5.0	87.0	11.7	50	4	0.59	1.47	33	9.3
Lepomis auritus	4.4	88.5	12.5	67	5	0.59	1.70	48	18.7
Lepomis cyanellus	3.9	97.2	16.5	93	5	0.67	1.82	56	15.9
Lepomis gibbosus	4.7	126.3	23.7	27	3	0.64	1.52	8	5.6
Lepomis gulosus	4.2	89.3	9.4	73	4	0.60	1.54	101	17.1
Lepomis macrochirus	4.2	78.3	8.3	100	3	0.52	1.19	12	5.0
Micropterus salmoides	4.4	101.6	16.4	73	4	0.66	1.71	160	13.2
Pomoxis nigromaculatus	3.4	126.9	19.0	55	5	0.48	1.46	6	4.0
Searchers									
Clinostomus funduloides	9.8	59.7	9.3	89	1	0	0	17	4.7
Hybopsis hypsinotus	6.7	56.6	5.3	100	5	0.35	1.12	7	2.6
Nocomis leptocephalus	9.7	80.9	8.3	67	4	0.48	1.36	13	3.3
Notemigonus crysoleucas	6.7	76.0	5.1	45	5	0.59	1.70	7	1.0
Notropis alborus	5.8	53.1	4.0	27	3	0.64	1.52	4	1.0
Notropis altipinnis	8.8	43.1	3.9	83	2	0.42	0.88	3	4.4
Notropis analostanus	9.9	49.0	4.9	44	4	0.56	1.35	5	3.5
Notropis ardens	11.1	54.7	5.3	80	2	0.42	0.88	10	4.4
Notropis cerasinus	9.2	60.7	6.3	100	3	0.34	0.88	11	6.6
Notropis chiliticus	12.0	51.5	5.7	100	3	0.56	1.37	11	4.8
Notropis procne	7.5	44.8	3.2	60	4	0.59	1.47	3	1.0
Phoxinus oreas	6.1	47.4	3.4	67	2	0.10	0.29	1	0.1
Semotilus atromaculatus	9.2	83.9	10.6	80	1	0	0	26	10.9

Prediction 2

Habitat specialiation, as measured by percentage of stations at which a species occurred, did not differ significantly between any of the predator categories (Table 3) as predicted.

Prediction 3

All statistically significant skews were positive (Table 3). Only one darter and three minnows showed insignificantly negative values for g_1. A 2 × 3 contingency table analysis of these results based on positive or negative sign of g_1 (ignoring statistical significance) gives $\chi^2 = 2.062$, 2 df, $P > 0.25$ that there are no differences among my three predator categories. Prediction 3 is, therefore, not supported.

Table 3. Means for Selected Features for Species Listed in Table 2*

Feature		Pursuers (darters and pickerels, n = 5)		Intermediate (sunfishes, n = 8)		Searchers (minnows, n = 13)
Red muscle ($\overline{x} \pm se$)		0.8% ± 0.49		4.3% ± 0.17		8.7% ± 0.54
Number of food categories	NS	3.0	NS	4.1	*	3.0
Diversity of food categories	NS	0.27	**	0.59	*	0.39
Equitability of food categories	NS	0.44	*	0.78	NS	0.60
Coefficient of variation in prey size	NS	84%	NS	53%	**	9%
Percentage of habitats found	NS	66%	NS	67%	NS	72%
Skew in prey-size distribution		3 positive 2 normal		8 positive 0 normal		10 positive 3 normal
Relative prey size	*	0.135	NS	0.118	*	0.062
Sexual dimorphism		3/3		1/8		5/6
Mouth size ratios	**	1.45	**	1.21	NS	1.21
Body size ratios	**	1.43	**	1.09	NS	1.13

*Notations in the left column after the features indicate the significance of differences between pursuers and searchers using two-tailed t-tests: NS, not significant; *, P < 0.05; **, P < 0.01. Similar notations between the values for either the pursuers and intermediate or the intermediate and searchers indicate the significance of the differences between the means for these two respective pairwise comparisons.

Prediction 4

Relative prey size varied among predator categories (Table 3). Although the direction of variation was as predicted, I do not interpret these results as support for the prediction. The result appears instead to be merely an artifact produced by the strong correlation between prey size and standard length of the fish ($r = 0.810$, 24 df, $P < 0.001$). My sample of pursuers included the largest fishes, the pickerels, and the intermediate predators, the sunfishes, were intermediate in size also. If the relative prey size of just the darters (0.083) from among the pursuers is compared with that for the similarly sized searchers (0.062), no significant difference is seen. Further, the mean relative prey size of the intermediate predators (0.118) is now significantly larger than that for the small pursuers.

Prediction 5

Carlander (1950; 1969; 1977) summarized the results of studies of differences in growth rates for many of the species used in this study. All three of my pursuers listed by Carlander show sexual dimorphism, as do five of the six of my searchers in which dimorphism has been studied. On the other hand, in only one of the eight sunfish species that I used has a consistent dimorphism been reported. There is an association between dimorphism and predator category (2 × 3 contingency $\chi^2 = 9.356$, $P < 0.01$), but it is the intermediate group rather than the most extreme searchers, as theory would predict, that shows the least sexual dimorphism. The searchers and pursuers show similarly high levels of dimorphism (Table 3). Therefore this prediction, too, is not supported by these fishes.

Prediction 6

Character difference ratios in mouth size and body size (mouth or body size of one species divided by mouth or body size of the next smaller species; see Schoener, 1965) were calculated for all sympatric members of each family of fishes (Table 3). They provide partial support for prediction 6, in that pursuers do show the highest mean ratios, although the searchers and intermediate predators do not differ as might be expected. MacArthur and Levins' (1964) expectation of coarse-grained differences in pursuing species was interpreted by Eckhardt (1979) to mean that character difference ratios ≥ 1.14 should be seen more frequently in pursuers than in searchers. Eckhardt apparently took this ratio from Schoener's (1965) ". . . somewhat arbitrary ratio of 1.15

Table 4. Character Difference Ratios for Different Predator Types*

Ratios	Pursuers	Intermediate	Searchers	Totals
Mouth-size				
≥ 1.15	5 (3.1)	9 (8.7)	9 (11.2)	23
< 1.15	0 (1.9)	5 (5.3)	9 (6.8)	14
Total	5	14	18	37
$\chi^2 = 4.236$; $P > 0.1$				
Standard-length				
≥ 1.15	4 (1.4)	2 (3.8)	4 (4.9)	10
< 1.15	1 (3.6)	12 (10.2)	14 (13.1)	27
Total	5	14	18	37
$\chi^2 = 8.104$; $P < 0.025$				

*Expected numbers are given in parentheses. See text for further explanation.

[which] was chosen as probably indicating a significant amount of separation by food size. . . ." Contingency table analyses of frequency of ratios ≥ 1.15 (Schoener, 1965) show no significant differences by predator type in mouth-size ratios but significant deviation from random in body-size ratios (Table 4). Again, the searchers and intermediate predators show very similar percentages of ratios on either side of this arbitrary cut-off figure.

DISCUSSION

The results presented here provide minimal support for only the last of the predictions of optimal foraging models (see Tables 1 and 3). Such an absence of concordance with theory is certainly not without precedent in the literature (e.g., Eckhardt, 1979) and may, in fact, indicate some basic flaws in the theory. Alternatively, the authors of the predictions themselves have offered numerous qualifiers and caveats that provide partial explanations for some of the deviations. I shall examine the predictions (Table 1) *seriatim* in an attempt to recognize causes for the discrepancies.

Predictions of differences in degree of prey specialization between different predator types depend on the availability of sufficient prey. If food is scarce, all predators are expected to generalize (Pyke et al., 1977). On the basis of the generally slower growth rates of many species of fishes in streams than in impoundments (Carlander, 1969), food may indeed be limiting at the study sites. The possibility of a limited food supply cannot wholly explain my results, however, since some significant differences in degree of specialization are seen. For instance, the intermediate predators, sunfishes, are more generalized in some measures of prey type than either searchers or pursuers. With regard to variability in prey size, MacArthur (1972) noted that larger species are more generalized in diet than smaller species, and my data showed just such a correlation. Moreover, Schoener (1969) noted that the size distributions of prey consumed become ever more similar with small predators. Similarities in prey-size distributions in darters and minnows might, therefore, be expected independently of their respective predatory behaviors, and thus lead to the similar coefficients of variation in prey size noted for these two groups. Because the predictions of optimal foraging are opposite from empirical generalizations related to predator size, my data permit a tentative acceptance of one alternative and rejection of the other. The data do not at all fit the predictions of optimal foraging theory; thus it seems that whatever selective forces exist relative to predator size alone overrode selection for simple prey maximization per unit time in these streams.

Habitat specialization complementary to food generalization is a rather general expectation of competition theory (Schoener, 1974). Given that all predator categories showed similar levels of food specialization, then similar levels of habitat specialization for all categories would be predicted by competition theory if these two dimensions are to remain equally complementary. Another possible explanation for the habitat specialization results is also in the literature; MacArthur and Pianka (1966) suggest that the degree of habitat specialization is related to food density. They predict that, in times of high food density, pursuers should be more habitat specialized than searchers. Obviously, the presumed absence of dense food can now be offered as a possible reason for my lack of association between predatory type and degree of habitat specialization. Note, however, that the prediction of MacArthur and Pianka is directly opposite to that of MacArthur and Levins (1964). A final possibility for the absence of differences in habitat specialization which also should be stated explicitly is that the mere presence or absence of a species at my rather large sampling stations (100 to 200 m of stream bed) provided too crude a measure of habitat specialization to show any correlations.

Skew of prey-size distributions has been traditionally measured as I measured it by numbers of prey items in various prey-size categories (Schoener, 1969). Still, a possible reason for the preponderance of positively skewed distributions for all species—especially pursuers—was the use of numbers of prey to generate prey-size distributions. Thus a pickerel that had eaten several small insects and only one large fish or crayfish would have a positively skewed prey-size distribution. If a volumetric basis had been used to weight the results, the negative skew predicted by theory may well have been seen.

Relative prey size followed the pattern of variation predicted by optimal foraging theory only if there were size differences among the different predatory categories. As soon as the size variation was factored out (i.e., when just darters and minnows were compared), the differences in relative prey size predicted by optimal foraging theory disappeared. Both these results are in accord with a theory that relative prey size is determined by predator size alone; the second result leads to a rejection of the hypothesis that relative prey size is determined by predatory behavioral type. Thus, as seems to have been the case in the prediction related to variability in prey size, whatever selective forces exist as a result of predator size alone may have overridden the considerations on which optimal foraging theory is based.

Along with his prediction, Schoener (1969) also provided several possible reasons why differences in degree of sexual dimorphism might not hold in any group of predators. First, he noted that, if pursuit and/or handling and swallowing times are a function of predator size, as they are in at least some fishes (Werner, 1974), then the prediction of dimorphism in pursuers no longer holds and searchers will be strongly monomorphic. My results do not support these alternative predictions either. Second, Schoener noted that the presence of competitors can cause dimorphism not to be seen. As dimorphism was seen here in two of three groups (Table 3), Schoener's competition caveat seems not to apply either. Third, Schoener noted that the presence of a large trophic apparatus decreases the probability of dimorphism. This final caveat provides a possible explanation for the low level of sexual dimorphism observed in the sunfishes. Relative mouth height (mouth height to standard length) and relative mouth width (mouth width to standard length) were both significantly larger in sunfishes than in either pursuers or searchers ($0.150 > 0.110 \approx 0.110$ and $0.139 > 0.089 \approx 0.085$, respectively). Thus dimorphism would be expected to be lowest among the sunfishes. Sexual dimorphism among the searchers still remains to be "explained," however. Sexual selection for large male body size for nest building or other reproductive behaviors among the species in my searcher category may have overridden any selective pressures as a result of foraging type and, therefore, may provide a possible explanation.

The prediction of coarse-grained differences in pursuers is the one most nearly supported by my data (Tables 3 and 4). The main discrepancy is the similarity between sunfishes and minnows; the sunfishes did not demonstrate intermediate character difference ratios.

In summation, two general factors, along with particular caveats, have thus far been suggested as being reponsible for the disagreement between my data and the theoretical predictions. These factors are differences in body size between the members of my searcher and pursuer groups (and the intermediate predators) and a limitation of the food supply in nature. In addition, several points made by Eckhardt (1979) bear restatement since they may apply to my study as well as his: (1) The time scales of foraging theory and a year-round field study such as this may be too different. (2) The predictions may be more qualified than indicated explicitly either in Table 1 or in the original papers. (3) The theory may apply only to unrealistic behavioral extremes rather than to the range of behavioral types covered by the species used here. (4) Competitive interactions among species of a single predatory type may obscure general foraging patterns for the group as a whole. Further studies in which attention is paid to the body-size and time-scale problems, especially, might be most profitable.

ACKNOWLEDGMENTS

I thank Ohio Wesleyan University for affording me the time off from my teaching responsibilities to undertake this study. E. E. Werner and W. K. Patton provided helpful comments on an earlier version of this paper.

REFERENCES

Boddoke, R., E. J. Slijper, and A. van der Stelt, 1959, Histological Characteristics of the Body Musculature of Fishes in Connexion with Their Mode of Life, *Akademie van Wetenshappen Koninkl. Nederl., Proceedings C,* 62: 576-588.

Carlander, K. D., 1950, *Handbook of Freshwater Fishery Biology,* 1st ed., W. C. Brown Co., Dubuque, IA.

———, 1969, *Handbook of Freshwater Fishery Biology,* Vol. 1, 3rd ed., Iowa State University Press, Ames.

———, 1977, *Handbook of Freshwater Fishery Biology,* Vol. 2, 1st ed., Iowa State University Press, Ames.

Eckhardt, R. C., 1979, The Adaptive Syndromes of Two Guilds of Insectivorous Birds in the Colorado Rocky Mountains, *Ecol. Monogr.,* 49: 129-149.

Gatz, A. J. Jr., 1973, Speed, Stamina, and Muscles in Fishes, *J. Fish. Res. Board Can.,* 30: 325-328.

———, 1979, Ecological Morphology of Freshwater Stream Fishes. *Tulane Stud. Zool. Bot.,* 21: 91-124.

Krebs, C. J., 1978, *Ecology: The Experimental Analysis of Distribution and Abundance,* 2nd ed., Harper & Row, Publishers, New York.

MacArthur, R. H., 1972, *Geographical Ecology,* Harper & Row, Publishers, New York.

————, and R. Levins, 1964, Competition, Habitat Selection, and Character Displacement in a Patchy Environment, *Proc. Nat. Acad. Sci.,* 51: 1207-1210.

————, and E. R. Pianka, 1966, On Optimal Use of a Patchy Environment, *Am. Nat.,* 100: 603-609.

Pianka, E. R., 1966, Convexity, Desert Lizards, and Spatial Heterogeneity, *Ecology,* 47: 1055-1059.

Pyke, G. H., H. R. Pulliam, and E. L. Charnov, 1977, Optimal Foraging: A Selective Review of Theory and Tests, *Quart. Rev. Biol.,* 52: 137-154.

Schoener, T. W., 1965, The Evolution of Bill Size Differences Among Sympatric Congeneric Species of Birds, *Evolution,* 19: 189-213.

————, 1969, Models of Optimal Size for Solitary Predators, *Am. Nat.,* 103: 277-313.

————, 1974, Resource Partitioning in Ecological Communities, *Science,* 185: 27-39.

Sokal, R. R., and F. J. Rohlf, 1969, *Biometry,* W. H. Freeman and Company, San Francisco.

Werner, E. E., 1974, The Fish Size, Prey Size, Handling Time Relation in Several Sunfishes and Some Implications, *J. Fish. Res. Board Can.,* 31: 1531-1536.

21. DYNAMICS OF LEAF PROCESSING IN A MEDIUM-SIZED RIVER

Robert W. Paul, Jr.*, Ernest F. Benfield, and John Cairns, Jr.
Department of Biology and Center for
   Environmental Studies
Virginia Polytechnic Institute and State University
Blacksburg, Virginia

ABSTRACT

Leaf processing in the middle reaches of the New River, Virginia, proceeds at rates equivalent to those reported for small streams. The initial stages of New River leaf procesing are mediated by microorganisms and enhanced by mechanical fragmentation, but macroinvertebrates important in initial leaf processing in small streams are unimportant in the section of New River studied. Sycamore (*Platanus occidentalis*) leaves decomposed at a significantly slower rate than did those of box elder (*Acer negundo*). Differences in processing rates were caused by differences in initial proportions and distributions of leaf chemical constituents and the rapidity of chemical constituent loss during incubation. Microbial numbers were greater and microbial activity higher on box elder leaves in comparison with sycamore; this was probably because the faster decomposing leaves were favored as a nutritional substrate. Seasonal reductions in chemical constituent losses, microbial activity, and processing rates occurred during winter. In spring, leaf chemical losses, microbial activity, and processing rates gradually increased; this suggests that seasonal temperature regulated all processing mechanisms studied.

---

*Current Address: Division of Natural Science and Mathematics, St. Mary's College of Maryland, St. Mary's City, MD.

INTRODUCTION

Particulate organic material of terrestrial origin frequently contributes significantly to energy budgets of running water systems (Teal, 1957). In small woodland streams as much as 99% of the total energy input may be allochthonous (Fisher and Likens, 1972; 1973). A substantial body of literature has accumulated on the decomposition and fragmentation (processing) of particulate allochthonous material (e.g., Kaushik and Hynes, 1971; Iversen, 1973; Petersen and Cummins, 1974), and the role of macroinvertebrates in detritus processing has recently been reviewed (Cummins, 1973; Cummins and Klug, 1979). A general model of organic matter processing in small streams has emerged, as summarized in Cummins (1977) and Cummins et al. (1973).

Leaves, the primary components of terrestrial litter, enter water bodies in large quantities during autumnal leaf abscission and heavy spring runoff (Hobbie and Likens, 1973; McDowell and Fisher, 1976). Immediately upon entering water, leaves undergo a rapid loss of water-soluble compounds through leaching. Rendered nutritionally depauperate within a few days by leaching (Krumholz, 1972; Nykvist, 1963), leaves are then transformed (conditioned) into substrates with nutritional value by colonizing microbes (Petersen and Cummins, 1974; Barlocher and Kendrick, 1975; Kostalos and Seymour, 1976). Fully conditioned leaf material is then fed upon by aquatic insect "shredders," which consume large leaf particles, and "collectors," which gather smaller detrital fragments (Cummins, 1974; Short and Maslin, 1977).

Leaves and fragments larger than 1 mm, coarse particulate organic matter (CPOM), become trapped (Young et al., 1978) and are processed near points of entry (Bilby and Likens, 1980); they are subsequently transported downstream as fine particulate organic matter (FPOM). Therefore, in-stream CPOM to FPOM ratios decline as stream order increases and processing nears completion (Sedell et al., 1978; Naiman and Sedell, 1979a; 1979b). Concomitant with these shifts in CPOM to FPOM ratios are apparent changes in processing speed as labile chemical constituents are rapidly used and a more refractory FPOM fraction is left (Vannote et al., 1980). As a result, CPOM probably plays a minor role relative to FPOM in the energy budgets of larger rivers.

Trophic analysis of macroinvertebrates inhabiting lotic systems shows that, as stream order increases, the predominance of shredders in first- and second-order streams gives way to collector-dominance in stream orders 4 to 6. Assemblages of detritus-processing macroinvertebrates in lotic ecosystems form a continuum of communities adapted for processing progressively smaller detrital fragments (Petersen and Cummins, 1974; Cummins, 1976; Vannote et al., 1980).

Although much information is available on many aspects of leaf processing in small woodland streams and lentic environments, only two studies have examined allochthonous detritus breakdown in a larger river (Mathews and Kowalczewski, 1969; Kowalczewski and Mathews, 1970). The contribution of leaves to the energy bases of larger streams, in relation to primary production, can only be inferred at present. It must be considerable, however, and is probably essential to the energy requirements of the biota inhabiting these systems.

The purpose of this study was to examine leaf processing in a medium-sized river, identify the agents responsible for processing, and compare the results with those from small woodland streams.

MATERIALS AND METHODS

Site Description

The New River at Glen Lyn (Giles County), Virginia was selected for study because it is a medium-sized river (approximately stream order 6) and is relatively free of environmental stress. The study site (km 153) is characterized by a series of extended riffles and pools with a tilted bedrock substrate overlain by loose cobble, gravel, and sand. Flow ranges over a 50-yr period are expected to be from 32 to 4361 m^3/s; mean annual flow is 139 m^3/s. The experimental site was 1 m deep, with a mean velocity of 18 cm/s.

Methods

Leaves of American sycamore (*Platanus occidentalis*) and box elder (*Acer negundo*) were collected at abscission and air-dried. Ten-gram packs of each species were sewn into 20- by 20-cm nylon bags (mesh size, 1 mm), and 10 bags of a single species were tied between 2 (1- by 1.5-m) poultry-wire screens (mesh size 2.5 cm) held in place on the river bed by concrete blocks.

A sequential leaf incubation scheme was used to simulate leaf material entering the New River at different times and to monitor leaf processing under different seasonal temperatures. An incubation experiment (run) was initiated by placing four screens of each leaf species at the study site.

One screen was retrieved on a weekly schedule for box elder and a biweekly schedule for sycamore until all screens in a particular run were removed. A single run required 4 weeks for box elder and 8 weeks for sycamore. Beginning October 25, 1976, successive runs were initiated at weekly intervals for box elder and biweekly intervals for sycamore. Between October 25, 1976 and March 28, 1977, twelve box elder runs and eight sycamore runs were completed. Continuous daily temperature and discharge were monitored throughout the study period.

Retrieved screens were returned to the laboratory, where remaining leaf material from eight of the ten bags was removed, washed, dried at 50°C, and weighed. Initial leaf weights were corrected for handling losses, and final leaf weights were adjusted for inorganic silt accumulations by ashing leaf material subsamples at 500°C. Hence, all final leaf weights reported here are on an ash-free dry weight (AFDW) basis. Assuming an exponential model (Petersen and Cummins, 1974), the percentage of leaf material remaining (AFDW) was then transformed (log_e) and regressed on both incubation time and accumulated day–degrees (Andrewartha and Birch, 1954) to compute processing coefficients.

Two randomly chosen leaf packs from each screen were selected from odd numbered runs and ground in a Wiley mill, and three 0.5-g subsamples were analyzed for soluble and fiber constituents (Goering and Van Soest, 1970). Packs not incubated in the river were treated similarly and served as controls.

In the field, two packs were immediately removed from retrieved screens. One leaf pack was opened, and a leaf disk (diameter, 2 cm) was cut from remaining leaf material and boiled for 10 min in 10 ml of 0.03 M Tris buffer to extract adenosine triphosphate (ATP) from leaf surface organisms. Both packs were then transported to the laboratory for microbial enumeration and activity. Two 2-cm leaf disks were cut from leaf material in the laboratory and homogenized in 10 ml of sterile distilled water. A 1-ml aliquot of each homogenate was serially diluted, and three replicate 0.1-ml aliquots were plated on plate count agar. In addition, six disks were cut from these leaves to measure microbial respiration and were placed in 15-ml Gilson flasks containing 5 ml of filtered (0.45 μm) river water. Flasks were fitted to two Gilson (1963) differential respirometers, covered with aluminum foil to exclude light, and incubated for 2 h at field temperature. After this period, five consecutive hourly readings of oxygen consumption were taken. The ATP analysis was performed with an ATP fluorimeter, and ATP content was determined by fitting experimental data to standard curves.

RESULTS

River Discharge and Thermal Regime

Mean daily New River discharge during the study was 165.8 m³/s, with a maximum discharge of 846 m³/s, occurring on March 14, 1977. New River temperature dropped from a maximum of 11.7°C on October 25, 1976, to near 0°C during mid-January and early February, 1977, and then rose to 11.1°C on March 28, 1978.

During the first 4 weeks of the study (box elder, run 1), leaf material accumulated 220.1 day–degrees, and during the first 8 weeks (sycamore, run 1), 350.3 day–degrees were accumulated by incubated samples. With progressively colder incubation temperatures, fewer day–degrees were accumulated by leaf material (Table 1).

Leaf Weight Loss

Box elder leaves exhibited the largest weight losses with incubation time. Over all box elder runs during the study, the mean percentages of leaf material remaining were 76.7, 68.6, 64.7 and 59.9% after incubation periods of 1, 2, 3 and 4 weeks, respectively. Although sycamore leaves showed a substantial weight reduction after the first 2 weeks of incubation over all runs (88.5% of initial weight remained), subsequent weight losses were less dramatic. After 4, 6, and 8 weeks, the mean percentages of sycamore leaf material remaining were 90.3, 86.5, and 82.7%, respectively; this indicates that the initial (11.5%) weight loss observed at 2 weeks was due primarily to leaching.

Processing coefficients (k), obtained by regressing $\log_e$ transformed weight-remaining data on days of incubation, permitted comparison of weight losses over 4-week combined incubation periods for box elder and over 8-week periods for sycamore (Table 1). The mean processing rate for all box elder runs was 0.0178, whereas the mean processing rate for sycamore leaves over all eight runs was substantially slower (0.0031). When tested by analysis of variance, both box elder and sycamore leaf packs showed a gradual but significant ($P \leq 0.05$) dimunition in processing speed as New River temperatures dropped during the study.

We obtained processing coefficients (k_t) based on cumulative incubation temperature by also regressing weight-remaining data on day–degrees accumulated (Table 1). The processing coefficient calculated for box elder leaves during run 1 was low when the largest number of day–degrees were

Table 1. Mean Percentage of Leaf Material Remaining After Final Incubation Period, Day–Degrees Accumulated by Incubated Leaf Packs, and Processing Coefficients Based on Incubation Time (k) and Cumulative Day–Degrees (k_t) for All Box Elder and Sycamore Runs*

	Box elder					Sycamore				
Placement date	Run number	Percentage of leaf remaining	Day-degrees accumulated	k	k_t	Run number	Percentage of leaf remaining	Day-degrees accumulated	k	k_t
Oct. 25	1	68.1	220.1	0.0154	0.0020	1	82.7	350.3	0.0026	0.0004
Nov. 1	2	50.7	187.8	0.0240	0.0036					
Nov. 8	3	60.6	161.1	0.0191	0.0033	2	78.2	254.7	0.0037	0.0008
Nov. 15	4	60.7	141.2	0.0166	0.0034					
Nov. 22	5	51.4	123.0	0.0235	0.0054	3	81.2	180.1	0.0037	0.0011
Nov. 29	6	56.2	101.6	0.0209	0.0056					
Dec. 6	7	56.4	89.9	0.0188	0.0058	4	82.5	125.4	0.0025	0.0010
Dec. 13	8	55.5	71.3	0.0187	0.0074					
Dec. 20	9		42.8	0.0136	0.0065	5	83.2	85.8	0.0042	0.0027
Dec. 27	10	65.7	40.8	0.0145	0.0098					
Jan. 3	11	66.6	34.8	0.0123	0.0102	6	78.7	99.1	0.0038	0.0022
Jan. 10	12	67.2	29.4	0.0121	0.0116					
Jan. 17						7	88.3	155.6	0.0023	0.0007
Jan. 24										
Jan. 31						8	86.9	264.7	0.0023	0.0003

*The k values are from Paul et al., 1978; reprinted by permission.

accumulated. With colder winter temperatures, box elder processing coefficients gradually increased as weight-remaining data were regressed on fewer day–degrees. Therefore, the last box elder run had the highest k_t value when the fewest day–degrees were accrued. Similarly, sycamore processing coefficients were initially low (run 1), increased in January, then declined dramatically as river temperatures increased in the early spring (runs 7 and 8).

Chemical Constituent Changes

The methods of Goering and Van Soest (1970) permitted the separation of all chemical compounds into soluble (all non-fiber compounds with the exception of cutin), fiber (hemicellulose, cellulose, and lignin), and cutin fractions. The soluble fraction, as used here, refers to the non-fiber fraction of leaf tissues and may or may not include leachate material, depending on the length of time the leaf tissue was incubated.

Soluble compounds comprised 62.0% of box elder leaves and 46.4% of sycamore leaves before incubation. The balance of initial chemical constituents was found to be 35.8% fiber and 2.3% cutin in box elder, whereas fiber and cutin made up 35.8% and 3.2% of the balance in sycamore, respectively. Soluble constituent losses from both species dominated leaf chemical changes, particularly during the first weeks of incubation (Figure 1). Very high soluble fraction losses were observed in box elder leaves during the first week of incubation; however, sycamore leaf packs did not show the same magnitude of soluble losses during the same incubation periods. Seasonal diminutions of soluble constituent losses occurred in both box elder and sycamore leaves as incubation temperatures dropped (Figure 1).

Changes in fiber content were not as pronounced as those observed for soluble constituents (Figure 1). Sycamore leaves, which had the greater initial proportion of fiber constituents, showed very little change in fiber content with incubation time, whereas box elder fiber was steadily degraded with incubation time.

Computing mean absolute fiber constituents over all runs permitted the comparison of chemical changes in both leaf species with incubation time and with leaf-pack weight losses (Figure 2). Of the fiber constituents, hemicellulose and cellulose were steadily degraded, but absolute amounts of lignin declined slightly or even increased with increased incubation in box elder leaves. The same general pattern of fiber degradation was seen in sycamore leaves, but reduction in absolute amounts of fiber in sycamore was not as pronounced as in box elder leaves. Absolute amounts of cutin either increased or remained unchanged with incubation time for both leaf species.

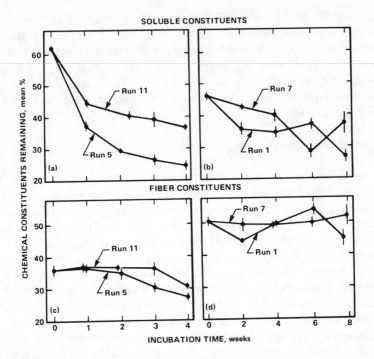

Figure 1. Mean percentages of soluble and fiber chemical constituents for selected box elder (a and c) and sycamore (b and d) leaf runs. Points represent the mean value of eight replicates, and vertical lines indicate one standard deviation above and below the mean.

Changes in percentages of chemical constituents reflected leaf-pack weight changes, indicating that leaf weight loss was mediated by chemical rather than physical processes. For example, box elder leaves incubated during the first week of run 1 had weight losses of 24% and corresponding chemical constituent losses of 23%. Of the 23% chemical constituent loss, 79% was attributable to soluble component loss, and the remaining 21% was lost from fiber constituents. Over all runs, weight losses in the first incubation period were, therefore, primarily attributable to the rapid leaching of soluble constituents and secondarily a result of relatively minor declines in fiber constituents. Soluble constituent losses were less pronounced after the first incubation period, and fiber components (hemicellulose and cellulose) became more important in leaf-pack weight loss. The other constituents (lignin and cutin) were poor indicators of processing speed because, on an absolute basis, their concentrations did not diminish with incubation time.

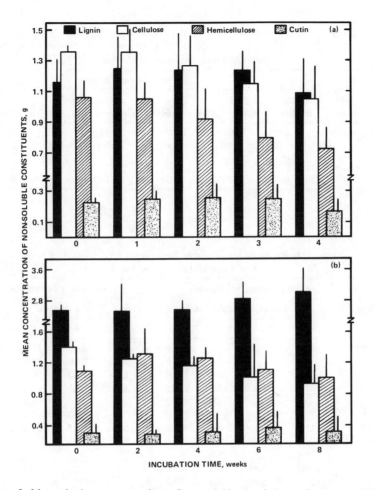

Figure 2. Mean absolute concentrations of non-soluble chemical constituents on ash-free dry weight basis, over all experimental box elder (a) and sycamore (b) runs. One standard deviation is indicated by narrow vertical lines.

To determine the degree to which leaf weight losses were explained by chemical constituent losses, we subjected all weight-remaining and chemical constituent data over all runs to a Spearman rank correlation analysis (Sokal and Rohlf, 1969). Resulting Spearman's coefficients of rank correlation (r_s) showed that leaf weight losses were most strongly correlated with soluble constituent losses from box elder leaves and most poorly correlated with lignin and cutin losses from both box elder and sycamore leaves (Table 2).

Table 2. Spearman's Coefficients of Rank Correlation (r_s) and Significance Probabilities (P) Correlating Leaf Weight and Leaf Chemical Constituents Remaining in Box Elder and Sycamore Leaf Packs

Leaf species		Chemical constituent					
		Soluble fraction	Total fiber	Hemicellulose	Cellulose	Lignin	Cutin
Box elder	(r_s)	0.92	0.69	0.73	0.55	0.19	0.33
	(P)	0.0001	0.0001	0.0001	0.0001	0.0265	0.0001
Sycamore	(r_s)	0.64	0.25	0.24	0.48	−0.13	−0.09
	(P)	0.0001	0.0189	0.0241	0.0001	0.2440	0.3871

Table 3. Mean Numbers of Viable Bacteria on Leaf Disks Cut From Box Elder and Sycamore Leaves

Run number	Box elder, 10^5 number				Sycamore, 10^5 number			
	Weeks of incubation				Weeks of incubation			
	1	2	3	4	2	4	6	8
1				27.0		38.0	23.5	455.0
2			67.0	75.0	22.0	25.0	39.5	30.5
3		65.0	52.0	173.0	25.5	62.0	6.1	10.3
4	33.0	27.0	93.5	134.5	39.5	23.5		42.5
5	0.9	39.0	128.0	495.0	8.0		35.5	
6	13.6	415.0	495.0	205.0	4.4	59.0	79.5	
7	111.5	390.0	230.0	415.0	14.5	99.5	32.0	
8	120.0	250.0	43.0	103.0	30.0	1.8		
9	125.0	36.5	95.0					
10	9.5	10.8	6.5	67.0				
11	39.5	5.9	38.5	28.5				
12	4.5	8.8	54.5					
$\bar{x}$	50.8	124.8	118.5	172.3	20.5	44.1	36.0	134.6

Microbial Numbers and Activity

Microbial Numbers

Mean numbers of microorganisms on leaf disks generally increased with incubation time for both leaf species (Table 3). Variability of microbial numbers and several missing data points, however, prevented the statistical comparison of successive runs to determine if statistically

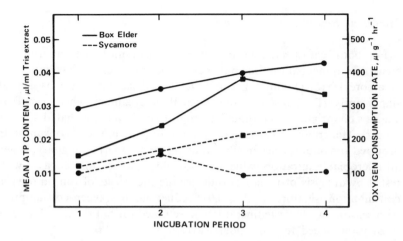

Figure 3. Mean ATP content (■) and mean oxygen consumption rate (●) per leaf disk dry weight over time for all box elder and sycamore leaf runs. Incubation periods were 1 week for box elder and 2 weeks for sycamore.

significant seasonal dimunition of numbers occurred. In general, leaf disks seemed to be poorly colonized during initial incubations, became increasingly well colonized during December (box elder runs 7 to 9, and sycamore runs 5 to 6), and then declined with colder January and February incubations. Box elder leaf disks supported approximately twice the microbial numbers of sycamore disks, despite much shorter incubation times.

Microbial Biomass

Estimation of microbial biomass of leaf disk surfaces by ATP content analysis revealed that, in general, microbial biomass, like microbial numbers, increased with incubation time for box elder and sycamore leaves (Figure 3). In relative terms, mean ATP content of box elder and sycamore leaves over all experimental runs indicated that box elder disks had higher ATP content than sycamore disks. The small number of samples and variability among samples, as was the case in microbial numbers analysis, prevented statistical comparison of ATP content among runs.

Microbial Respiration

For both leaf species, mean oxygen consumption over all runs increased with incubation time but declined after 6 weeks of incubation for sycamore leaves (Figure 3). Box elder leaf disks consumed oxygen at significantly (P ≤ 0.001) faster rates than sycamore when tested by analysis of variance, even though sycamore disks were incubated 4 weeks longer. Generally, a seasonal dimunition in oxygen consumption occurred for both species incubated under colder thermal regimes. When all microbial respiration values for box elder and sycamore leaves were tested by analysis of variance to determine the effects of run on oxygen consumption, significant (P ≤ 0.01) reductions in oxygen consumption with seasonally colder temperatures were noted for both species but were most pronounced for box elder leaves.

DISCUSSION

Leaf Substrate Changes

Processing coefficients based on incubation time (k) were within the range of values reported in the literature (Petersen and Cummins, 1974). Chemically similar to white oak (Suberkropp et al., 1976), sycamore leaves had processing rates close to those reported for white oak and other species classified by Petersen and Cummins (1974) as slow decomposers. Sycamore processing rates, as reported here, were low with respect to those reported by Benfield et al. (1977) for sycamore in an Appalachian pastureland stream. However, sycamore leaves exposed to thermal perturbation in the New River (Paul et al., 1978) were processed at rates similar to those reported by Benfield et al. (1977). The overall processing coefficient for box elder (0.0178) was five times larger than the sycamore coefficient and almost twice as large as the coefficients reported for other Aceraceae (Petersen and Cummins, 1974).

Differences in weight loss between leaf species may be caused, in part, by the fact that box elder is a fast decomposing species, whereas sycamore is a slow decomposer, in the sense of Petersen and Cummins (1974), under all temperature regimes examined here. The faster decomposition rate of box elder was not unexpected; Suberkropp et al. (1976) and Triska and Sedell (1976) demonstrated that species-specific processing rates are related to initial proportions of refractory materials. Initial fiber concentrations in sycamore and in box elder leaves strongly suggested that the greater proportion of more refractory fiber in sycamore would slow decomposition relative to box elder.

To further examine the role of chemical constituent changes in overall leaf weight losses, we calculated separate leaching and fiber processing coefficients, and the differences in weight loss patterns between species became apparent. Regression of weight-remaining data on the initial incubation period produced leaching coefficients for box elder (0.0368) and sycamore (0.0083). Fiber processing coefficients, calculated from weight remaining data after the initial incubation period, were 0.0100 for box elder and 0.0032 for sycamore. These coefficients emphasize the fact that, not only were soluble constituents lost from box elder leaves at significantly faster rates than from sycamore, but also box elder fiber constituents were more rapidly degraded than those of sycamore.

Differences observed in processing speeds of sycamore and box elder were caused by the initial proportions of chemical constituents in these species. Overall, box elder leaves showed the greater susceptibility to chemical change of the two species, primarily because of higher proportions of soluble constituents relative to sycamore. Although cellulose and hemicellulose contents in both species were initially equivalent, sycamore contained 15% more fiber and 14% more lignin than box elder leaves. Cutin is apparently very resistant to degradation and may have inhibited chemical degradation of sycamore leaf surfaces. Poor correlations between leaf weight losses and lignin–cutin changes over time (Table 2) suggest that the higher proportions of cutin and lignin in sycamore render these leaves resistant to decomposition, and that lignin and cutin were poor predictors of decomposition speed.

Increases in absolute lignin concentrations of both box elder and sycamore leaves with incubation time were probably the result of lignin degradation products complexing with other chemical constituents and being analyzed as lignin. Lignin content has been used to predict leaf decomposition speed in various environments (e.g., Cromack and Monk, 1975; Triska et al., 1975). Our data do not support this idea, but rather agree with the conclusions of Suberkropp et al. (1976), who questioned whether lignin values obtained by these previous investigators were lignin or "lignin–nitrogen complexes." Other chemical constituents (soluble components, hemicellulose, or cellulose), as implied by the correlations in Table 2, may be better predictors of decomposition speed.

Temperature played a major role in determining the processing speeds of leaf material incubated in this study. In November, up to 83% of the total weight lost by box elder leaves occurred within the first week of incubation, but, in January, only 57% of the total weight lost from box elder leaves occurred within the first week of incubation. Since all weight losses during the first week of box elder and first 2 weeks of sycamore incubations had a chemical basis, it is very likely that colder incubation temperatures resulted in slower leaching rates. Although Petersen and Cummins (1974) asserted that leaching is not influenced by temperature, contrary evidence has been presented by Nykvist (1963) and more recently by Short and Ward (1980).

Reductions in decomposition speed with lower temperature regimes have been noted in the laboratory (Hynes and Kaushik, 1969; Kaushik and Hynes, 1971) and in the field (Iverson, 1975; Reice, 1974; Suberkropp et al., 1975). However, evidence has appeared that temperature is not limiting to leaf weight loss in the field (Hart and Howmiller, 1975; Sedell et al., 1975; Triska et al., 1975; Suberkropp et al., 1976). It has been suggested that differences in the intensity of macroinvertebrate attack (Hart and Howmiller, 1975), leaf chemistry (Suberkropp et al., 1976), or a combination of the two (Sedell et al., 1975) may override the effects of temperature on processing speed. All these studies compared processing rates in different streams having different thermal regimes and did not use consecutive placement of leaf material to study processing speed changes in the same stream as we did in this study. We reported (Paul et al., 1978) that an artificially elevated thermal regime in the New River not only produced increased leaf decomposition rates, but also offset the effects of winter on processing speed; this clearly indicates that temperature dominated processing events. Because the only difference between runs in this study was thermal regime, temperature overrode all other factors in determining leaf processing speeds, in our opinion.

The discrepancy in processing coefficients based on increasing incubation temperature (k_t) and those based on decreasing incubation time (k) during progressively colder incubations is an interesting anomaly. Petersen and Cummins (1974) found that microbial and invertebrate colonization of ash and aspen leaf packs required fewer calendar days and 218 and 305 day–degrees, respectively, during October–November incubations but needed a greater number of calendar days and 65 day–degrees for both species to achieve equivalent colonization during November–January incubations. These investigators explained shorter day–degree conditioning time in January by pointing out that high quality leaves were scarce during mid-winter and that leaves placed in November did not function as valid natural leaf pack analogues because they were apparently selected by invertebrates over slow decomposing leaves. Our results are not readily explainable by this contention, but rather suggest that increasing processing coefficients (k_t) are explainable when the dependent variables used in regression (temperature and time) are examined. Obviously, processing coefficients would become larger when weight-remaining data are regressed on 30 day–degrees as opposed to 140 day–degrees and would give the false impression that processing efficiency was increasing when in actuality it was not. Processing coefficients based on incubation time, however, provided a better assessment of weight loss data under lower thermal regimes because incubation times were exactly the same for all runs, whereas regressions based on temperature were not equivalent.

Macroinvertebrates were collected in all retrieved litter bags, but species normally credited with ability to shred leaves (e.g., Tipulidae, Limnephilidae, Lepidostomatidae, and Pteronarcidae) were absent from all samples. The bulk of macroinvertebrates encountered were collectors (e.g., Hydropsychidae, Baetidae, Heptageniidae, and various Chironomidae), some scrapers (e.g., Gastropoda, Ephemerellidae, and Elmidae), and a few predators (e.g., Megaloptera, Perlidae, and Odonata). Macroinvertebrates at the New River study site apparently had little effect on reducing whole leaves to fragments although many of the species found are probably heavily dependent on fine particles as food sources, part of which certainly originate from leaf fragments. Their presence on leaf litter bags suggests that macroinvertebrates were selecting litter bags and leaf material as refuges rather than as food (Benfield et al., 1977; Reice, 1978; Winterbourne, 1978; Short and Ward, 1980).

The initial steps of New River leaf processing, based on weight loss and chemical constituent changes, were largely caused by chemical processing of leaves followed by mechanical fragmentation by water currents. Microbes apparently mediated leaf processing in the New River and probably fostered mechanical fragmentation.

Leaf Surface Activity

Paul et al. (1977), using scanning electron microscopy (SEM), demonstrated that leaf material incubated in the New River was rapidly colonized (within 10 days) by fungi and bacteria. Degradation of sycamore epidermal tissue was visually evident in samples retrieved after 14 weeks of incubation. Although aquatic hyphomycetes were present on sycamore leaves incubated in the New River, their occurrence was not as ubiquitous as observed on leaves incubated in Augusta Creek (Suberkropp and Klug, 1974).

Later comparison of both sycamore and box elder colonization in the New River by SEM (Paul, 1978) revealed that both sycamore and box elder leaves showed progressively heavier colonization by microbes with incubation time, but box elder midvein areas were much more heavily covered than corresponding areas of sycamore leaves. Fungi were the dominant microbial colonizers in these studies; this is in agreement with previous reports of fungal dominance of colonization (Triska, 1970; Kaushik and Hynes, 1971; Suberkropp and Klug, 1974; 1976).

In our study, evidence for microbial presence (microbial numbers and biomass) and activity (respiration) support the SEM evidence that box elder leaves were more heavily colonized than sycamore. Presumably, the heavier colonization of box elder leaves resulted in the faster and more complete mobilization of box elder chemical constituents; however, this cause and effect relationship was not tested and can only be inferred for this study.

Clearly lower microbial numbers and metabolic activity during the colder incubation period contributed to the slower processing rates of both leaf species during winter. Since microbial numbers, biomass, and respiration rates on both species were substantially higher under thermal elevation, and because of reduction in leaf weight loss and changes in chemical constituents during these incubations (Paul, 1978), the logical conclusion that microbial activity dimunitions were the result of declining winter temperatures is not unreasonable.

Leaf Processing in Medium-Sized Rivers

The river continuum concept (Vannote et al., 1980) is rapidly becoming the paradigm for integrating predictable and observable biological features of lotic systems. Using the results of our study and those conducted in other stream orders, we can make generalizations on leaf processing in medium-sized rivers, relative to the river continuum concept.

Small headwater streams appear to be very effective in retaining and processing the bulk of entering CPOM (e.g., Malmqvist et al., 1978; Sedell et al., 1978; Young et al., 1978; Naiman and Sedell, 1979a; 1979b; 1980). Upstream processing inefficiencies (leakage) (Vannote et al., 1980) and fluctuations in stream discharge (Winterbourne, 1976) result in FPOM transport into the middle reaches of the river continuum, where heterotrophic communities shift to process more refractory FPOM more efficiently.

The New River seems to be typical of the river continuum scenario. Newbern (1978) found that, of the total organic material reaching our study site, the vast majority was dissolved organic matter (DOM) and FPOM. Additionally, macroinvertebrate communities of the middle reaches of the New River are capitalizing on upstream processing inefficiencies by processing FPOM. Certainly CPOM inputs into medium-sized rivers are small in comparison with those into headwaters, primarily because the ratios of river shoreline to width and shoreline to discharge volume diminish downstream. However, CPOM enters medium-sized rivers through direct riparian leaf fall, blow-in, or transport during flood discharges. The paucity of headwater processing mechanisms (absence or near absence of CPOM-processing macroinvertebrates) in the New River segment studied suggests that CPOM processing efficiency would be discontinuous and the leaf processing rates would be much slower than its headwater counterparts.

Our study does not support this contention for the section of New River studied. We found, instead, that leaf processing proceeded in the New River at rates approximately equivalent to those of small streams, despite a lack of leaf-shredding macroinvertebrates. Apparently the activity of leaf surface microorganisms, under the control of temperature, enhanced mechanical fragmentation. This processing mechanism, under the thermal regimes examined here, was as efficient in the initial stages of leaf processing as the microbial–macroinvertebrate mechanisms of headwater streams. This mechanism may represent functional redundancy to insure complete processing of CPOM "leaking" out of headwater streams or directly entering medium-sized rivers before downstream export as FPOM.

Equivalent CPOM processing efficiencies of medium-sized rivers relative to headwater streams are not particularly surprising when we consider three parameters important in processing—temperature, stream power, and biological activity. In our study and the work of Suberkropp et al. (1975), Paul et al. (1978), and Short and Ward (1980), temperature overrode all other factors in determining leaf processing speeds. Maximum diel temperature ranges (ΔT max) are greater in medium-sized rivers than in headwater streams, and biological diversity may stabilize energy flow and mitigate the influence of high physical variance in medium-sized lotic systems (Vannote et al., 1980). As a result, leaf litter microbes may be exposed to favorable optimum temperatures at specific points of diel temperature cycles as Vannote and coworkers (1980) suggest. Because of extreme daily and seasonal temperature fluctuations in medium-sized rivers, microbial elaboration of extracellular enzymes may be species specific and sequential and may result in more efficient processing of refractory leaf material in comparison with headwater streams, where thermal regimes are more uniform.

Unit stream power (Leopold et al., 1964) has been linked to POM transport by Naiman and Sedell (1979a; 1979b) and may be important in the physical processing of POM, as other investigators have suggested (Boling et al., 1975; Benfield et al., 1977; Triska and Buckley, 1978). The New River probably possesses a unit stream power comparable with the McKenzie River (Naiman and Sedell, 1979a; 1979b; 1980), a seventh-order stream, and the potential for physical processing is reasonably high (Paul, 1978). Distinguishing between chemical and physical processing was impossible however, in our study, and has not, to our knowledge, been attempted in leaf processing studies to date.

In medium-sized rivers, macroinvertebrates appear to play a minor role in the initial stages of leaf processing, as the river continuum concept proposes. The role of microorganisms in mediating leaf processing and mechanical fragmentation probably has been underestimated, however. Leaf processing in the New River proceeds at rates approximately equivalent to those of small streams, perhaps because there is higher microbial activity in the New River than in smaller streams. Seasonal temperature changes and thermal perturbation had a profound effect on leaf processing in the New River, regulating all parameters associated with leaf degradation except physical processing. The nature and extent of physical processing, as well as the physiological capabilities of leaf processing microbes, need further study.

ACKNOWLEDGMENTS

This investigation was supported in part by U. S. Energy Research and Development Administration, contract no. E-(40-1)-4939. We are also indebted to Ellen Corson for the typing of preliminary and final manuscripts.

REFERENCES

Andrewartha, H. G., and L. C. Birch, 1954, *The Distribution and Abundance of Animals,* pp. 165-166, University of Chicago Press, Chicago.

Barlocher, F., and B. Kendrick, 1975, Leaf-Conditioning by Microorganisms, *Oecologia,* 20: 359-362.

Benfield, E. F., D. R. Jones, and M. F. Patterson, 1977, Leaf Pack Processing in a Pastureland Stream, *Oikos,* 29: 99-103.

Bilby, R. E., and G. E. Likens, 1980, Importance of Organic Debris Dams in the Structure and Function of Stream Ecosystems, *Ecology,* 61: 1107-1113.

Boling, R. H., Jr., E. D. Goodman, J. A. VanSickle, J. O. Zommer, K. W. Cummins, R. C. Petersen, and S. R. Reice, 1975, Toward a Model of Detritus Processing in a Woodland Stream, *Ecology,* 56: 141-151.

Cromack, K., Jr., and C. D. Monk, 1975, Litter Production, Decomposition, and Nutrient Cycling in a Mixed Hardwood Watershed and White Pine Watershed, in F. G. Howell, J. B. Gentry and M. M. Smith (Eds.), *Mineral Cycling in Southeastern Ecosystems,* ERDA Symposium Series, CONF-740513, pp. 609-624, NTIS, Springfield, VA.

Cummins, K. W., 1973, Trophic Relations of Aquatic Insects, *Ann. Rev. Entomol.*, 18: 183-206.

_____, 1974, Structure and Function of Stream Ecosystems, *Bioscience*, 24: 631-641.

_____, 1976, The Use of Macroinvertebrate Benthos in Evaluating Environmental Damage, in R. K. Sharma, J. D. Buffington, and J. T. McFadden (Eds.), *Biological Significance of Environmental Impacts*, pp. 139-149, NTIS, Springfield, VA.

_____, 1977, From Headwater Streams to Rivers, *Am. Biol. Teacher*, 39: 305-312.

_____, and M. J. Klug, 1979, Feeding Ecology of Stream Invertebrates, *Annu. Rev. Ecol. Systemat.*, 10: 147-172.

_____, R. C. Petersen, F. O. Howard, J. C. Wuycheck, and V. I. Holt, 1973, The Utilization of Leaf Litter by Stream Detritivores, *Ecology*, 54: 336-345.

Fisher, S. G., and G. E. Likens, 1972, Stream Ecosystem: Organic Energy Budget, *Bioscience*, 22: 33-35.

_____, and G. E. Likens, 1973, Energy Flow in Bear Brook, New Hampshire: An Integrative Approach to Stream Ecosystem Metabolism, *Ecol. Monogr.*, 43: 421-439.

Gilson, W. E., 1963, Differential Respirometer of Simplified and Improved Design, *Science*, 141: 531-532.

Goering, H. K., and P. J. Van Soest, 1970, *Forage Fiber Analysis (Apparatus, Reagents, Procedures and Some Applications)*, Agricultural Handbook No. 379, pp. 1-20, Agricultural Research Division, U.S. Department of Agriculture, Washington, D.C.

Hart, S. D., and R. P. Howmiller, 1975, Studies of the Decomposition of Allochthonous Detritus in Two Southern California Streams, *Verh. Internat. Verein. Limnol.*, 19: 1665-1674.

Hobbie, J. E., and G. E. Likens, 1973, Output of Phosphorus, Dissolved Organic Carbon, and Fine Particulate Carbon from Hubbard Brook Watersheds, *Limnol. Oceanogr.*, 18: 734-742.

Hynes, H. B. N., and N. K. Kaushik, 1969, The Relationship Between Dissolved Nutrient Salts and Protein Production in Submerged Autumnal Leaves, *Verh. Internat. Verein. Limnol.*, 17: 95-103.

Iversen, T. M., 1973, Decomposition of Autumn-Shed Beech Leaves in a Springbrook and Its Significance for the Fauna, *Arch. Hydrobiol.*, 72: 305-312.

_____, 1975, Disappearance of Autumn Shed Beech Leaves Placed in Bags in Small Streams, *Verh. Internat. Verein. Limnol.*, 19: 1687-1692.

Kaushik, N. K., and H. B. N. Hynes, 1971, The Fate of Dead Leaves That Fall into Streams, *Arch. Hydrobiol.*, 68: 465-515.

Kostalos, M., and L. R. Seymour, 1976, Role of Microbial Enriched Detritus in the Nutrition of *Gammarus minus*, *Oikos*, 28: 512-516.

Kowalczewski, A., and C. P. Mathews, 1970, The Leaf Litter as a Food Source of Aquatic Organisms in the River Thames, *Polskie Archiwum Hydrobiol.*, 17: 133-134.

Krumholz, L. A., 1972, *Degradation of Riparian Leaves and Recycling of Nutrients in a Stream Ecosystem,* Research Report 57, Water Resources Institute, University of Kentucky, Louisville.

Leopold, L. B., M. G. Wolman, and J. P. Miller, 1964, *Fluvial Processes in Geomorphology,* p. 522, W. H. Freeman and Company, San Francisco.

McDowell, W. H., and S. G. Fisher, 1976, Autumnal Processing of Dissolved Organic Matter in a Small Woodland Stream Ecosystem, *Ecology,* 57: 561-569.

Malmqvist, B., L. M. Nilsson, and B. S. Svensson, 1978, Dynamics of Detritus in a Small Stream in Southern Sweden and Its Influence on the Distribution of the Bottom Animal Communities, *Oikos,* 31: 3-16.

Mathews, C. P., and A. Kowalczewski, 1969, The Disappearance of Leaf Litter and Its Contribution to Production in the River Thames, *J. Ecol.,* 57: 543-552.

Naiman, R. J., and J. R. Sedell, 1979a, Characterization of Particulate Organic Matter Transported by some Cascade Mountain Streams, *J. Fish. Res. Board Can.,* 36: 17-31.

———, and J. R. Sedell, 1979b, Benthic Organic Matter as a Function of Stream Order in Oregon, *Arch. Hydrobiol.,* 87: 404-422.

———, and J. R. Sedell, 1980, Relationships Between Metabolic Parameters and Stream Order in Oregon, *Can. J. Fish. Aquat. Sci.,* 37: 834-847.

Newbern, L. A., 1978, Detritus Transport in the New River, Virginia, M.S. thesis, p. 40, Virginia Polytechnic Institute and State University, Blacksburg.

Nykvist, N., 1963, Leaching and the Decomposition of Water-Soluble Organic Substances from Different Types of Leaf and Needle Litter, *Stud. For. Suec.,* 3: 1-29.

Paul, R. W., Jr., 1978, Leaf Processing and the Effects of Thermal Perturbation on Leaf Degradation in the New River, Virginia, Ph.D. thesis, Virginia Polytechnic Institute and State University, Blacksburg.

———, D. L. Kuhn, J. L. Plafkin, J. Cairns, Jr., and J. G. Croxdale, 1977, Evaluation of Natural and Artificial Substrate Colonization by Scanning Electron Microscopy, *Trans. Am. Microsp. Soc.,* 96: 506-519.

———, E. F. Benfield, and J. Cairns, Jr., 1978, Effects of Thermal Discharge on Leaf Decomposition in a River Ecosystem, *Verh. Internat. Verein. Limnol.,* 20: 1759-1766.

Petersen, R. C., and K. W. Cummins, 1974, Leaf Processing in a Woodland Stream, *Freshwater Biol.,* 4: 343-368.

Reice, S. R., 1974, Environmental Patchiness and the Breakdown of Leaf Litter in a Woodland Stream, Ecology, 55: 1271-1282.

———, 1978, Role of Detritivore Selectivity in Species Specific Litter Decomposition in a Woodland Stream, *Verh. Internat. Verein. Limnol.,* 20: 1396-1400.

Sedell, J. R., R. J. Naiman, K. W. Cummins, G. W. Minshall, and R. L. Vannote, 1978, Transport of Particulate Organic Material in Streams as a Function of Physical Processes, *Verh. Internat. Verein. Limnol.,* 20: 1366-1375.

———, F. J. Triska, and N. S. Triska, 1975, The Processing of Conifer and Hardwood Leaves in Two Coniferous Forest Streams: I. Weight Loss and Associated Invertebrates, *Verh. Internat. Verein. Limnol.,* 19: 1617-1627.

Short, R. A. and P. W. Maslin, 1977, Processing of Leaf Litter by a Stream Detritivore: Effect on Nutrient Availability to Collectors, *Ecology,* 58: 935-938.

———, and J. V. Ward, 1980, Leaf Litter Processing in a Regulated Rocky Mountain Stream, *Can. J. Fish. Aquat. Sci.,* 37: 123-127.

Sokol, R. R., and F. J. Rohlf, 1969, *Biometry,* pp. 538-540, W. H. Freeman and Company, San Francisco.

Suberkropp, K. F., and M. J. Klug, 1974, Decomposition of Deciduous Leaf Litter in a Woodland Stream, *Microb. Ecol.,* 1: 96-103.

———, and M. J. Klug, 1976, Fungi and Bacteria Associated with Leaves During Processing in a Woodland Stream, *Ecology,* 57: 707-719.

———, M. J. Klug, and K. W. Cummins, 1975, Community Processing of Leaf Litter in Woodland Streams, *Verh. Internat. Verein. Limnol.,* 19: 1653-1658.

———, G. L. Godshalk, and M. J. Klug, 1976, Changes in the Chemical Composition of Leaves During Processing in a Woodland Stream, *Ecology,* 57: 720-727.

Teal, J. M., 1957, Community Metabolism in a Temperate Cold Spring, *Ecol. Mongr.,* 27: 283-302.

Triska, F. J., 1970, Seasonal Distribution of Aquatic Hyphomycetes in Relation to the Disappearance of Leaf Litter from a Woodland Stream. Ph.D. thesis, University of Pittsburgh, Pittsburgh, PA.

———, and B. M. Buckley, 1978, Patterns of Nitrogen Uptake and Loss in Relation to Litter Disappearance and Associated Invertebrate Biomass in Six Streams of the Pacific Northwest, U.S.A., *Verh. Internat. Verein. Limnol.,* 20: 1324-1332.

———, and J. R. Sedell, 1976, Decomposition of Four Species of Leaf Litter in Response to Nitrate Manipulation, *Ecology,* 57: 783-792.

———, J. R. Sedell, and B. Buckley, 1975, The Processing of Conifer and Hardwood Leaves in Two Coniferous Streams: II. Biochemical and Nutrient Changes, *Verh. Internat. Verein. Limnol.,* 19: 1628-1639.

Vannote, R. I., G. W. Minshall, K. W. Cummins, J. R. Sedell, and C. E. Cushing, 1980, The River Continuum Concept, *Can. J. Fish. Aquat. Sci.,* 37: 130-137.

Winterbourne, M. J., 1976, Fluxes of Litter Falling into a Small Beech Forest Stream, *N. Z. J. Mar. Freshwater Res.,* 10: 399-416.

———, 1978, An Elevation of the Mesh Bag Method for Studying Leaf Colonization by Stream Invertebrates, *Verh. Internat. Verein. Limnol.,* 20: 1557-1561.

Young, S. A., W. P. Kovalak, and K. A. Del Signore, 1978, Distances Traveled by Autumn-Shed Leaves Introduced into a Woodland Stream, *Am. Midl. Nat.,* 100: 217-222.

22. THE ROLE OF SHREDDERS IN DETRITAL DYNAMICS OF PERMANENT AND TEMPORARY STREAMS

J. M. Kirby, J. R. Webster, and E. F. Benfield

Biology Department
Virginia Polytechnic Institute and State University
Blacksburg, Virginia

ABSTRACT

It has been suggested that leaf-shredding insects have an important role in breakdown of leaf detritus and the production of particulate organic matter (POM) in streams. This role was evaluated by comparing detrital dynamics in three permanent and three temporary tributaries of Guys Run, Rockbridge County, Virginia. In general, the streams with fastest leaf breakdown, highest low-flow POM concentrations, and largest average POM particle sizes were found to have the greatest shredder and total insect densities. It was concluded that this is further evidence for the importance of shredders in woodland streams.

INTRODUCTION

Leaves form the major trophic base of many streams, especially small headwater streams draining forested watersheds in the eastern deciduous forest of North America. Leaves entering these streams are broken down by a combination of physical, chemical, and biological factors, including mechanical breakage, leaching of water-soluble compounds, microbial decomposition, and invertebrate feeding. Our concern in this chapter is the importance of invertebrate feeding on leaf breakdown rates and generation of small particles of detritus. Detritivores have relatively low assimilation efficiencies (e.g., Berrie, 1976), and much of the leaf material ingested by detritivores (or shredders, Cummins, 1973) is returned to the stream greatly reduced in size. This fine particulate organic matter (FPOM) provides an energy source for other components of the stream fauna, which feed on deposits of FPOM or filter FPOM from suspension (Short and Maslin, 1977; Wallace et al., 1977). The importance of shredder activity in stream detrital dynamics has been difficult to quantify, however (Anderson and Sedell, 1979). The importance of shredders in leaf breakdown and FPOM generation has been evaluated by two general techniques: breakdown rates in streams with different shredder abundances have been compared, and detritus use budgets have been calculated from shredder ingestion or production data. Hart and Howmiller (1975), Iversen (1975), and Sedell et al. (1975) compared leaf breakdown in two or more streams and, in each case, suggested that differences in breakdown rates were caused by differences in the invertebrate fauna. Also, Petersen and Cummins (1974) found slower leaf breakdown rates in channels where shredders were excluded than in those where they were present.

Fisher and Likens (1973) found that macroconsumers accounted for only a small portion of the energy flow in Bear Brook. However, Cummins (1971) estimated that detritivores in a woodland stream ingested almost 32% of gross large particulate organic matter (LPOM) input on a daily basis, and Webster and Patton (1979) calculated that annual detritivore ingestion in a small stream at Coweeta Hydrologic Laboratory was approximately 80% of leaf fall. In another study, Grafius and Anderson (1979) found that, though *Lepidostoma quercina* production was only a small portion of the total secondary production in Berry Creek, this shredder produced sufficient FPOM to support 25 to 50% of the simuliid production in the stream. Finally, on the basis of laboratory experiments, Cummins et al. (1973) concluded that shredders have an important influence on energy flow in detritus-dominated stream ecosystems.

In this study we compared rates of leaf breakdown and POM transport in three permanent and three temporary streams draining small forested watersheds. Since the periodic dry period of temporary streams has been shown to greatly reduce the number of invertebrates in such streams (e.g., Clifford, 1966; Williams and Hynes, 1976), we anticipated that shredder influences on leaf breakdown and POM production should be evident in a comparison of these two types of streams.

DESCRIPTION OF STUDY AREA

Transport of particulate organic matter and leaf breakdown rates were measured in six first-order tributaries of Guys Run, a tributary of the Calfpasture River (James River Basin, Rockbridge County, Virginia; 79°39′ W longitude, 38°58′ N latitude) (Figure 1). Most of the 19-km^2 watershed of Guys Run is located within the Goshen Wildlife Mangement Area. Overstory vegetation is primarily oak, hickory, maple, and pine, with an understory of dogwood, rhododendron, and mountain laurel. The six tributaries of Guys Run used in this study are typical of the low-order, low-nutrient streams of the southern Appalachian Mountains. Glade Brook, Beckney Hollow, and Three Dwarf Run have channel flow throughout the year, with highest flows in winter and lowest flows in summer. Dry Branch, Tower Branch, and Grave Branch are temporary streams, with no channel flow during late summer and autumn. Stream lengths, gradients, and watershed areas for the six streams are given in Table 1.

METHODS

Leaf breakdown rates were examined in the six study streams by using nylon mesh bags (10 by 10 cm, with 1-cm openings) filled with approximately 5 g of dried and weighed red maple (*Acer rubrum*) leaves. Thirty bags were placed in each stream on January 20, 1979, and five bags were recovered from each site on six dates, beginning on February 10, 1979, and ending on July 5, before the temporary streams dried up. Retrieved leaf bags were placed on ice and transported to the laboratory, where invertebrates were recovered and preserved for identification. The leaf material was air-dried to constant weight and weighed. Subsamples (1 g) from each bag were ashed for 15 min at 500°C to determine ash-free dry weight (AFDW) of the remaining leaf material. Breakdown rates (Jenny et al., 1949; Olson, 1963) were calculated by regressing log-transformed percent weight remaining against exposure time. Breakdown rates were compared by analysis of covariance (Sokal and Rohlf, 1969).

Particulate organic matter was collected from the six study streams on seven dates beginning in January 1979 and ending in January 1980. Large POM size fractions were collected by pouring measured volumes of water through a 20-μm plankton net. Water was also collected in carboys to obtain samples of smaller particles. All POM samples reported in this

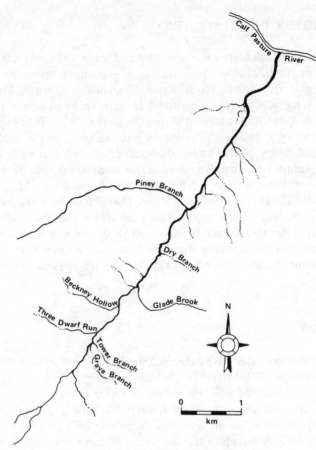

GUYS RUN DRAINAGE

Figure 1. Map of Guys Run study area.

Table 1. Stream Length, Gradient, and Watershed Areas for Three Permanent and Three Temporary Streams in the Guys Run Watershed

Stream	Stream length, m	Stream gradient, m/m	Watershed area, km^2
Permanent streams			
Glade Brook	400	0.22	1.13
Beckney Hollow	1350	0.08	0.83
Three Dwarf Run	550	0.06	0.38
Temporary streams			
Dry Branch	549	0.15	0.33
Tower Branch	555	0.13	0.94
Grave Branch	1173	0.05	0.36

paper were collected during non-storm periods. Samples were analyzed with wet filtration system (Gurtz et al., 1980). Measured volumes of water or resuspended net samples were filtered with suction through a series of stainless-steel screens into the following size classes: >234 μm, 105 to 234 μm, 43 to 105μm, and 25 to 43 μm. Material collected on the screens was resuspended and collected on preashed, preweighed Gelman A/E glass-fiber filters. An aliquot of material passing through the 25-μm screen was filtered through a glass-fiber filter to provide a 0.5 to 25 μm size fraction. All samples were oven-dried for 24 h at 50°C, desiccated (24 h), weighed, ashed 15 min at 500°C, rewetted to restore water of hydration (Weber, 1973)], redried, desiccated, and weighed. We calculated POM concentrations (mg AFDW/l) and average particle sizes (based on weight) from these weights.

RESULTS AND DISCUSSION

Breakdown of red maple leaf packs in the six study streams is shown in Figure 2. Glade Brook and Beckney Hollow, two of the permanent streams, had breakdown rates significantly faster than the other four streams (Table 2). Leaf breakdown in Three Dwarf Run, the smallest of the three permanent streams, was more similar to that in the temporary streams. Of the three temporary streams, leaf breakdown in Dry Branch was more rapid than breakdown in Grave Branch and faster, but not statistically significantly so, than in Tower Branch. There were no significant correlations between breakdown rates and stream gradients, steam velocity, or watershed area. Leaf breakdown rates of red maple measured in this study were all slower than the rate of 0.0298/day found by Thomas (1970) for a small stream in Tennessee but similar to the 0.0062/day breakdown rate measured in Augusta Creek by Petersen and Cummins (1974).

Shredder and total insect density on leaf packs increased with time, especially after about 70 days, once the leaves became conditioned (Figure 3). The dominant shredders found on the leaves were two Trichoptera (*Lepidostoma* sp. and *Pycnopsyche* sp.), three Plecoptera (*Leuctra* sp., *Peltoperla* sp., and *Pteronarcys proteus*), and a Diptera (*Tipula* spp). Densities of shredders and total insects reflected leaf breakdown rates. Two permanent streams, Glade Brook and Beckney Hollow, which had the fastest breakdown rates, had highest insect densities. Three Dwarf Run (permanent) and Grave Branch (temporary), the two streams with the slowest leaf breakdown rates, had consistently low insect densities. In Tower Branch and Dry Branch (both temporary),

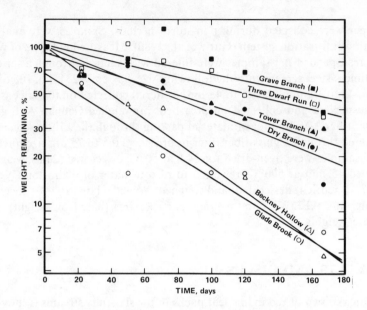

Figure 2. Leaf breakdown in the six study streams. Each point represents the average of five samples. Lines were fit by regression.

Table 2. Red Maple Leaf Breakdown Rates in the Six Study Streams

Stream	N	r^2	Breakdown rate/day	95% Confidence interval	Analysis of* covariance
Beckney Hollow	35	0.72	0.0175	±0.0078	
Glade Brook	35	0.66	0.0149	±0.0075	
Dry Branch	35	0.65	0.0098	±0.0050	
Tower Branch	35	0.53	0.0077	±0.0051	
Three Dwarf Run	35	0.46	0.0059	±0.0045	
Grave Branch	34	0.33	0.0045	±0.0047	

*Vertical bars indicate rates that were not significantly different ($\alpha = 0.05$).

which had intermediate breakdown rates, shredder densities were low at the end of the study, but peaks of shredder abundance (primarily *Leuctra* sp. in both cases) were noted after 90 days of exposure in Tower Branch and 125 days in Dry Branch.

On dates when comparisons were possible, POM concentrations were generally higher in the permanent than in the temporary streams (Figure 4). Concentrations in the permanent streams were highest in summer,

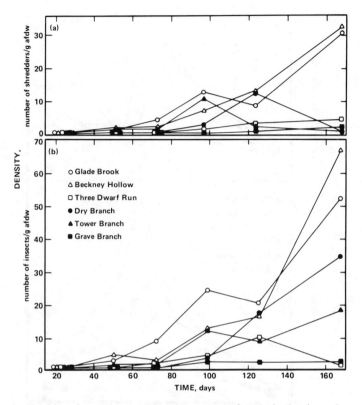

Figure 3. Densities of shredders and total insects on leaf packs in the six study streams. Each point is the mean from five samples.

lowest in winter, and intermediate in spring and autumn. During the peak period (May to September), POM concentrations were consistently highest in Beckney Hollow, lowest in Three Dwarf Run, and intermediate in Glade Brook. Among the temporary streams, Dry Branch generally had the highest POM concentration. Trends in the average POM particle sizes were not as clear, but followed the same general pattern. The average particle size was usually largest in Beckney Hollow and Glade Brook (both permanent) and smallest in Tower Branch and Grave Branch (both temporary). Three Dwarf Run (permanent) and Dry Branch (temporary) were intermediate.

The POM concentrations were generally lower than concentrations measured in other small streams in the southern Appalachian Mountains (Gurtz et al., 1980). Higher POM concentrations were also found in small streams in Pennsylvania (Sedell et al., 1978), Michigan (Wetzel and

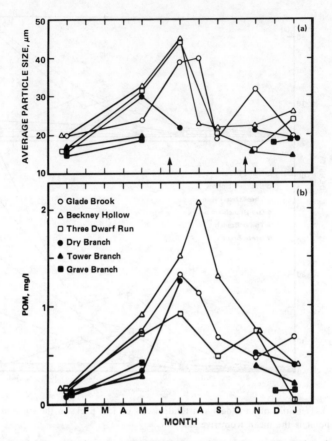

Figure 4. Particulate organic matter (POM) concentrations and average particle sizes in the six study streams. Temporary streams were dry from early June to mid October, as indicated by arrows on part a of the figure.

Manny, 1977), and Mississippi (de la Cruz and Post, 1977). Concentrations more similar to ours (annual average 1 mg/l) were reported from Hubbard Brook streams (Bormann et al., 1969; Fisher and Likens, 1973; Bilby and Likens, 1979) and streams in the western United States (Maciolek, 1966; Maciolek and Tunzi, 1968; Sedell et al., 1978; Naiman and Sedell, 1979). Our average particle size was generally smaller than the 37 to 65 μm found in Hugh White Creek, a second-order stream in North Carolina (Gurtz et al., 1980), but larger than that (5 to 12 μm) reported by Naiman and Sedell (1979) for first- and second-order streams in Oregon. The average particle sizes we observed were well below the particle sizes found in shredder feces (J. O'Hop, University of Georgia, personal communication).

The results of this study suggest that shredders have a significant influence on the rates of leaf breakdown and POM production in tributaries of Guys Run. In two permanent streams, Glade Brook and Beckney Hollow, we observed the fastest leaf breakdown, greatest low-flow POM concentrations, largest average POM particles sizes, and the greatest abundances of shredders on leaf packs. In contrast, in the two temporary streams with the longest dry periods, Tower Branch and Grave Branch, we found significantly slower breakdown, lower POM concentrations, smaller average POM particle sizes, and lower shredder densities. This suggests that in streams where shredders are abundant their feeding accelerates weight loss from leaves, and the relatively large particles in their feces become a major portion of the POM carried by the streams, resulting in a larger average POM particle size.

It might be argued that shredder densities are not relevant, but that greater POM concentrations, larger particle sizes, and faster leaf breakdown are results of flows and greater stream power in Beckney Hollow and Glade Brook since it is well known that as stream power increases during storms, POM transport increases (e.g., Bilby and Likens, 1979). If greater POM transort in the permanent streams was caused by their greater stream power, we would expect the greatest POM transport and largest particle sizes in winter when flows were high, and lower transport and smaller particle sizes during low summer flows. We observed just the opposite, however. In summer, even with lower flows, daily transport at base flow was higher than daily transport at winter base flows. For example, in July 1979, daily transport in Beckney Hollow was 436 g/day at a flow of 3.3 l/s. In January 1980, at a flow of 23.8 l/s transport in the same stream was only 235 g/day.

Two of the streams we studied failed to fit our preconceived ideas about temporary and permanent streams. Three Dwarf Run, one of the permanent streams, had slower leaf breakdown, lower POM concentrations, and smaller particle sizes than the other permanent streams. In these respects, Three Dwarf Run was more like the temporary streams. Dry Branch, a temporary stream, illustrated the opposite situation by having detrital characteristics more like a permanent stream. Shredder and total insect densities in both of these streams were also the opposite of what we anticipated. Shredder and total insect densities were generally high in Dry Branch but very low in Three Dwarf Run. We are unable to explain why these two streams failed to fit the permanent–temporary pattern, but the correlations between detrital characteristics and shredder densities further support our argument for the importance of shredders.

ACKNOWLEDGMENTS

We thank Larry Hornick, Bob Sinsabaugh, Steve Golladay, Sandra Gay, and Peter Wallace for their assistance in gathering data presented here. This research was supported by a grant from the USDA Forest Service, Southeastern Forest Experimental Station. Access to the study area was provided by the Virginia Commission of Game and Inland Fisheries.

REFERENCES

Anderson, N. H., and J. R. Sedell, 1979, Detritus Processing by Macroinvertebrates in Stream Ecosystems, *Ann. Rev. Entomol.*, 24: 351–377.

Berrie, A. D., 1976, Detritus, Micro-Organisms and Animals in Fresh Water, in J. M. Anderson and A. Macfadyen (eds.) *The Role of Terrestrial and Aquatic Organisms in Decomposition Processes*, pp. 323–337, Blackwell Scientific Publications, Oxford, England.

Bilby, R. E., and G. E. Likens, 1979, Effect of Hydrologic Fluctuations on the Transport of Fine Particulate Organic Carbon in a Small Stream, *Limnol. Oceanogr.*, 24: 69–75.

Bormann, F. H., G. E. Likens, and J. S. Eaton, 1969, Biotic Regulation of Particulate and Solution Losses from a Forest Ecosystem, *BioScience*, 19: 600–611.

Clifford, H. F., 1966, The Ecology of Invertebrates in an Intermittent Stream, *Invest. Indiana Lakes Streams*, 7: 57–98.

Cummins, K. W., 1971, *Predicting Variations In Energy Flow Through a Semi-controlled Lotic Ecosystem*, Technical Report No. 19, Institute of Water Research, Michigan State University, East Lansing.

———, 1973, Trophic Relations of Aquatic Insects, *Ann. Rev. Entomol.*, 18: 183–206.

———, 1974, Structure and Function of Stream Ecosystems, *BioScience*, 24: 631–641.

———, R. C. Petersen, F. O. Howard, J. C. Wuycheck, and V. I. Holt, 1973, The Utilization of Leaf Litter by Stream Detritivores, *Ecology*, 54: 336–345.

de la Cruz, A. A., and H. A. Post, 1977, Production and Transport of Organic Matter in a Woodland Stream, *Arch. Hydrobiol.*, 80: 227–238.

Fisher, S. G., and G. W. Likens, 1973, Energy Flow in Bear Brook, New Hampshire: An Integrative Approach to Stream Ecosystem Metabolism, *Ecol. Monogr.*, 43: 421–439.

Grafius, E., and N. H. Anderson, 1979, Population Dynamics, Bioenergetics, and Role of *Lepidostoma quercina* Ross (Trichoptera: Lepidostomatidae) in an Oregon Woodland Stream, *Ecology*, 60: 433–441.

Gurtz, M. E., J. R. Webster, and J. B. Wallace, 1980, Seston Dynamics in Southern Appalachian Streams: Effects of Clear-Cutting, *Can. J. Fish. Aquat. Sci.*, 37: 624–631.

Hart, S. D., and R. P. Howmiller, 1975, Studies on the Decomposition of Allochthonous Detritus in Two Southern California Streams, *Verh. Internat. Verein. Limnol.*, 19: 1665–1674.

Iversen, T. M., 1975, Disappearance of Autumn Shed Leaves Placed in Bags in Small Streams, *Verh. Internat. Verein. Limnol.*, 19: 1687–1692.

Jenny, H., S. P. Gessel, and F. T. Bingham, 1949, Comparative Study of Decomposition Rates of Organic Matter in Temperate and Tropical Regions, *Soil Sci.*, 68: 417–432.

Maciolek, J. A., 1966, Abundance and Character of Microseston in a California Mountain Stream, *Verh. Internat. Verein. Limnol.*, 16: 639–645.

———, and M. G. Tunzi, 1968, Microseston Dynamics in a Simple Sierra Nevada Lake-Stream System, *Ecology*, 49: 60–75.

Naiman, R. J., and J. R. Sedell, 1979, Characterization of Particulate Organic Matter Transported by Some Cascade Mountain Streams, *J. Fish. Res. Board Can.*, 36: 17–31.

Olson, J. S., 1963, Energy Storage and the Balance of Producers and Decomposers in Ecological Systems, *Ecology*, 44: 322–332.

Petersen, R. C., and K. W. Cummins, 1974, Leaf Processing in a Woodland Stream, *Freshwater Biol.*, 4: 343–368.

Sedell, J. R., R. J. Naiman, K. W. Cummins, G. W. Minshall, and R. L. Vannote, 1978, Transport of Particulate Organic Material in Streams as a Function of Physical Processes, *Verh. Internat. Verein. Limnol.*, 20: 1366–1375.

———, F. J. Triska, and N. S. Triska, 1975, The Processing of Conifer and Hardwood Leaves in Two Coniferous Forest Streams. I. Weight Loss and Associated Invertebrates, *Verh. Internat. Verein. Limnol.*, 19: 1617–1627.

Short, A. S., and P. E. Maslin, 1977, Processing of Leaf Litter by a Stream Detritivore: Effect on Nutrient Availability to Collectors, *Ecology*, 58: 935–938.

Sokal, R. R., and F. J. Rohlf, 1969, *Biometry*, W. H. Freeman and Company, San Francisco.

Thomas, W. A., 1970, Weight and Calcium Losses from Decomposing Tree Leaves on Land and in Water, *J. Appl. Ecol.*, 7: 237–242.

Wallace, J. B., J. R. Webster, and W. R. Woodall, 1977, The Role of Filter Feeders in Flowing Waters, *Arch. Hydrobiol.*, 79: 506–532.

Weber, C. I. (Ed.), 1973, *Biological Field and Laboratory Methods for Measuring the Quality of Surface Water and Effluents*, Report EPA 670/4-73-001, Environmental Protection Agency, Cincinnati.

Webster, J. R., and B. C. Patten, 1979, Effects of Watershed Perturbation on Stream Potassium and Calcium Budgets, *Ecol. Monogr.*, 49: 51–72.

Wetzel, R. G., and B. A. Manny, 1977, Seasonal Changes in Particulate and Dissolved Organic Carbon and Nitrogen in a Hardwater Stream, *Arch. Hydrobiol.*, 80: 29–39.

Williams, D. D., and H. B. N. Hynes, 1976, The Ecology of Temporary Streams. I. The Faunas of Two Canadian Streams, *Int. Rev. Gesamten. Hydrobiol.*, 61: 761–787.

23. MICROBIAL ACTION IN DETRITAL LEAF PROCESSING AND THE EFFECTS OF CHEMICAL PERTURBATION

James F. Fairchild, Terence P. Boyle, and Everett Robinson-Wilson

> Columbia National Fisheries Research Laboratory
> United States Department of the Interior
> Fish and Wildlife Service
> Columbia, Missouri

John R. Jones

> School of Forestry, Fisheries, and Wildlife
> University of Missouri
> Columbia, Missouri

ABSTRACT

A laboratory study was conducted to determine the effects of microbial action in detrital leaf processing and to quantify the effects of an antimicrobial agent on microbial functions. Preleached disks from leaves of *Acer saccharum* were incubated in flasks of stream water on a rotary shaker at 20°C for 56 days. Treatments consisted of a control, killed control, and 1.0 and 10.0 mg/l sodium pentachlorophenol (PCP). Leaf disks were periodically measured for changes in dry weight, ash weight, ash-free dry weight (AFDW), and nitrogen. Microbial measurements included respiration, adenosine triphosphate, and ^{14}C acetate incorporation in lipids. Changes in AFDW of leaf disks in the control indicated that microbes processed 21% of leaf material during the 56-day study. The PCP treatments did not differ significantly from the control in loss of

437

AFDW from leaf disks. Significant differences were noted among several microbial characteristics, however; this possibly indicates structural differences caused by the PCP treatments. Chemical perturbation could alter the structure of microbial communities and therefore adversely affect leaf processing, invertebrate productivity, and fishery resources.

INTRODUCTION

During the last 15 years a conceptual scheme has been developed concerning the processing of organic matter in small streams (Minshall, 1967; Cummins, 1974). Basic to this sequence of events is the microbial colonization of detrital material to form what is known as the matter–microbial complex (Cummins et al., 1972). This processing unit is an important link in the transfer of fixed solar energy from the terrestrial watershed to the heterotrophic stream community (Fisher and Likens, 1973).

Using detritus both nutritionally and as a site of activity, bacteria and fungi oxidize both dissolved and particulate allochthonous organics (Cummins, 1974). Microbial production increases the net protein content of the decomposing matter (Kaushik and Hynes, 1971). Invertebrates, in turn, shred and scrape the colonized organic matter to derive nutrition (Cummins, 1974). Thus microbial populations provide an improved food source for detritivores, while also reducing the organic load of streams.

The relative importance of microbial populations in processing particulate organic matter, however, warrants further study (Anderson and Sedell, 1979). The contribution of microbial populations is difficult to determine in the presence of invertebrates, epiphytic algae, and physical abrasion, which combine in complex systems of synergistic and antagonistic processes. Therefore we conducted a laboratory study to help corroborate the importance of microbial processing of leaves and to examine trends among parameters known to be associated with detrital leaf processing. In addition, we sought to determine the extent to which an environmental contaminant such as pentachlorophenol (PCP) might disrupt microbial leaf processing within streams. Manufactured at the rate of 50 million lb/yr, PCP is used in the wood-products industry to retard decay of wood and is also used to prevent algal and microbial growth in cooling towers (Cirelli, 1978). The magnitude of production and use confirm the potential for the entry of PCP into aquatic systems.

MATERIALS AND METHODS

Leaves of sugar maple (*Acer saccharum*) were collected at abscission during autumn 1979. The leaves were pressed, lyophilized, and stored frozen. Before experimental treatment began, 6-mm disks were punched from leaves of similar size, shape, and texture. The leaf disks were preleached at 20°C for 36 h in well water to remove readily soluble organics.

Incubation water was obtained on March 23, 1980, from Clifty Creek, a relatively undisturbed oak–hickory watershed overlying limestone bedrock near Hayden township, Maries County, Missouri. Initial water conditions were: temperature, 17°C; conductivity, 305 μmho/cm; pH, 8.0; alkalinity, 135 mg/l as $CaCO_3$; and Kjeldahl nitrogen, 0.73 mg/l as N.

On March 24, 1980, 60 leaf disks and 300 ml of unfiltered stream water were incubated in 500-ml Erlenmeyer flasks on a rotary shaker at 20 $\pm$ 2°C in the dark. Treatments consisted of an untreated biotic control, killed control (100 mg/l $HgCl_2$), low PCP (1.0 mg/l Na salt of PCP), and high PCP (10.0 mg/l Na-salt of PCP). Triplicate flasks within each treatment were sampled on days 1, 3, 7, 14, 28, and 56 in the following manner.

Temperature, oxygen, conductivity, pH, and alkalinity were determined by conventional means (American Public Health Association, 1975). Water for determinations of dissolved organic carbon (DOC) was filtered through a 0.2-μm Nucleopore membrane filter, acidified to pH 2 with 1.0 N H_2SO_4, and frozen until analysis. The DOC samples were measured with a Technicon Autoanalyzer II Industrial System DOC Cartridge (Technicon Industrial Systems, 1976).*

Upon sampling, leaf disks were lyophilized for 48 h and frozen until analysis. At the time of analysis, leaf disks were lyophilized for an additional 24 h and then stored in a desiccator. Leaf disks were weighed to the nearest 0.1 μg to determine dry weights, ashed at 550°C, and reweighed to determine ash weights. Ash-free dry weight (AFDW) was determined by subtraction.

Nitrogen concentrations in leaf disks and water were determined colorimetrically after Kjeldahl digestion (Kopp and McKee, 1979).

Adenosine triphosphate (ATP) samples were extracted in cold acid by a method modified from Karl and LaRock (1975). Ten-disk subsamples,

---

*Mention of products or brand names does not imply Government or University endorsement.

15-ml H_2O filtrates (0.2-μm Nucleopore membrane), and triplicate spikes and blanks were each immersed in 5 ml of 0.6 N H_2SO_4 at 4°C for 30 min (Knauer and Ayers, 1977). The extract was diluted with 2 ml of MOPS diluent, neutralized with NaOH to pH 7.2, and adjusted to 10 ml in volume with MOPS diluent (E. I. DuPont DeNemours and Co., 1970). The extracts were assayed immediately for ATP content with a DuPont 760 Luminescent Biometer. Recoveries of spiked samples (4.46 ng ATP) averaged 64% during the study; sample values were subsequently corrected to reflect this recovery.

Microbial activity in leaf disks and water was estimated by relative rates of lipid biosynthesis determined by ^{14}C acetate uptake and incorporation into lipids (White et al., 1977). Ten leaf disks, 50 ml of flask water, and 12 μCi [1-^{14}C] of acetate (specific activity, 59 mCi/mmole) were incubated for 2 h in a 125-ml Erlenmeyer flask on a rotary shaker in the dark at 20 ± 2°C. After incubation the water was filtered through a 0.4-μm Nucleopore membrane filter. Disks and filters were subsequently extracted according to White et al. (1977). The chloroform extract was decanted into a scintillation vial and evaporated to dryness. The vials were filled with 10 ml of scintillation fluid (Beckman Fluoralloy dry mix in toluene) and counted in a Beckman Model LS-3133T Scintillation Counter. Recovery was determined by triplicate extraction of spiked ^{14}C stearic acid. Recoveries of spiked samples averaged 94%, and sample values were subsequently corrected for recovery.

Seven days before respiratory measurements, triplicate flasks were stoppered and allowed to incubate on the darkened shaker table. On the seventh day, 500 μl of the flask airspace was removed with a syringe via a septum in the stopper and injected into a gas chromatograph equipped with a thermal conductivity detector. Carbon dioxide was eluted with a stainless steel column (2 m by 0.32 cm ID) packed with Carbosieve "B" (80/100 mesh) maintained at 90°C. Carbon dioxide evolution was calculated from a regression line obtained from CO_2 analysis of standards.

Concentrations of carbon dioxide dissolved in the water of the stoppered flask were calculated from an algorithm of Stumm and Morgan (1970). Total-flask respiration was calculated as CO_2 (evolved) plus CO_2 (dissolved).

Leaf disk samples were periodically removed and examined by scanning electron microscopy for their characteristic microbial flora and degree of colonization. Disks were fixed in 1% osmium tetroxide for 1 h and stored in absolute ethanol at 5°C. Upon analysis, the disks were critical-point dried, sputter coated with gold, and scanned for dominant microflora with a Joel JSM-1 Scanning Electron Microscope.

In analyzing the data, we plotted variables by treatment through time and conducted a one-way analysis of variance of treatment means using orthogonal contrasts (Chew, 1977) on each sampling day. Differences were considered statistically significant at $p \leqslant 0.05$.

RESULTS

Water Chemistry

The water chemistry of the biotic control, low PCP, and high PCP treatments was stable through time, with no significant differences caused by the treatment. The measurements and their ranges were: temperature, 18 to 22°C; conductivity, 260 to 283 μmho/cm; pH, 8.25 to 8.44; dissolved oxygen, 8.2 to 8.7 mg/l; and total alkalinity, 120 to 132 mg/l as $CaCO_3$. The water chemistry in the killed control, however, differed from these treatments in conductivity, pH, and alkalinity caused by the $HgCl_2$. The killed control measurements and their ranges were: temperature, 18 to 22°C; conductivity, 291 to 307 μmho/cm; pH, 8.06 to 8.26; dissolved oxygen, 8.3 to 8.7 mg/l; and total alkalinity, 94 to 124 mg/l as $CaCO_3$. Although most measurements in the killed control were uniform through time, the alkalinity decreased.

Microbial Biomass

Ratios of cellular carbon to ATP are fairly constant for a wide variety of microorganisms (Holm-Hansen and Booth, 1966). Properly employed, the ATP concentration of detrital matter provides a rapid, sensitive estimate of microbial biomass (Bancroft et al., 1976).

ATP-Disks

The ATP concentrations of microbes associated with leaf disks increased from 2.0 to 12.5 ng ATP/disk in the biotic control and low PCP treatments through the course of the experiment (Figure 1). The trend in these treatments amounted to a 525% increase in ATP content of leaf disks during the 8 weeks. These two treatments paralleled each other through time, with no significant differences between them. The high PCP treatment, however, remained significantly lower than the biotic control on every sampling date (Figure 1). On the average, the ATP content of leaf disks in the high PCP treatment was only 38% of that in

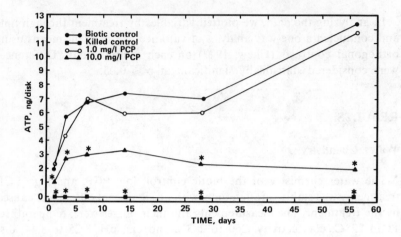

Figure 1. Changes in ATP content of leaf disks over time by treatment. The asterisk (*) indicates significantly different from biotic control at $p \leq 0.05$.

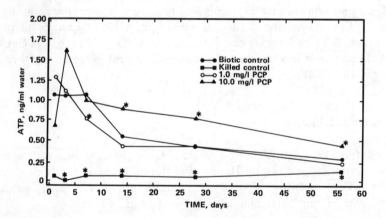

Figure 2. Changes in ATP content of water over time by treatment. The asterisk (*) indicates significantly different from biotic control at $p \leq 0.05$.

the biotic control. The ATP concentration of leaf disks in the killed control remained at minimally detectable levels (Figure 1).

ATP-Water

The ATP concentration of water and its suspended microbes decreased through time in the biotic control, low PCP, and high PCP treatments (Figure 2). The ATP in biotic control and low PCP treatments decreased nearly 65% in 56 days. In general, these two treatments did not differ significantly in ATP of suspended microbes. The ATP in water in the high PCP treatment also decreased after day 3 but was significantly higher than the biotic control after day 7 (Figure 2). After day 7 the high PCP treatment contained an average of 78% more ATP in water than the biotic control or low PCP treatments. Minimal ATP was detected in water of the killed control treatment (Figure 2).

ATP-Total

When the ATP values are separated into three categories (ATP associated with the 60 leaf disks, ATP associated with the 300 ml of water, and total flask ATP), it is clear that ATP was distributed differently in the various treatments through time (Figure 3). If ATP is used as an estimate of biomass, it seems that the microbial communities in the biotic control and low PCP treatments shifted from suspended forms to a leaf disk association as time progressed. Microbial biomass was highest in these two treatments on day 56, and nearly 85% of the total ATP was contained in the leaf disk compartment of these two treatments (Figure 3).

In the high PCP treatment, however, the total microbial biomass increased rapidly through day 3 but declined during the remainder of the study and was equally distributed between the leaf disks and the water on day 56 (Figure 3). On this day the high PCP treatment was 70% lower in total ATP than the biotic control or low PCP treatments. The ATP data suggest that the high PCP treatment affected the microbial community by exerting a toxic chemical effect. The high PCP treatment altered the total biomass and distribution of the microbial community, or, alternatively, the high PCP treatment may have caused a population shift to a group of organisms with an ATP concentration different from that of organisms in the biotic control or low PCP treatments.

Scanning electron micrographs (SEM) verified the colonization of leaf disks by microbes (Figure 4). The SEM photos of the ventral leaf surface at day 0 indicated that the leaf disks entered the study with little colonization (Figure 4a). On days 10 and 20 the microbial community nearly obscured the surface of the leaf disks in the biotic control (Figures 4b and 4c). In contrast, the disks in the killed control were never colonized; on day 20 the stomates were clearly visible, with only a few abiotic precipitates present on the disk surface (Figure 4d).

Microbial Lipid Synthesis

The incorporation of ^{14}C acetate into microbial lipid is known to parallel α–D mannosidase and respiratory activity of microbes in the environment. Under experimental conditions such as these, this measure-

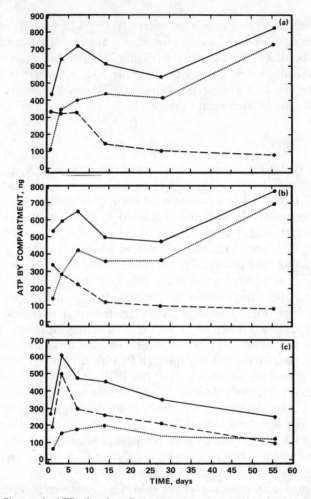

Figure 3. Changes in ATP of various flask compartments over time for (a) biotic control, (b) low PCP (1.0 mg/l), and (c) high PCP treatments (10.0 mg/l). Solid lines denote total flask ATP; dotted lines, ATP associated with 60 leaf disks; and dashed lines, ATP associated with 300 ml of flask water.

ment can be used as a relative measure of heterotrophic microbial activity (White et al., 1977).

^{14}C Lipid-Disks

The uptake and incorporation of ^{14}C acetate into microbial lipids associated with leaf disks increased rapidly in the biotic control during the

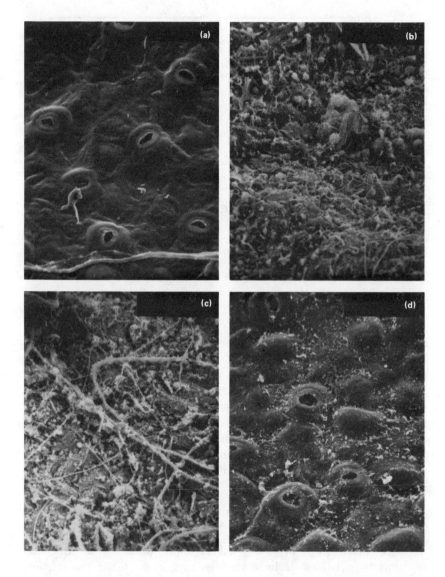

Figure 4. Scanning electron micrographs of the ventral surfaces of leaf disks. (a) Day 0. (b) Day 10, biotic control. (c) Day 20, biotic control. (d) Day 20, killed control. All photos were magnified 1000X.

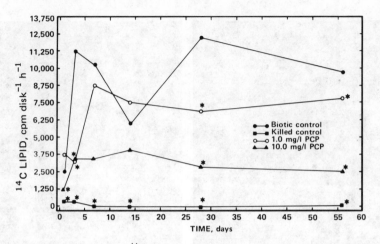

Figure 5. Changes in rates of ^{14}C lipid synthesis in the microbial community associated with leaf disks. Plots represent ^{14}C lipid synthesis over time by treatment. The asterisk (*) indicates significantly different from biotic control at $p \leqslant 0.05$.

first three days of the study (Figure 5). The rate decreased in the biotic control on days 7 and 14 but increased at days 28 and 56 to about 11,000 cpm disk^{-1} h^{-1}. The rates of ^{14}C lipid synthesis by disk microbes in the low and high PCP treatments increased during the first 7 days of the study (8750 and 3400 cpm disk^{-1} h^{-1}, respectively) (Figure 5). The rates remained stable through day 56 but were significantly lower than in the biotic control on days 28 and 56. On day 56 the rates of ^{14}C lipid synthesis were 20% less in the low PCP treatment and nearly 70% less in the high PCP treatment than in the biotic control. Leaf disks in the killed control did not exhibit any appreciable uptake and incorporation of ^{14}C acetate during the experiment (Figure 5).

^{14}C Lipid-Water

Uptake and incorporation of ^{14}C acetate into lipids by the suspended microbial populations declined through time in the biotic control, low PCP, and high PCP treatments (Figure 6). Rates in the biotic control and low PCP treatments declined 80% in 56 days to 700 cpm ml^{-1} water h^{-1}. In contrast, the rate in the high PCP treatment declined only 35% in 56 days to approximately 1500 cpm ml^{-1} water h^{-1} (Figure 6). The day-56 value was significantly greater in the high PCP treatment than in the biotic control and corresponds to the significantly greater ATP concentration (water) in the suspended microbial community. There was little

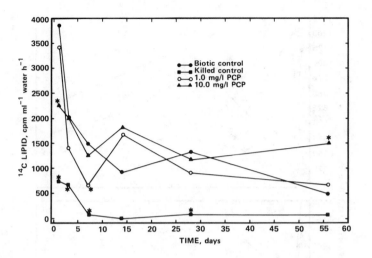

Figure 6. Changes in rates of ^{14}C lipid synthesis by the suspended microbial community. Plots represent ^{14}C lipid synthesis over time by treatment. The asterisk (*) indicates significantly different from biotic control at $p \leqslant 0.05$.

incorporation of ^{14}C acetate by suspended microbes in the killed control (Figure 6).

^{14}C Lipid-Total

By partitioning the ^{14}C lipid into the total associated with the 60 leaf disks, the total associated with the 300 ml water, and total flask ^{14}C lipid, we can demonstrate that the biotic control, low PCP, and high PCP treatments differed in the distribution of microbial activity. Microbial activity in the biotic control and low PCP treatments shifted from the water to the disk locus through time (Figure 7). On day 56 approximately two-thirds of the total activity was associated with leaf disks in these two treatments. In the low PCP treatment, however, ^{14}C lipid synthesis by disk microbes was significantly lower on days 28 and 56 (Figure 5). Furthermore, the trends in ^{14}C lipid synthesis in the low PCP treatment did not parallel fluctuations within the biotic control (Figure 7). The fluctuations in the biotic control data may indicate succession in the microbial community (White et al., 1977). Similar, parallel fluctuations were not observed in the low PCP treatment; this may indicate that chemical perturbation prevented such succession.

The distribution of microbial activity in the high PCP treatment also seemed to differ from that in the biotic control. Primarily disk associated

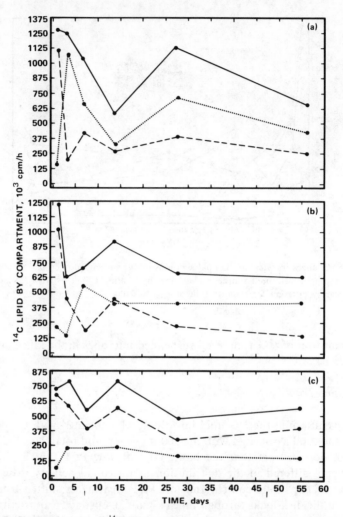

Figure 7. Changes in rates of ^{14}C lipid synthesis in the various flask compartments over time for (a) the biotic control, (b) low PCP (1.0 mg/l), and (c) high PCP (10.0 mg/l) treatments. Solid lines denote total flask synthesis; dotted lines, synthesis associated with 60 leaf disks; and dashed lines, synthesis associated with 300 ml flask water.

in the biotic control, ^{14}C lipid synthesis was mostly in the aqueous compartment of the high PCP treatment (Figure 7). This trend was similar to that shown for ATP and suggests that the high PCP treatment altered the production, distribution, or succession of the microbial community.

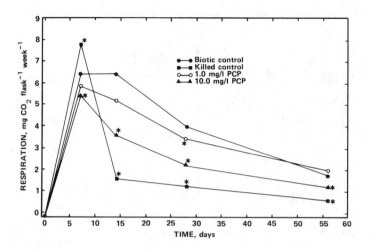

Figure 8. Changes in whole-flask respiration over time by treatment. The asterisk (*) indicates significantly different from biotic control at $p \leqslant 0.05$.

Respiration

Since heterotrophic organisms use fixed organic carbon as their primary energy source, and carbon that is not incorporated or egested is liberated as respiratory CO_2, respiration levels can provide a relative estimate of rates of leaf decomposition caused by microbial activity.

The respiration rate declined among all treatments through time (Figure 8). In the biotic control and low PCP treatments, it declined from about 6 mg CO_2 flask^{-1} week^{-1} at week 1 to about 2 mg CO_2 flask^{-1} week^{-1} by week 8. This amounted to a 66% decrease over the course of the study. On most sampling days these two treatments were not significantly different from one another. Respiration in the high PCP treatment decreased 80% in 56 days and was significantly lower than that in the biotic control on each sampling day (Figure 8). The respiration data in the high PCP treatment supported the ATP and ^{14}C lipid synthesis results in indicating that the high PCP treatment may have altered the function as well as the structure of the microbial community. Evolution of CO_2 was also detected in the killed control (Figure 8). Because all other variables indicated that these flasks were sterile, we assumed that the CO_2 evolved in the killed control was probably chemically liberated by the presence of $HgCl_2$.

The decline in respiratory rates during the study may have been caused by depletion of labile organic carbon (Cummins et al., 1972; Wetzel and Manny, 1972). This hypothesis is reinforced by the dissolved organic

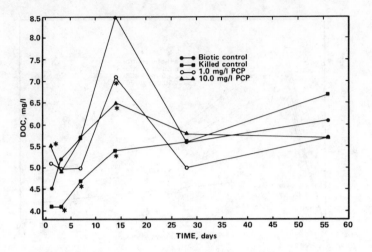

Figure 9. Changes in dissolved organic carbon (DOC) concentrations over time by treatment. The asterisk (*) indicates significantly different from biotic control at $p \leqslant 0.05$.

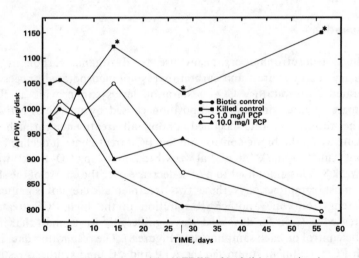

Figure 10. Changes in ash-free dry weight (AFDW) of leaf disks over time by treatment. The asterisk (*) indicates significantly different from biotic control at $p \leqslant 0.05$.

carbon (DOC) data (Figure 9). Respiration was highest on days 7 and 14, the same days on which the DOC levels were highest. Both respiration and DOC declined during the rest of the study as the more readily metabolized elements of the carbon pool were consumed (Figures 8 and 9). Because these were whole-flask respiration estimates, it is not known whether the carbon was oxidized from particulate or dissolved sources.

Processing

Changes in the ash-free dry weight (AFDW) of detrital material are often used as a measure of processing. The AFDW of leaf disks decreased through time in the biotic control, low PCP, and high PCP treatments by 213, 193, and 176 μg/disk, respectively (Figure 10). This decrease amounted to 18 to 21% of AFDW of leaf disks in these treatments. Neither of the PCP treatments were significantly different from the biotic control on any sampling date. Leaf disks in the killed control treatment actually increased in AFDW by over 15% (Figure 10). This increase may have been caused by adsorption (high dry weight measurement) and subsequent volatilization (low ash weight measurement) of mercury during the ashing procedure.

Although the respiration data indicated significant differences in processing between the biotic control and high PCP treatments, AFDW of leaf disks did not. We attributed this in part to the inherent variability in the leaf disks, which may have hindered detection of significant differences in processing. This variability may imply that AFDW of leaf disks is not a highly sensitive indicator of chemical perturbation of leaf processing. In addition, microbes may contribute to loss of organic carbon from leaves in ways other than metabolic oxidation to CO_2. For instance, microbial activity may contribute to further carbon losses from detritus by micro-fragmentation or a form of biotic leaching. Under certain conditions, leaf leachates can form flocculent precipitates with divalent cations that abiotically remove DOC from the water column (Lush and Hynes, 1973). We did not have the analytical capabilities to analyze small particulate organic matter. Lack of information concerning this carbon pool may have caused some error in our processing measurements and may explain why we did not detect significant differences in AFDW of leaf disks in the high PCP treatment.

Nitrogen

The nitrogen concentration of leaf litter is known to increase during microbial decomposition as a result of two processes—an increase in protein as a result of microbial growth on the leaf surface and a simultaneous decrease in organic carbon caused by microbial respiration (Howarth and Fisher, 1976). An increasing nitrogen concentration in detritus indicates that the decomposing matter is increasing in value as an invertebrate food source (Kaushik and Hynes, 1971).

In our 56-day study, the nitrogen concentration of leaf disks increased 18 to 30% in the biotic control, low PCP, and high PCP treatments

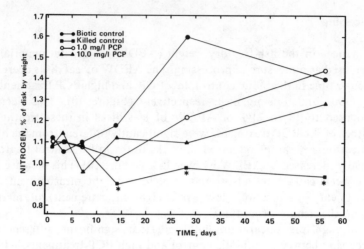

Figure 11. Changes in nitrogen concentration of leaf disks over time by treatment. Nitrogen values are expressed as a percent of disk weight. The asterisk (*) indicates significantly different from biotic control at $p \leqslant 0.05$.

(Figure 11). The two PCP treatments were not significantly different from the biotic control on any sampling day. Nitrogen concentrations of leaf disks in the killed control decreased by 14% in 56 days. This trend may be a result of the leaching of soluble organic nitrogen or may represent a mathematical artifact of the observed increase in dry weight of leaf disks in the killed control. Nitrogen concentrations of leaf disks were significantly lower in the killed control than in the biotic control on days 14, 28, and 56 (Figure 11) because the killed control lacked a microbial community.

The ATP content of leaf disks indicated that microbial biomass associated with leaf disks was significantly lower in the high PCP treatment than in the biotic control (Figure 1). This significant difference was not reflected in the nitrogen concentration of leaf disks because of two inherent problems. First, the variability in weights of leaf disks may have hindered precise calculation of nitrogen concentration. Second, the leaf disks were the major source of nitrogen in the experimental flasks. As microbial protein developed on the leaf disks, nitrogen was merely translocated from the disk to the microbes; thus the disk-microbe complex in the biotic control and PCP treatments contained similar absolute measurements of nitrogen. Although the nitrogen may have been distributed differently between the leaf disks and the microbes in these treatments, it was not discernible in the nitrogen concentration of the disk-microbe complex.

DISCUSSION

Microbial Processing of Leaf Disks

Leaf disks lost 21% of AFDW in the biotic control over the course of the experiment. Leaf processing is known to be the result of four physical and biological factors: (1) abiotic leaching, (2) physical abrasion, (3) microbial metabolism, and (4) invertebrate feeding activity (Cummins, 1974). Invertebrates were intentionally excluded from our system. Lack of changes in ash weights through time indicated that physical abrasion was not a factor in weight loss, and, since leaf disks did not decrease in dry weight during the first 7 days of the experiment, little abiotic leaching occurred. Therefore we attributed the majority of processing to microbial metabolism of leaf disk carbon.

Our estimate of 21% microbial processing of leaves is similar to a 14% estimate obtained with maple leaf disks in a recirculating laboratory stream maintained at 18°C for 5 weeks (Howarth and Fisher, 1976) and to a 24% processing estimate from a laboratory study with unleached elm leaves incubated in streamwater at 21°C for 56 days (Hynes et al., 1974).

There are relatively few estimates of rates of microbial processing of particulate matter in natural streams. Sedell et al. (1975) indicated that leaves decayed slowly by microbial metabolism, but that decomposition increased dramatically after invertebrates entered the leaf packs. These investigators measured over 50% decay of red alder, vine maple, big-leaf maple, and conifer leaves in less than 70 days in a Cascade Mountain stream and attributed this rapid processing to invertebrate feeding. This high percentage suggests that invertebrates mediate most of the leaf decomposition in streams.

However, we cannot ignore the importance of microbial populations in processing leaves in streams. Invertebrates are known to prefer leaves that have been adequately conditioned by microbes (Boling et al., 1975) and select leaves that are most rapidly colonized by microbes (Sedell et al., 1975). Thus microbes play a major role in invertebrate processing activity. Invertebrate activity in itself may increase the ability of microbes to decompose particulate organics. By feeding action, invertebrates reduce the particle size of organic matter; this in turn increases the total surface area available to microbial colonization and metabolism (Cummins, 1974). By excluding invertebrates, we may have reduced the particle-size diversity of organic matter available to microbial decomposition. This reduction may result in an underestimate of the ability of microbial populations to oxidize particulate organic matter. We further recognize that our experimental system may have been nutrient limited; this would lead to a conservative estimate of microbial processing of leaves. Other researchers have demonstrated increased microbial leaf processing after nutrient enrichment (Howarth and Fisher, 1976; Hynes et al., 1974), and recent work in our laboratory has demonstrated similar results.

We believe that these results have confirmed that microbes can process particulate organic materials to a significant extent. Other research has shown that bacteria are the principal processors of dissolved organic matter derived from terrestrial inputs, leaching, and invertebrate activities (Cummins et al., 1972; Lock and Hynes, 1976). Thus it is evident that microbes probably serve a substantial purpose in processing organic matter within streams. The functional diversity of microbial and invertebrate activity probably plays a significant role in reducing the organic load of streams, thus maximizing heterotrophic production and protecting downstream aquatic resources from organic enrichment.

Potential for Perturbation

Data for the killed control indicated that, in extreme instances, chemical perturbation could completely disrupt microbial leaf processing. The PCP treatments did not significantly affect processing, as judged by the loss of AFDW from leaf disks. Microbial parameters indicated, however, that chemical perturbation by PCP could affect the biomass, distribution, or succession of the microbial community.

Such changes in the microbial community could conceivably disrupt the processing of organic matter in streams. Suberkropp and Klug (1976) determined that fungi are the dominant microbes involved in processing leaf material and that bacteria mainly process fungal excretions and dissolved organic leachates. In addition, Kostalos (1980) recently demonstrated that fungi associated with detritus may be the primary source of nutrition for adult amphipods and other shredding and scraping invertebrates. Therefore a structural change in the microbial community that adversely affects fungal communities could alter not only fungal decomposition of leaves but also invertebrate nutrition and production. A decrease in invertebrate production could eventually have serious fishery implications as well.

We tested only one compound; other types of compounds, which could affect aquatic microbial communities in different ways, should be investigated. Of further interest is the toxicity of environmental contaminants to invertebrate communities. We believe that chemical perturbation might be most damaging at this level since invertebrates are frequently quite sensitive to contaminants and may not be as resilient as microbes. Reduction in invertebrate numbers or diversity could have a multitude of effects on the processing of organic matter and on fishery resources.

ACKNOWLEDGMENTS

We extend our gratitude to Paul Heine and Phil Lovely for their fine technical assistance. We also wish to thank Darrell Kinden and Preston Stogsdill, University of Missouri, Columbia, School of Veterinary Medicine, for their expertise in scanning electron microscopy.

REFERENCES

American Public Health Association, 1975, *Standard Methods for the Examination of Water and Wastewater*, 14th ed., Washington, DC.

Anderson, N. H., and J. R. Sedell, 1979, Detritus Processing by Macroinvertebrates in Stream Ecosystems, *Ann. Rev. Entomol.*, 24: 351-377.

Bancroft, K., E. A. Paul, and W. J. Wiebe, 1976, The Extraction and Measurement of Adenosine Triphosphate from Marine Sediments, *Limnol. Oceanogr.*, 21: 473-480.

Boling, R. H., E. D. Goodman, J. A. Van Sickle, J. O. Zimmer, K. W. Cummins, R. C. Peterson, and S. R. Reice, 1975, Toward a Model of Detritus Processing in a Woodland Stream, *Ecology*, 56: 141-151.

Chew, Victor, 1977, *Comparisons Among Treatment Means in an Analysis of Variance*, Agricultural Research Service, Report No. ARS/H/6, U.S. Government Printing Office, Washington, DC.

Cirelli, D. P., 1978, Patterns of Pentachlorophenol Usage in the United States of America—An Overview, *Pentachlorophenol: Chemistry, Pharmacology, and Environmental Toxicology*, pp. 13-18, Plenum Press, New York.

Cummins, K. W., 1974, Structure and Function of Stream Ecosystems, *Bioscience*, 24: 631-641.

_____, J. J. Klug, R. G. Wetzel, R. C. Peterson, K. F. Suberkropp, B. A. Manny, J. C. Wuycheck, and F. O. Howard, 1972, Organic Enrichment with Leaf Leachate in Experimental Lotic Ecosystems, *BioScience*, 22: 719-722.

E. I. DuPont DeNemours & Co., 1970, *DuPont 760 Luminescent Biometer Instruction Manual*, Instrument Products Division, Wilmington, DL.

Fisher, S. G., and G. E. Likens, 1973, Energy Flow in Bear Brook, New Hampshire: An Integrative Approach to Stream Ecosystem Metabolism, *Ecol. Monogr.*, 43: 421-439.

Holm-Hansen, O., and C. R. Booth, 1966, The Measurement of Adenosine Triphosphate in the Ocean and Its Ecological Significance, *Limnol. Oceanogr.*, 11: 510-519.

Howarth, R. A., and S. G. Fisher, 1976, Carbon, Nitrogen, and Phosphorus Dynamics During Leaf Decay in Nutrient-Enriched Stream Microecosystems, *Freshwater Biol.*, 6: 221-228.

Hynes, H. B. N., N. K. Kaushik, M. A. Lock, D. L. Lush, Z. S. J. Stocker, R. R. Wallace, and D. D. Williams, 1974, Benthos and Allochthonous Organic Matter in Streams, *J. Fish. Res. Board Can.*, 31: 545-553.

Karl, D. M., and P. A. LaRock, 1975, Adenosine Triphosphate Measurements in Soil and Marine Sediments, *J. Fish. Res. Board Can.,* 32: 599-607.

Kaushik, N. K., and H. B. N. Hynes, 1971, The Fate of Dead Leaves that Fall into Streams, *Arch. Hydrobiol.,* 68: 465-515.

Knauer, G. A., and A. V. Ayers, 1977, Changes in Carbon, Nitrogen, Adenosine Triphoshate, and Chlorophyll α in Decomposing *Thalassia testudinum* Leaves, *Limnol. Oceanogr.,* 22: 408-414.

Kopp, J. F., and G. D. McKee, 1979, *Methods for Chemical Analysis of Water and Wastes,* EPA-600/4-79-020, Office of Research and Development, Environmental Protection Agency, Cincinnati, OH.

Kostalos, M. S., 1980, Comparison of the Role of Microbial Enriched Detritus in the Nutrition of Adult and Juvenile *Gammarus minus* (Amphipoda), *Bulletin of the Ecological Society of America,* 61: 151, abstracts of papers presented at the Aug. 3-7, 1980, meeting with AIBS at the University of Arizona, Tucson.

Lock, M. A., and H. B. N. Hynes, 1976, The Fate of "Dissolved" Organic Carbon Derived from Autumn-Shed Maple Leaves (*Acer saccharum*) in a Temperate Hard-Water Stream, *Limnol. Oceanogr.,* 21: 436-443.

Lush, D. L., and H. B. N. Hynes, 1973, The Formation of Particles in Freshwater Leachates of Dead Leaves, *Limnol. Oceanogr.,* 18: 968-977.

Minshall, G. W., 1967, The Role of Allochthonous Detritus in the Trophic Structure of a Woodland Springbrook Community, *Ecology,* 48: 139-149.

Sedell, J. R., F. J. Triska, and N. S. Triska, 1975, The Processing of Conifer and Hardwood Leaves in Two Coniferous Forest Streams: I. Weight Loss and Associated Invertebrates, *Verh. Internat. Verein. Limnol.,* 19: 1617-1627.

Stumm, W., and J. J. Morgan, 1970, *Aquatic Chemistry: An Introduction Emphasizing Chemical Equilibria in Natural Waters,* pp. 118-159, Wiley-Interscience, New York.

Suberkropp, K., and M. J. Klug, 1976, Fungi and Bacteria Associated with Leaves During Processing in a Woodland Stream, *Ecology,* 57: 707-719.

Technicon Industrial Systems, 1976, *Dissolved Organic Carbon in Water and Wastewater,* Technicon Industrial Method No. 451-76W, Tarrytown, NY.

Wetzel, R. G., and B. A. Manny, 1972, Decomposition of Dissolved Organic Carbon and Nitrogen Compounds from Leaves in an Experimental Hard-Water Stream, *Limnol. Oceanogr.,* 17: 927-931.

White, D. C., R. J. Bobbie, S. J. Morrison, D. K. Oosterhof, C. W. Taylor, and D. A. Meeter, 1977, Determination of Microbial Activity of Estuarine Detritus by Relative Rates of Lipid Biosynthesis, *Limnol. Oceanogr.,* 22: 1089-1099.

$\frac{24/30}{75}^{b}$

24. NITRATE-NITROGEN MASS BALANCES FOR TWO ONTARIO RIVERS

Alan R. Hill

Department of Geography
York University
Toronto, Ontario

ABSTRACT

Analysis of nitrogen transport revealed an average daily loss of 45 ± 9 kg NO_3–N (160 mg N/m^2) in Duffin Creek and 35 ± 21 kg NO_3–N (60 mg N/m^2) in the Nottawasaga River during low flows between May and October. Laboratory experiments suggested that most of the nitrate disappearance was caused by denitrification and nitrate reduction in stream sediments. Contrasts in nitrate removal rates in the two rivers probably resulted mainly from differences in sediment characteristics. The Nottawasaga River sediments had very low percentages of organic carbon and silt in comparison to sediments from pool reaches in Duffin Creek. Nitrate losses during stream transport rose rapidly to high levels in May and fell to lower levels in November as water temperature declined. Laboratory experiments suggested that denitrification and nitrate reduction may still occur, although at very low rates, during the winter months when water temperatures in the Nottawasaga River and Duffin Creek are 0 to 2°C. Denitrification was considered to be negligible during major flood events. The removal of nitrate represented <5% of the annual export of nitrogen from the two rivers. In low flow conditions, however, between May and October, nitrate losses corresponded to 50% of the average daily input of total nitrogen and 75% of the nitrate input in Duffin Creek. The Nottawasaga River had a much smaller assimilative capacity, removing only about 13% of the daily nitrate input.

INTRODUCTION

Nitrogen transformations, including such dissipative reactions as denitrification, have been investigated mainly in lakes (Keeney, 1973; Brezonik, 1977). There have been relatively few studies of nitrogen budgets and sink processes in rivers. This neglect has resulted from a tendency to regard rivers as inert pipelines rather than as functioning ecosystems with respect to nutrient transport (Likens, 1975).

Approximate nitrogen budgets based on a combination of measured and estimated fluxes revealed apparent losses of inorganic nitrogen in some rivers during low summer flows (Owens et al., 1972). The investigators attributed these losses primarily to bacterial denitrification in sediments. Several recent studies indicated that denitrification in anaerobic sediments may play a major role in removing nitrogen from streams. Van Kessel (1977b) measured a NO_3–N loss of 913 mg m^{-2} day^{-1} during a 20-day period for a small canal receiving treated sewage. Laboratory experiments with undisturbed water–sediment profiles indicated that the disappearance of nitrate was caused mainly by denitrification in the sediment (Van Kessel, 1977a). Field observations revealed considerable losses of nitrate during transport in two small headwater streams in Ontario, Canada (Kaushik et al., 1975; Robinson et al., 1978a). Laboratory studies showed rapid removal of nitrate from well-oxygenated water overlying sediments from these streams (Sain et al., 1977). Subsequently $^{15}NO_3$–N has been used to show that the losses were caused by denitrification (Chatarpaul and Robinson, 1979; Chatarpaul et al., 1980).

The studies cited do not provide the detailed field measurements of nitrogen transport in streams which are needed to assess the importance of denitrification and other sink processes in relation to seasonal and annual river nitrogen budgets. I have recently analyzed the magnitude of denitrification in relation to the annual export of nitrogen from Duffin Creek, Ontario (Hill, 1979). My data suggest that the removal of NO_3–N, primarily by denitrification, in the downstream reaches of the river represents 5 to 6% of the annual export of total nitrogen from this river basin. During low summer flows, however, nitrate sink processes removed approximately 75% of the daily NO_3–N input to the downstream reaches of the river (Hill, 1981).

A greater understanding of nitrogen flux and cycling in rivers requires field studies of nitrogen mass balances in a wide variety of rivers. In this paper I analyze nitrate sink processes in relation to nitrogen transport in the Nottawasaga River, Ontario. The data are compared with my measurements from Duffin Creek to examine the magnitude, seasonal variability, and relative importance of nitrate removal processes in these two river ecosystems.

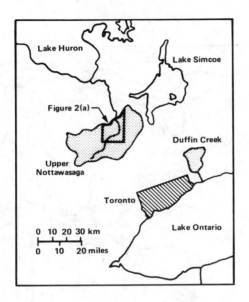

Figure 1. Location of study areas.

THE STUDY AREA

Nottawasaga River

The upper Nottawasaga River basin (1230 km²) is located about 50 km north of Toronto (Figure 1). Research was restricted to the Alliston sand plain area of the basin, which is drained by the Nottawasaga River and its main tributaries, the Boyne River and Inisfil Creek (Figure 2a). The sand plain, which is underlain by a shallow groundwater table, is an area of intensive cash farming of potatoes, corn, and sod. Many of these crops are irrigated during the summer with water pumped from the major rivers. Alliston is served by a sewage treatment plant that discharges effluent into the Boyne River about 2 km upstream from station 33.

The Nottawasaga River is a hard-water sixth-order stream, ranging in width from 10 to 20 m and having a mean annual discharge of 7.9×10^5 m³/day at station 34. The channels of the Nottawasaga River and its major tributaries are frequently incised to a depth of 10 to 20 m below the surface of the sand plain. The river between station 32 and the junction with the Inisfil Creek consists of alternating riffles and pools. Gravel riffles are generally absent in other reaches, and the river bed consists of unstable medium-to-coarse sandy sediments. Benthic algae occur on the

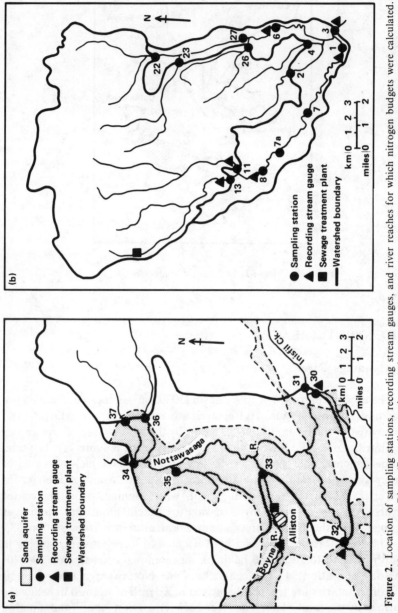

Figure 2. Location of sampling stations, recording stream gauges, and river reaches for which nitrogen budgets were calculated. (a) Nottawasaga River. (Sampling stations on four small sand plain streams are not shown). (b) Duffin Creek.

gravel riffles between station 32 and the junction with Inisfil Creek; elsewhere the river channels are largely devoid of benthic algae and macrophytes.

Duffin Creek

Duffin Creek is a shallow, hard-water, sixth-order stream that drains an area of 262 km² about 30 km east of Toronto (Figure 1). Land use for the drainage basin includes about 49% agricultural land, 21% forest, and 18% abandoned farm land in various stages of old-field succession. Stouffville, a small urban center is served by a tertiary sewage treatment plant that discharges effluent into the west branch of Duffin Creek (Figure 2b).

The mean annual discharge of the river is 2.2×10^5 m³/day at the Pickering stream gauge (station 3). Both branches of the river are composed of riffles and pools, and the average channel width increases downstream from 4 m at stations 11, 13, and 22 to 30 m at station 3. Riffle substrates consist of gravel, ranging in size from a few millimeters up to 50 cm or more in diameter, and comprise approximately 50% of the total stream bed area. Calcareous pool sediments, varying between silt loam and sandy loam in texture, occupy areas between riffles. The river channel is generally unshaded and flows across a floodplain through unused pastures. Aquatic macrophytes are almost entirely absent from the river and the major component of the flora consists of benthic algae dominated by *Cladophora sp.,* which form a mat on the gravel riffles.

MATERIALS AND METHODS

The construction of a nutrient mass balance for river reaches involves measuring a number of input and output vectors. During low flows, the nutrient load of the river as it enters and leaves a particular reach constitutes the major input and output terms in the budget. In some river reaches additional inputs may be represented by small tributary streams and groundwater entering the main channel. The analysis of a mass balance for flood events was not attempted because it was impossible, in view of the number of sites involved, to measure adequately the rapid changes in chemistry and discharge which occur during storm runoff.

River Inputs and Exports

River discharge and nitrogen concentrations were measured at 13 stations in the Nottawasaga basin and at 14 stations on Duffin Creek (Figure 2). Government stream-flow recording gauges provided discharge records at a number of sampling stations. Current meter measurements coupled with cross-sectional areas were used to calculate discharge at times of stream water sampling for the remaining stations. The quantity of river water used for irrigation was estimated from water extraction permits submitted by individual landowners to the Ontario Ministry of Environment.

Groundwater Inputs

In the Nottawasaga study area the shallow Alliston sand plain aquifer represents an important nitrate input to the river. The concentration of NH_4-N and NO_3-N was measured at 60 seepage sites adjacent to the main river channels. The quantity of groundwater entering the river reaches was evaluated by using data from six small streams that drain portions of the sand plain. These streams are sustained by groundwater during summer low flows and their discharge, divided by drainage area, provides an estimate of groundwater flux per square kilometer. This estimate was then applied to areas of the sand plain which drain directly to the Nottawasaga River and its major tributaries, Inisfil Creek and the Boyne River. The input derived by this procedure agrees closely with the groundwater flux that can be calculated as the difference between daily inputs of water from major rivers and the sand plain tributaries and the export of water measured at station 34, together with estimates of withdrawals for irrigation.

Recent hydrogeologic studies indicate that there are no major inputs of groundwater in the downstream reaches of Duffin Creek (Sibul et al., 1977; Ostry, 1979). Small terraces composed of 1 to 3 m of gravel occur in the river valleys however, and contain perched groundwater tables that cause small increases in stream flow in some reaches during May and late October. No attempt was made to sample the concentration of nitrogen forms in seepages from these areas.

Water Chemistry

Water samples were collected in acid-washed polyethylene bottles and transported to the laboratory on ice. Nitrogen analysis was usually run within 2 days of sampling. An automated cadmium reduction method was used to measure $NO_3 + NO_2$–N (American Public Health Association, 1976). Tests were also made separately for NO_2–N in the water samples. Since concentrations of NO_2–N were very low, the $NO_3 + NO_2$–N value is henceforth referred to as NO_3–N. Water samples were analyzed for NH_4–N with an automated indophenol blue method (Technicon Industrial Systems, Method No. 98-70W). Measurements of all nitrogen forms were also available for the sampling stations on Duffin Creek from government water-quality monitoring studies (Ontario Ministry of Environment, 1973-1978).

Sediment Samples

A coring device was used to remove sediment samples from the river bed at ten sites in each study area. Composite samples for the 0 to 4 and 4 to 10-cm sediment depths were obtained by combining six samples collected at intervals across the river channel at each site. Ammonium–nitrogen and $NO_3 + NO_2$–N were determined by extracting moist sediment samples with $2M$ KCl followed by colorimetric analysis according to the procedures outlined for water samples. Organic nitrogen in sediments was determined with a block digestor, then the ammonia was determined on the Auto Analyzer (Issac and Johnson, 1976). Organic carbon was estimated by the Walkley–Black method (Allison, 1965). Sediment pH was measured with a glass electrode on a 1 : 2 soil and water mixture.

Laboratory Systems

The possible role of sediments in nitrate removal from stream water was investigated in a laboratory experiment. Intact cores were removed from the river bed at ten locations in each drainage basin. A stainless steel column of 5.8 cm OD and 0.1 cm in wall thickness was pushed into the sediment; its top end was closed with a rubber stopper; and the column was then removed, together with a core about 10 cm deep. The core was gently pushed out of the metal column into a tall glass cylinder with the same internal diameter as the metal coring device. Gravel samples were also collected in polypropylene pans from stream riffles at five sites on Duffin Creek. Additional details of riffle sampling procedures are given elsewhere (Hill, 1981).

To each glass cylinder, 500 ml of a solution containing 5 mg/l nitrate–N as KNO_3 was added to give a solution depth of about 15 cm over the sediment. To each pan, 4000 ml of the same solution was added to cover the gravel to a depth of 5 cm. Cylinders and pans containing only the solution were used as controls. Air bubbles were introduced near the water–sediment interface to reproduce the turbulence and aeration characteristic of river conditions. The cylinders and baths were incubated in the dark at 20°C, a temperature similar to the average summer water temperature of the rivers. Nitrate concentrations in the solutions were analyzed at intervals over a period of 6 days.

Material Mass Balances

The data used to examine mass balances were collected on days between late May and October. Discharge and nitrogen concentrations were measured only during periods of low stream flow, characterized by discharge fluctuations of less than 10% at stations 3 and 34 in the 24 to 36 h before sample collection.

Analysis of the mass balance for the Alliston sand plain area of the Nottawasaga River was based on a total of 21 days of observation in 1977 to 1979. The daily input for each day was derived from the summation of the individual fluxes at stations 30, 31, 32, 33, 36, and 37 on the major streams; the sand plain tributaries (e.g., station 35); and the groundwater contribution. The daily output from this study area consisted of the flux at station 34 and the estimated removal of water and nitrogen for irrigation (Figure 2a).

A mass balance was calculated for the downstream reach of Duffin Creek on the basis of data collected on a total of 18 days from 1973 to 1975 and 1978. The summation of the individual stream discharges and nitrogen fluxes at stations 2, 4, 11, 13, 22, 23, and 26 were compared with the output of water and nitrogen at station 3 for each day (Figure 2b).

RESULTS AND DISCUSSION

The water budget showed that average daily inputs and outputs were in balance for the Nottawasaga River, whereas Duffin Creek had a small but significant daily gain of about 8% (Table 1). Most of the daily water input to the downstream reaches of Duffin Creek was contributed by stations 22 (40%), 11 (27%), and 13 (14%). In the Alliston sand plain area stations 32 and 33 accounted for 48% and 24%, respectively, of the average daily input of water, and the other major streams contributed an additional 20%. Approximately 8% of the daily input was derived from groundwater, either in direct seepage to the main river channels (3%) or in sand plain tributary streams (5%).

Nitrate Budgets During Low Summer Flows

Low concentrations of NH_4-N (<0.08 mg/l) and trace levels of NO_2-N occurred at all sampling locations in both drainage basins. Mean concentrations of organic nitrogen varied between 0.25 and 0.61 mg/l for the Duffin Creek stations. Nitrate levels were generally in the range 0.10 to 1 mg N/l at most stream sampling stations. However, mean concentrations of NO_3-N at stations 13 and 33, downstream from sewage treatment plants, were 2.63 and 1.32 mg/l respectively. High NO_3-N concentrations (ranging from 3 to 40 mg/l) occurred in some small tributaries and groundwater seeps on the heavily fertilized Alliston sand plain.

The NO_3-N mass balance for the Nottawasaga River showed a loss on all 21 days of observation. Individual daily deficits ranged from less than 1 kg to a maximum of 70 kg, and the average daily NO_3-N mass balance indicated a loss of 35 ± 21 kg, which is equivalent to about 60 mg N/m^2 of stream bed (Table 1). In Duffin Creek significant NO_3-N losses were recorded on all 18 days of observation. The average daily loss for the 18 days was 45 ± 9 kg NO_3-N, which is equivalent to approximately 160 mg N/m^2 of stream bed (Table 1). Individual reaches on the east branch of

Table 1. Mean Daily Input and Output of Water and NO_3-N in the Nottawasaga River and Duffin Creek During Low Flows from Late May to October*

	Input	Output	Net gain or loss	Gain or loss as % of input
Nottawasaga River				
Flow, l/s				
Total	3310 ± 561	3364 ± 595	$+54 \pm 141$	2
Major streams	3058 ± 555			
Sand plain streams	147 ± 8			
Groundwater	105			
NO_3 – N, kg/day				
Total	267 ± 38	232 ± 45	-35 ± 21**	13
Major streams	172 ± 37			
Sand plain streams	16 ± 2			
Groundwater	79			
Duffin Creek				
Flow, l/s	996 ± 170	1078 ± 249	$+82 \pm 127$*	8
NO_3 – N, kg/day	60 ± 11	15 ± 9	-45 ± 9**	76

*Means are $\pm$ standard deviation. Mean daily gain or loss significantly different from zero at the 5% level is indicated by * and at the 1% level, by ** (a pair comparison t-test was used).

the river had average losses ranging from 40 to 110 mg N m^{-2} day^{-1}, whereas much higher losses (300 and 480 mg N m^{-2} day^{-1}) occurred downstream from the sewage treatment plant between stations 7 and 8 and between stations (11 + 13) and 8, respectively (Figure 2b). Analysis of the transport of organic nitrogen in Duffin Creek indicated that small daily gains and losses occurred with equal frequency and the average daily difference between inputs and outputs was less than 3%.

Mechanisms for Nitrate Removal in River Reaches

A variety of processes may be responsible for the occurrence of nitrate losses during river transport. Nitrate removal as a result of temporary immobilization during decomposition of organic matter is of minor significance during the summer in southern Ontario (Kaushik and Hynes, 1971; Kaushik et al., 1975). Nitrate assimilation by riparian tree roots in the streambed is also likely to be a negligible factor (Meyer, 1978). Aquatic macrophytes and benthic algae are unimportant in the Nottawasaga study area. Analysis of the role of benthic algae on gravel riffles in Duffin Creek suggests that nitrate uptake by the algae can account for only approximately 20% of the observed daily nitrate loss (Hill, 1981). The two-way exchange of water between a stream and a subsurface groundwater reservoir can cause element losses from river reaches (Rigler, 1979). Cation mass balances and hydrogeologic data indicate an absence of such water exchanges in the Nottawasaga River and Duffin Creek (Hill, 1982).

Laboratory experiments revealed disappearance of NO_3–N from solutions overlying sediment cores, with almost total loss occurring in some samples within 6 days despite the presence of high dissolved oxygen levels in the solutions. The control cylinders and pans containing no sediment did not exhibit any change in NO_3–N concentration during the experimental period. Less than 5% of the nitrate disappeared from solutions overlying sediment samples that had been sterilized with formaldehyde; this indicates that nitrate removal was caused by biological mechanisms.

In Duffin Creek average rates of NO_3–N loss over a 6-day period ranged from 80 to 250 mg/m^2 of sediment surface daily for twenty pool sediment cores and from 10 to 60 mg/m^2 for nine riffle gravel samples. The highest loss rates for the gravels were recorded in samples taken from small slackwater areas of the riffles where a thin veneer of silt, 5 to 10 mm thick, covered the surface of the gravel. The twelve sediment cores collected from the Nottawasaga River and its major tributaries exhibited a daily loss ranging from 10 to 90 mg/m^2 of sediment surface over a 6-day period. The medium-to-coarse sands that occupy 80 to 90% of the stream bed area of the Nottawasaga study site have a mean nitrate loss rate (38 ± 27 mg N m^{-2} day^{-1}) which is only slightly higher than the average for the gravel riffles in Duffin Creek (26 ± 15 mg N m^{-2} day^{-1}). The pool sediments that cover approximately 50% of the bed area in Duffin Creek have a mean loss rate of 150 ± 53 mg N m^{-2} day^{-1}, which is almost four times greater than the Nottawasaga sediments.

The removal of nitrate from solutions overlying sediments has been attributed primarily to bacterial denitrification. (Sain et al, 1977; Van Kessel, 1977b). Laboratory nitrate losses may provide an overestimate of denitrification, however, since some assimilatory nitrate reduction and respiratory reduction to the level of ammonia may also occur (Keeney, 1973; Brezonik, 1977). Labeled $^{15}NO_3$–N experiments with sediments from Canagagigue Creek, about 70 km west of Toronto, indicated that 98% of the nitrate removed resulted from denitrification (Chatarpaul and Robinson, 1979). Van Kessel (1977c) also indicated that assimilatory reduction was responsible for less than 5% of NO_3–N losses in laboratory tests with sediments from a canal. A much greater proportion of immobilization and nitrate reduction to ammonia, ranging from 20 to 60% of the nitrate loss, has been recorded for lake sediments (Keeney et al., 1971; Chen et al., 1972b; 1979). Chatarpaul and Robinson (1979) suggested that these higher immobilization rates may result from the technique of adding the labelled nitrogen directly to the sediment rather than to the solution overlying the sediment. The latter technique was used in the studies that recorded negligible assimilatory reduction.

The rates of nitrate disappearance observed in the laboratory are of the same order of magnitude as the field losses recorded for the Nottawasaga River and Duffin Creek study reaches. Nevertheless, a precise comparison cannot be made between the laboratory and field data. The initial NO^3–N concentration of the solutions used in the laboratory experiment was considerably greater than the concentration in the rivers; consequently the laboratory experiment may overestimate denitrification rates in the rivers. (Van Kessel, 1977a). In other respects the laboratory experiment may seriously underestimate the real rate of denitrification in streams. Nitrate transport into sediments from the overlying water is a major determinant of denitrification rate. The turbulence and mixing caused by stream flow and the presence of benthic animals create much higher rates of nitrate transport into sediment under natural conditions (Van Kessel, 1977b).

Spatial Variations in Nitrate Removal Rates

Analysis of nitrate sink processes indicates that denitrification may be a major factor in nitrate removal in the Nottawasaga River and Duffin Creek. However, nitrate mass balances and my laboratory experiments suggest that the two rivers differ considerably with respect to denitrification and nitrate reduction rates in stream sediments during low summer flows.

Denitrification is affected by a number of environmental factors. The process requires an absence of oxygen, although not a highly reducing environment (National Academy of Sciences, 1978). It is influenced by temperature and pH, proceeding slowly at low temperatures and in acidic conditions (Keeney, 1973; Terry and Nelson, 1975; Van Kessel, 1977a). Carbon, although required as an energy source by denitrifers, does not appear to affect denitrification rates over a range from approximately 1 to 30% organic carbon content (Robinson, et al., 1978b). However, there must be some critical level of carbon below which denitrification rates will decrease. A positive correlation is evident between denitrification rates and nitrate concentrations in the overlying water (Van Kessel, 1977a). Stirring of sediment by benthic fauna aids rapid diffusion of nitrate into sediments and can have a positive effect on denitrification (Chatarpaul et al., 1979; 1980).

The wide variation in daily rates of nitrate removal (40 to 480 mg N/m^2) in the six downstream reaches of Duffin Creek results mainly from differences in stream nitrate concentration (Hill, 1981). Stream sediments are similar in all reaches; however, mean NO_3–N concentration varies from 0.5 to 1.0 mg/l upstream from station 7A, whereas NO_3–N levels are usually 0.2 to 0.4 mg/l in the other reaches of the river. The average NO_3–N concentration in the Nottawasaga River and its major tributaries is 0.5 to 1.0 mg/l, which is similar to or higher than levels in Duffin Creek. Despite the high nitrate concentration, the field and laboratory data indicate that nitrate removal rates in the Nottawasaga River are considerably lower than in Duffin Creek. A comparison of the two rivers suggests that they are similar in water retention times, temperature, and pH but differ in many sediment characteristics.

The Nottawasaga river sediments are medium-to-coarse sands with a much lower content of silt and organic carbon than the Duffin Creek pool sediments (Table 2). Differences in organic carbon and nitrogen are particularly evident in the 0- to 4-cm zone. Duffin Creek sediments in this zone contain about six times more organic carbon than those in the Nottawasaga River. Denitrification in stream sediments occurs mainly in the upper few centimeters of the river bed (Sain et al., 1977; Van Kessel,

Table 2. Characteristics of the Sediment at 0 to 4 and 4 to 10 cm Depth in Pool Reaches of Duffin Creek and the Nottawasaga River*

	Duffin Creek		Nottawasaga River	
	0-4 cm (N = 10)	4-10 cm (N = 10)	0-4 cm (N = 10)	4-10 cm (N = 10)
pH	7.9 ± 0.2	8.1 ± 0.2	8.1 ± 0.2	8.2 ± 0.1
NH_4-N, µg/g	9.0 ± 7.7	1.8 ± 2.5	1.8 ± 2.1	0.5 ± 1.1
NO_3 + NO_2-N, µg/g	0.07 ± 0.08	0.06 ± 0.09	0.16 ± 0.02	0.13 ± 0.13
Organic nitrogen, %	0.110 ± 0.07	0.015 ± 0.007	0.017 ± 0.016	0.011 ± 0.006
Organic carbon, %	1.40 ± 0.91	0.20 ± 0.10	0.22 ± 0.23	0.11 ± 0.08
C : N ratio	12.7 : 1	13.3 : 1	12.9 : 1	10.0 : 1
Gravel, %	16.1 ± 16.2	23.9 ± 25.6	4.6 ± 4.1	9.4 ± 16.8
Sand, %	53.2 ± 18.4	72.1 ± 24.4	91.0 ± 6.2	88.1 ± 16.3
Silt, %	25.1 ± 17.7	3.2 ± 1.9	3.8 ± 3.4	1.7 ± 1.4
Clay, %	5.6 ± 4.1	0.5 ± 0.3	0.6 ± 0.5	0.8 ± 0.4

*Means ± standard deviation (air-dry basis).

Table 3. Mean Daily Input and Output of Water and NO_3-N in the Nottawasaga River
and Duffin Creek
during Low Flows from Late November to March*

	Input	Output	Net gain or loss	Gain or loss as % of input
Nottawasaga River				
Flow, l/s	4078 ± 934	4319 ± 1124	+241 ± 249*	6
NO_3-N, kg/day	527 ± 97	535 ± 105	+ 8 ± 14	2
Duffin Creek				
Flow, l/s	1413 ± 255	1540 ± 306	+127 ± 93*	9
NO_3-N, kg/day	123 ± 16	137 ± 36	+ 14 ± 23	11

*Means ± standard deviation. Mean daily gain or loss significantly different from zero at the 5% level is indicated by * (a pair comparison t-test was used).

1977a); consequently the low rates of denitrification in the Nottawasaga River may result from insufficient levels of organic carbon in the surface sediment. In addition, the coarse texture of these sediments may reduce the number of anoxic microsites for denitrification near the surface.

Temporal Variations in Nitrate Removal Rates

Nitrate loss during stream transport may also occur during low flows in winter. Nitrate inputs and outputs were measured on 9 days in the Nottawasaga study area and on 5 days in Duffin Creek between late November and early March (Table 3). Stream discharge could not be measured accurately with a current meter at several stations because of a thick ice cover. In these cases daily discharge was estimated from regression analysis between nearby continuously gauged stations and the ungauged streams.

During winter low flows, small NO_3-N losses (1 to 12 kg) occurred on 3 days, and gains of 5 to 29 kg were observed on 6 days in the Nottawasaga River. A similar pattern of nitrate gains and losses was also evident in Duffin Creek. Mean daily nitrate outputs exceeded inputs by 2 and 11% in the Nottawasaga River and Duffin Creek, respectively (Table 3). These differences between inputs and outputs were not statistically significant, however; this indicates a balanced budget.

A more detailed analysis of the seasonal dynamics of nitrate sinks was undertaken in the late spring and early autumn of 1979 for the river reach between stations 7 and 8 on Duffin Creek (Figure 2b). Nitrate removal from the river reach became evident in early May as water temperatures

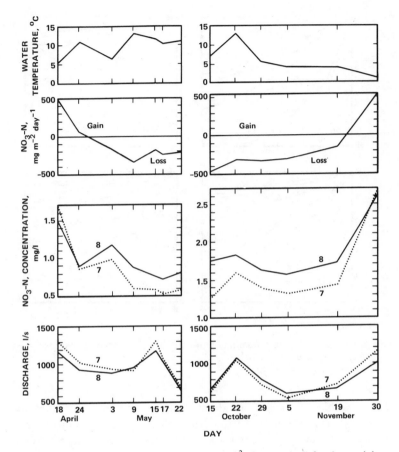

Figure 3. Variations in daily nitrate losses (mg N/m² of stream bed) for the reach between stations 7 and 8 on Duffin Creek during late spring and autumn 1979.

rose and stream discharge declined (Figure 3). A decrease in rates of nitrate removal also occurred in late October and early November.

These data suggest that denitrification and other sink processes decline during the autumn and increase in importance during May in Ontario rivers. The field data indicate an absence of nitrate removal between December and April. However, this evidence cannot be used to infer that nitrate sink processes cease during these months. The inaccuracy of discharge measurements on ice-covered streams and nitrate inputs from unsampled ephemeral tributaries create errors in the mass balance which preclude the detection of small losses of nitrate resulting from biological utilization under cold temperature conditions.

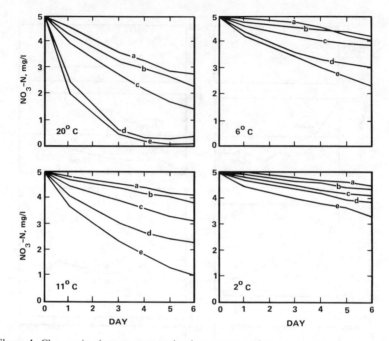

Figure 4. Changes in nitrate concentration in water overlying stream sediments incubated at temperatures between 2 and 20°C.

The possible occurrence of denitrification and nitrate reduction in sediments during the period from November to April in Ontario rivers was also examined by a laboratory experiment. The disappearance of nitrate was observed in twelve sediment–water profiles incubated at temperatures of 20, 11, 6, and 2°C. From December to March the water temperature in the Nottawasaga River and Duffin Creek fluctuates between 0 and 2°C. The laboratory data revealed that nitrate was lost from water in contact with sediments even at these low temperatures, although nitrate removal was only approximately 20% of the rate in summer water temperatures, which average 20°C (Figure 4). River water temperatures range from 4 to 11°C during November, April, and early May. Water–sediment profiles incubated at 6 and 11°C exhibited rates of nitrate loss that were 35 and 45%, respectively, of the nitrate removal at 20°C. These results suggest that river sediments may remove nitrate from stream water during the winter and spring, although at a much lower rate than is characteristic of the months from May to October.

Fluctuations in denitrification rates may also be influenced by river discharge. During major flood events the considerable disturbance and scouring of sediments probably creates aerated conditions that inhibit denitrification. Increases in stream velocity during smaller storm runoff events may still be sufficient to transport the upper few centimeters of sediment, although the stream bed at greater depths remains undisturbed. Since denitrification occurs mainly in the 0- to 5-cm sediment layer, this partial disturbance may reduce denitrification rates.

It is difficult to measure the effects of storm flows on denitrification and other sink processes in the field. However, the data collected for the reach between stations 7 and 8 on Duffin Creek provide some evidence of the effects of small discharge peaks in relation to nitrate removal rates (Figure 3). Higher discharges on May 15 and October 22 were not associated with increases in suspended sediment load, and there was only a small decrease in nitrate loss per square meter of stream bed in comparison to days with lower discharge. These two discharge peaks were approximately three times greater than average summer low flows in this reach; this suggests that nitrate sink processes may still be important during moderate stream flows.

Disturbance of stream sediments was simulated in the laboratory using beakers with 2 cm of sediment covered by 500 ml of solution containing 5 mg NO_3-N/l. The sediments were magnetically stirred at 20°C for 6 days. Analysis of stirred sediments from several sites on Duffin Creek revealed a 1- to 2-day lag phase, followed by an increase in nitrate; concentrations in the overlying solution rose to 10 mg N/l in some cases. Cessation of the disturbance resulted in a decline in nitrate concentration after a delay of 1 to 2 days. Subsamples of the sediment cores incubated in a nonstirred state showed rapid disappearance of nitrate over a 6-day period. The increase in nitrate levels in disturbed sediments is probably caused by rapid nitrificaiton. Similar results have been reported for stirred sediments from hardwater lakes (Chen et al., 1972a). These data suggest that denitrification is probably negligible in river channels during major flood events.

Importance of Nitrate Sink Processes in Relation to River Nitrogen Budgets

The mass balance for Duffin Creek during summer low flows indicates an average daily removal of 45 ± 9 kg NO_3-N (Table 1). In contrast, the Kjeldahl nitrogen load does not change significantly during stream transport. Bacterial denitrification and other minor nitrate sink processes remove approximately 50% of the average daily input of total nitrogen and 75% of the nitrate input received by the downstream river reaches (Hill, 1981). In the Alliston sand plain area of the Nottawasaga River the average daily removal of NO_3–N is 35 ± 21 kg, which represents approximately 13% of the nitrate input (Table 1). These nitrate losses, although substantial, probably underestimate the actual daily nitrate removal. The mass balance technique cannot measure the denitrification of nitrate generated within bottom sediments. Nitrate produced by nitrification in the thin aerobic surface layer of sediment may undergo denitrification in microsites with reducing conditions in this zone or after diffusion to deeper anaerobic sediments (Chatarpaul et al., 1979; 1980; Chen et al., 1972a).

Almost 80% of the annual nitrogen exports from Duffin Creek and the Nottawasaga River occur between November and April when denitrification and other nitrate sink processes are of minor importance because of cold temperatures and periods of high river discharge. In rivers that are not influenced by major point pollution sources, storm periods account for most of the annual nitrogen export. Approximately 65% of the nitrate is exported during 20% of the year when mean daily discharges exceeded 3 m^3/s in Duffin Creek and 14 m^3/s in the Nottawasaga River.

These element transport dynamics imply that nitrate removal by denitrification and other sink processes is relatively small in relation to annual nitrogen flux in most rivers. The average daily summer low flow loss of 35 kg NO_3–N in the Nottawasaga River would result in the removal of about 4600 kg NO_3–N between May and October. From the laboratory data on rates of denitrification in relation to water temperature, it is estimated that an additional 1000 kg is removed during low flows between November and April, for a total removal of 5600 kg. The average annual export of NO_3–N from the Nottawasaga basin at station 34 for the 1977–1979 period was about 410,000 kg (Hill, in preparation), and therefore nitrate removal during stream transport represents only about 1% of the annual nitrate export in this river. Similarly, nitrate losses in the downstream reaches of Duffin Creek represent only 5 to 6% of the annual export of total nitrogen (Hill, 1979).

This study focused on various aspects of nitrogen transport and transformations. The two rivers examined revealed considerable differences in the magnitude of nitrate sink processes during low flows. The high nitrate assimilative capacity of some rivers, as exemplified by Duffin Creek, is of considerable ecological significance because of the resulting reduction in nitrate concentration in stream water and the smaller export of nitrate to lakes and rivers downstream. Although a relatively small percentage of the annual nitrogen load of most rivers is transported during low discharges, the considerable time duration of these conditions enhances the ecological importance of nitrate sink processes. In southern Ontario low flows usually occur on 130 to 150 days between May and

October. In areas of the world that lack a cold winter, denitrification and other sink processes may be important in low flow conditions throughout the year.

Further research is required to evaluate nitrogen flux and cycling in rivers. The recent development of the acetylene blockage technique (Knowles, 1979; Yoshinari and Knowles, 1976) provides a relatively simple and inexpensive procedure for measuring denitrification rates, as well as evaluating the relative importance of NO_3-N immobilization and permanent loss by denitrification. The ecological effects of nitrate removal during low flows should be examined in relation to the functioning of downstream ecosystems. Increasing nitrogen inputs from agriculture, sewage treatment plants, and other sources are characteristic of many rivers; however, little is known about the capacity of streams to assimilate additional inputs. Studies are needed to evaluate the effects of increased nitrogen inputs on denitrification and other nitrate sink processes in rivers.*

ACKNOWLEDGMENTS

This work was supported by grants from the Natural Sciences and Engineering Research Council of Canada. I thank N. K. Kaushik and J. B. Robinson for their helpful advice during this study. I am also grateful to K. Sanmugadas, W. Scott, and T. Irvine for laboratory assistance; D. McLeish and L. Nordstrom for field assistance, and the Cartographic staff in the Department of Geography for the figures.

REFERENCES

Allison, L. E., 1965, Organic Carbon, in C. A. Black (Ed.), *Methods of Soil Analysis,* pp. 1367-1378, American Society of Agronomy, Madison, WI.

American Public Health Association, 1976, *Standard Methods for the Examination of Water and Wastewater,* 14th ed. New York.

Brezonik, P. L., 1977, Denitrification in Natural Waters. *Prog. Water Technol.,* 8: 373-392.

---

*Stream sediments from Duffin Creek and the Nottawasaga River have been incubated in the laboratory and monitored periodically for N_2O production using the acetylene blockage technique. Results indicate that >90% of the nitrate removal from solutions overlying the sediments can be attributed to denitrification.

Chatarpaul, L. and J. B. Robinson, 1979, Nitrogen Transformations in Stream Sediments: ^{15}N Studies, in C. D. Litchfield and P. L. Seyfried (Eds.), *Methodology for Biomass Determinations and Microbial Activities in Sediments*, pp. 119-127, Publication ASTM STP 673, Philadelphia.

———, J. B. Robinson, and N. K. Kaushik, 1979, Role of Tubificid Worms on Nitrogen Transformations in Stream Sediment, *J. Fish Res. Board Can.*, 36: 673-678.

———, J. B. Robinson, and N. K. Kaushik, 1980, Effects of Tubificid worms on Denitrification and Nitrification in Stream Sediment, *Can. J. Fish. Aquat. Sci.*, 37: 656-663.

Chen, R. L., D. R. Keeney, and J. G. Konrad, 1972a, Nitrification in Sediments of Selected Wisconsin Lakes, *J. Environ. Qual.*, 1: 151-154.

———, D. R. Keeney, D. A. Graetz, and A. J. Holding, 1972b, Denitrification and Nitrate Reduction in Wisconsin Lake Sediments, *J. Environ. Qual.*, 1: 158-162.

———, D. R. Keeney and L. J. Sikora, 1979, Effects of Hypolimnetic Aeration on Nitrogen Transformations in Simulated Lake Sediment-Water Systems, *J. Environ. Qual.*, 8: 429-433.

Hill, A. R. 1979, Denitrification in the Nitrogen Budget of a River Ecosystem, *Nature (London)*, 281: 291-292.

———, 1981, Nitrate-Nitrogen Flux and Utilization in a Stream Ecosystem During Low Summer Flows, *Can. Geog.*, 25: 225-239.

———, 1982, Phosphorus and Major Cation Mass Balances for Two Rivers During Low Summer Flows, *Freshwater Biol.* 12: 293-304.

Issac, R. A., and W. C. Johnson, 1976, Determination of Total Nitrogen in Plant Tissue, Using a Block Digester, *Journal AOAC*, 59: 98-100.

Kaushik, N. K. and H. B. N. Hynes, 1971, The Fate of Dead Leaves that Fall Into Streams, *Arch. Hydrobiol.*, 68: 465-515.

———, J. B. Robinson, P. Sain, H. R. Whiteley and W. N. Stammers, 1975, A Quantitative Study of Nitrogen Loss from Water of a Small Spring-Fed Stream, *Proc. Can. Symp. Water Pollut. Res., Toronto*, 10: 110-117.

Keeney, D. R., 1973, The Nitrogen Cycle in Sediment-Water Systems, *J. Environ. Qual.*, 2: 15-29.

———, R. L. Chen, and D. A. Graetz, 1971, Denitrification and Nitrate Reduction in Sediments: Importance to the Nitrogen Budget of Lakes, *Nature (London)*, 233: 66-67.

Knowles, R., 1979, Denitrification, Acetylene Reduction and Methane Metabolism in Lake Sediment Exposed to Acetylene, *Appl. Environ. Microbiol.*, 38: 486-493.

Likens, G. E., 1975, Nutrient Flux and Cycling in Freshwater Ecosystems, in F. G. Howell, J. B. Gentry, and M. H. Smith (Eds.), *Mineral Cycling in Southeastern Ecosystems*, ERDA Symposium Series, CONF-740513, pp. 314-348, NTIS, Springfield, VA.

Meyer, J. L., 1978, Transport and Transformation of Phosphorus in a Forest Stream Ecosystem, PhD. Thesis, Cornell University, Ithaca, NY.

National Academy of Sciences, 1978, *Nitrates: An Environmental Assessment*, Washington, DC.

Ontario Ministry of Environment, 1973-1978, *Water Quality Data for Ontario Lakes and Streams,* Vols. 8-14, Toronto.

Ostry, R. C. 1979, The Hydrogeology of the IFYGL Duffin Creek Study Area, Water Resources Report 5C, Ontario Ministry of Environment, Toronto.

Owens, M., J. H. N. Garland, I. C. Hart, and G. Wood, 1972, Nutrient Budgets in Rivers, *Symp. Zool. Soc. London,* 29: 21-40.

Rigler, F. H., 1979, The Export of Phosphorus from Dartmoor Catchments: A Model to Explain Variations of Phosphorus Concentrations in Streamwater, *J. Mar. Biol. Assoc. U.K.,* 59: 659-687.

Robinson, J. B., N. K. Kaushik, and L. Chatarpaul, 1978a, *Nitrogen Transport and Transformations in Canagagigue Creek,* International Joint Commission. Windsor, Ontario.

———, H. R. Whiteley, W. Stammers, N. K. Kaushik, and P. Sain, 1978b, The Fate of Nitrate in Small Streams and Its Management Implications, in R. C. Lohr et. al. (Eds.), *Best Management Practices for Agriculture and Silviculture,* pp. 247-259, Ann Arbor Science Publications, Ann Arbor, MI.

Sain, P., J. B. Robinson, W. N. Stammers, N. K. Kaushik, and H. R. Whiteley, 1977, A Laboratory Study of the Role of Stream Sediment in Nitrogen Loss from Water, *J. Environ. Qual.,* 6: 274-278.

Sibul, U., K. T. Wang, and D. Vallery, 1977, Ground-Water Resources of the Duffin-Rouge River Drainage Basins, Water Resources Report 8, *Ontario Ministry of Environment,* Toronto.

Terry, R. E., and D. W. Nelson, 1975, Factors Influencing Nitrate Transformations in Sediments, *J. Environ. Qual.,* 4: 549-554.

Van Kessel, J. F., 1977a, Factors Affecting the Denitrification Rate in Two Water-Sediment Systems, *Water Res.,* 11: 259-267.

———, 1977b, Removal of Nitrate from Effluent Following Discharge on Surface Water, *Water Res.,* 11: 533-537.

———, 1977c, the Immobilization of Nitrogen in a Water-Sediment System by Denitrifying Bacteria as a Result of Nitrate Respiration, *Prog. Water Technol.,* 8: 155-160.

Yoshinari, T., and R. Knowles, 1976, Acetylene Inhibition of Nitrous Oxide Reduction by Denitrifying Bacteria, *Biochem. Biophys. Res. Comm.,* 69: 705-710.

25. CHARACTERIZATION OF THE KINETICS OF "DENITRIFICATION" IN STREAM SEDIMENTS

**W. N. Stammers,* J. B. Robinson,† H. R. Whiteley* and
N. K. Kaushik†**

> *School of Engineering and †Department of
> Environmental Biology
> University of Guelph
> Guelph, Ontario

ABSTRACT

Inferences have been drawn from experiments, involving sediment columns with supernatant nitrate solutions, on the kinetic nature of the denitrification process which occurs in stream sediments. Work reported in this paper examines the utility of continuous and lumped models of nitrate transport as bases for the investigation of both transient and steady-state responses to pulse and constant supernatant nitrate solution input concentrations to sediment columns. The measured system output was supernatant nitrate concentration in the case of the pulse input and nitrate flux into the sediment in the case of the constant supernatant nitrate solution concentration. The transient response in all cases was found to be inappropriate for the investigation of denitrification process kinetics. However, the steady-state response to a constant supernatant nitrate solution concentration was found appropriate for testing hypotheses about process kinetics using both lumped and continuous transport models. Results indicate that the denitrification process may be kinetically described as a zero-order process.

INTRODUCTION

An understanding of the kinetics of denitrification, nitrification, mineralization, and immobilization is necessary for both the development of a detailed dynamic model of stream nitrogen transport and the successful budgeting of nitrogen in small streams. Despite recent efforts at characterizing these processes (e.g., ^{15}N studies of denitrification and nitrification in stream sediments by Chatarpaul et al., 1980), insufficient information on these processes is available at this time to construct a detailed dynamic model for nitrogen transport and transformation. However, since the nitrate form of nitrogen is the principle form of nitrogen transport in streams, a useful model of nitrate transport in streams (Stammers et al., 1978) can be developed by combining all biochemical processes in the stream sediments which use or produce nitrate. This combined process, which has been shown to remove nitrate from stream water (Kaushik et al., 1975), is defined as denitrification for the purposes of this paper.

Characterizing denitrification kinetics requires an experiment that can be satisfactorily modelled analytically to provide the basis for formulating

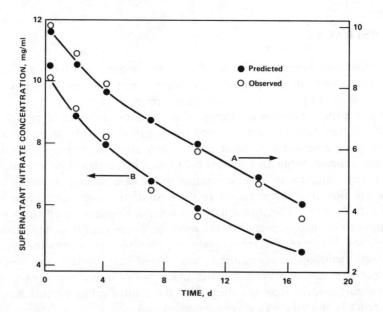

Figure 1. Measured and predicted supernatant nitrate concentration for zero- (A) and first-order (B) kinetics. Zero-order kinetics: column depth, 10 cm; $\lambda = 0.723 \times 10^{-5}$ mg l^{-1} sec^{-1}; and D = 7.25×10^{-5} cm^2/sec. First-order kinetics: Column depth, 5 cm; $\lambda = 0.254 \times 10^{-5}$ mg l^{-1} sec^{-1}; and D = 8.45×10^{-5} cm^2/sec. Temperature for both was 10°C.

and testing a hypothesis concerning process kinetics. The type of experiments we performed to provide this opportunity were in some respects similar to the sediment/water-column studies used by Patrick and his associates (Patrick and Reddy, 1976). In this type of experiment, water-saturated sediment columns are prepared and a supernatant nitrate solution is added so that nitrate diffuses from the supernatant solution into the sediment. Complete details of column preparation, incubation, and sediment characterization were reported by Sain et al. (1977). The sediment used is carbon rich, with 55 to 60% loss on ignition and a C/N ratio of approximately 18. The time-dependent behavior of the supernatant solution for typical experiments of this type is shown in Figure 1. The following section shows that neither a distributed nor a lumped model of nitrate diffusion and denitrification can be used in conjunction with the measured transient or steady-state behavior of the supernatant nitrate solution as a means of nitrate-sink characterization. It is also demonstrated that if the concentration of the supernatant nitrate solution is maintained constant, then under certain conditions a hypothesis can be developed using both distributed and lumped models.

MODEL AND HYPOTHESIS DEVELOPMENT

Time-Variant Supernatant Nitrate Solution Concentration

The transport equation for nitrate diffusion in the sediment is:

$$\frac{\partial C(x,t)}{\partial t} = D \frac{\partial^2 C(x,t)}{\partial x^2} + \lambda \qquad 0 < x < \infty \qquad (1)$$

where $C(x,t)$ is sediment nitrate concentration at position x and time t, D is the sediment nitrate diffusion coefficient, and λ is the net denitrification sink strength at position x and time t. The coordinate x is measured positively from the sediment surface into the sediment, and time t is specified from the time of addition of the supernatant nitrate solution. The dependence of λ on x and t is, in general, implicit through explicit dependence on $C(x,t)$. The nature of this explicit dependence is unknown.

For experiments of the type reported by Sain et al. (1977), in which at time zero a supernatant nitrate solution of known concentration is added to the sediment surface, the resulting supernatant concentration transient behavior, which is measurable, must be related to the denitrification sink strength, λ. The supernatant nitrate concentration, $C_l(t)$, is embedded in the nitrate flux boundary condition that must prevail at the sediment surface:

$$L_l \frac{dC_l(t)}{dt} = D \frac{\partial C(o,t)}{\partial x} \qquad t > 0 \tag{2}$$

where L_l is the depth of supernatant solution above the sediment. The relationship sought requires that the transport equation be solved subject to this boundary condition and an appropriate initial condition. Such a solution can only be obtained by making an assumption about the dependence of λ on the nitrate concentration. The only hypothesis available from this approach is one that would allow a choice, among several assumed alternative kinetic models for λ, of the model providing the best fit between measured and predicted values of supernatant nitrate concentration. This approach was followed by Philips et al. (1978).

A lumped model can be developed from the transport equation. Integrating Equation 1 over the interval $0 < x < \delta$, where δ is the depth of penetration of diffusing nitrate in the sediment, yields:

$$\frac{d[\delta\overline{C}(t)]}{dt} = [C(\delta,t) - \overline{C}(t)] \frac{d\delta}{dt} + q(t) - q_\delta(t) + \int_0^\delta \lambda dx \tag{3}$$

Here, $\overline{C}(t)$ is an average sediment nitrate concentration defined by

$$\overline{C}(t) = \frac{1}{\delta} \int_0^\delta C(x,t)dx \tag{4}$$

$q(t)$ is the flux of nitrate into the sediment from the supernatant solution, and $q_\delta(t)$ is the flux of nitrate into the boundary δ. The measurable quantity $C_l(t)$ is related to $q(t)$ by

$$q(t) = -L_l \frac{dC_l(t)}{dt} \tag{5}$$

Since δ is time dependent and parametrically dependent on $C_l(0)$ (Sain et al., 1977), Equation 3 cannot be used to relate $C_l(t)$ to λ without assuming, for example, that the product term $\delta\overline{C}$ is a constant and that the first and third terms on the right-hand side of Eq. 3 can be ignored. Such assumptions are equivalent to ignoring diffusive transport of nitrate in the sediment completely, and thus treating the sediment and supernatant solution as a stirred-tank reactor.

Time-Invariant Supernatant Nitrate Solution Concentration

When the concentration of the supernatant solution is maintained constant by continued addition of nitrate, the experimental observation is the flux of nitrate out of the supernatant solution into the sediment rather than the supernatant concentration as in the previously described situation.

The transport equation for nitrate diffusion in the sediment is again Equation 1, with a constant supernatant nitrate solution concentration as the boundary condition. Again, using the transient behavior of the system requires that an assumption be made about the kinetics of the denitrification process to solve the differential equation and relate the measured nitrate flux into the sediment to this solution. The criticism previously stated is also appropriate to this situation.

In contradistinction to the time-variant case, the steady-state behavior can be used. For large times, with the system in equilibrium with the constant-concentration supernatant nitrate solution, the transport equation for nitrate diffusion in the sediment is:

$$D \frac{d^2 C_\infty(x)}{dx^2} + \lambda = 0 \qquad 0 < x < \delta \tag{6}$$

where $C_\infty(x)$ is the steady-state nitrate concentration distribution in the sediment and δ is the time-invariant depth of nitrate penetration in the sediment. Equation 6 can be integrated over $0 < x < \delta$, yielding

$$D \frac{dC_\infty(\delta)}{dx} - D \frac{dC_\infty(0)}{dx} + \int_0^\delta \lambda dx = 0 \tag{7}$$

In this equation, the first term is zero and the second term is the steady-state nitrate flux, q_∞, into the sediment. Therefore

$$q_\infty = - \int_0^\delta \lambda dx \tag{8}$$

Sediment column experiments using different values of C_l will, in general, yield different values of q_∞ since the equilibrium value of δ will undoubtedly increase with increasing values of C_l. In addition, λ will depend implicitly on C_l if it is concentration dependent. The variation of q_∞ with C_l is given by

$$\frac{dq_\infty}{dC_l} = \int_0^\delta \frac{\partial \lambda}{\partial C_l} dx + \lambda(\delta, C_l) \frac{d\delta}{dC_l} \tag{9}$$

Previous experiments described by Sain et al. (1977) demonstrated that, for C_l values in the neighborhood of 5 to 10 mg/l, sediment columns 1 to 2 cm in length behaved like finite columns so that full penetration of nitrate occurred even in the early stages of the transient buildup of nitrate in the sediment. For columns exhibiting this behavior, δ is independent of C_l and Equation 9 reduces to

$$\frac{dq_\infty}{dC_l} = \int_0^\delta \frac{\partial \lambda}{\partial C_l} dx \tag{10}$$

A hypothesis may be formulated about the concentration dependence of λ on the basis of whether or not dq_∞/dC_l equals zero. For dq_∞/dC_l equal to zero, the integral of the right-hand side of Equation 10 is also zero, and thus, if λ is a nondecreasing function of C_l, $\partial\lambda/\partial C_l$ is identically zero. In this case denitrification process kinetics must be concentration independent. The lumped model (Equation 3) provides exactly the same basis and conclusion.

RESULTS AND DISCUSSION

Typical results for the time-varying supernatant nitrate solution concentration are presented in Figure 1. Parameters, assuming zero-order and first-order kinetics for the denitrification process, are given in Table 1. The minimum sums of squares of deviations between observed and predicted supernatant concentrations are also provided. The minimum sum of squares support first-order kinetics as the better of the two alternatives.

Results of the sediment column experiments in which the supernatant nitrate solution is maintained at a constant concentration are shown in Figure 2 for columns 2 cm long. The data are presented as cumulative mass inflow of nitrate into the sediment as a function of time. The steady-state flux, q_∞, is thus the slope of the curve at large times. For C_l values of 5 and 9 mg/l, q_∞ is independent of C_l, and thus $\partial\lambda/\partial\overline{C}_l$ equals zero and λ is nitrate-concentration independent. The denitrification process rate for the experimental results presented is 1.63×10^{-3} mg NO_3 m^{-2} sec^{-1}.

The steady-state transport equation (Equation 6) can be readily solved

Table 1. Optimal Denitrification Sink Parameters and Diffusion Coefficients with Corresponding Residual Sums of Squares*

Column depth, cm	Sink parameter	Diffusion coefficient, 10^5 cm^2/sec	Residual Sums of square
Zero-Order Kinetics ($\lambda : 10^5$ mg l^{-1} sec^{-1})			
5	0.403	12.90	0.94
10	0.723	7.28	0.58
First-Order Kinetics ($\lambda : 10^5$ sec^{-1})			
1	0.783	2.73	0.27
2.5	0.461	3.35	0.07
5	0.254	8.45	0.42

*All experiments at 10°C.

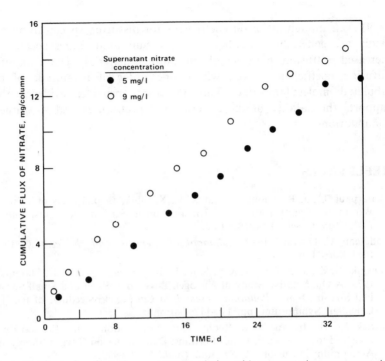

Figure 2. Nitrate flux from the supernatant solution with constant nitrate concentration overlying 2 cm sediment column at 25°C.

and used to provide a relationship between λ, δ, C_l, and D which is of some interest. A relationship for q_∞ derived from the solution is:

$$q_\infty = -\frac{\lambda\delta}{2} + \frac{\epsilon C_l D}{\delta} \tag{11}$$

where ϵ is sediment porosity. Equation 11, combined with the relation

$$q_\infty = -\lambda\delta \tag{12}$$

which states that the flux must be equal to the total denitrification sink strength at equilibrium, yields:

$$D = -\frac{\lambda\delta^2}{2\epsilon C_l} \tag{13}$$

The diffusion coefficient D (computed for experimental values of 2 cm for δ, 5 mg/l for C_l, -1.63×10^{-3} mg m^{-2} sec^{-1} for λ, and 0.8 for ϵ) is 4×10^{-5} cm^2/sec. A comparison can be made with the bulk-fluid molecular diffusion coefficient for potassium nitrate, 1.87×10^{-5} cm^2/sec (Hodgman,

1979) even though diffusion coefficients for dissolved species in porous media are generally somewhat less than bulk-fluid values because of increased diffusion path length in porous material. The computed diffusion coefficient is well within one order of magnitude of the tabulated molecular value. This agreement provides evidence that supports the validity of the experimental procedures and subsequent computations.

REFERENCES

Chatarpaul, L., J. B. Robinson, and N. K. Kaushik, 1980, Effects of Tubificid Worms on Denitrification and Nitrification in Stream Sediments, *Can. J. Fish. Aquat. Sci.,* 37(4): 656-663.

Hodgman, C. D. (Ed.), 1979, *Handbook of Chemistry and Physics,* CRC Press Inc., Boca Raton, FL.

Kaushik, N. K., J. B. Robinson, P. Sain, H. R. Whiteley, and W. N. Stammers, 1975, A Quantitative Study of Nitrogen Loss from Water of a Small Spring-Fed Stream, *Water Pollution Research in Canada,* Proceedings of the 10th Canadian Symposium, pp. 110-117, Toronto.

Patrick, W. H., Jr., and K. R. Reddy, 1976, Nitrification–Denitrification Reactions in Flood Soils and Water Bottoms; Dependence on Oxygen Supply and Ammonium Diffusion, *J. Environ. Qual.,* 5: 469-472.

Phillips, R. E., K. R. Reddy, and W. H. Patrick Jr., 1978, The Role of Nitrate Diffusion in Determining the Order and Rate of Denitrification in Flooded Soil. II. Theoretical Analysis and Interpretation, *J. Soil Sci. Soc. Am.,* 42(2): 272-278.

Sain, P., J. B. Robinson, W. N. Stammers, N. K. Kaushik, and H. R. Whiteley, 1977, A Laboratory Study on the Role of Stream Sediment in Nitrogen Loss from Water, *J. Environ. Qual.,* 6(3): 274-278.

Stammers, W. N., J. B. Robinson, H. R. Whiteley, N. K. Kaushik, and P. Sain, 1978, Modeling Nitrate Transport in a Small Upland Stream, *Water Pollution Research in Canada,* Proceedings of the 13th Canadian Symposium, pp. 161-174, Hamilton, Ontario.

AUTHOR INDEX

487

SUBJECT INDEX